T0235910

Communications in Computer and Information Science   1112

Commenced Publication in 2007
Founding and Former Series Editors:
Phoebe Chen, Alfredo Cuzzocrea, Xiaoyong Du, Orhun Kara, Ting Liu,
Krishna M. Sivalingam, Dominik Ślęzak, Takashi Washio, Xiaokang Yang,
and Junsong Yuan

Editorial Board Members

More information about this series at http://www.springer.com/series/7899

Brendan Eagan · Morten Misfeldt ·
Amanda Siebert-Evenstone (Eds.)

Advances
in Quantitative Ethnography

First International Conference, ICQE 2019
Madison, WI, USA, October 20–22, 2019
Proceedings

 Springer

Editors
Brendan Eagan ⓘ
University of Wisconsin–Madison
Madison, WI, USA

Morten Misfeldt ⓘ
University of Copenhagen
Copenhagen, Denmark

Amanda Siebert-Evenstone ⓘ
University of Wisconsin–Madison
Madison, WI, USA

ISSN 1865-0929   ISSN 1865-0937 (electronic)
Communications in Computer and Information Science
ISBN 978-3-030-33231-0   ISBN 978-3-030-33232-7 (eBook)
https://doi.org/10.1007/978-3-030-33232-7

This Springer imprint is published by the registered company Springer Nature Switzerland AG
The registered company address is: Gewerbestrasse 11, 6330 Cham, Switzerland

Preface

This volume contains the proceedings of the First International Conference on Quantitative Ethnography (ICQE). We have attempted to bring together scholars from a wide range of fields interested in the use of quantitative techniques to address issues more traditionally dealt with using qualitative methods. As the name suggests, we hope this event will be the first in a series of conferences—and in writing this preface as part of the final preparation for the meeting, the scope and quality of the work being presented at the conference suggests that this is a realistic ambition.

We had 53 paper submissions, which was more than we originally anticipated. Papers were reviewed using a double-blind review process, with two sub-reviews and one meta-reviewer from the conference's international Program Committee. The contributions were of high quality overall, and we accepted 23 as full papers and 9 as short papers for inclusion in these conference proceedings. The contributions come from a diverse range of fields and perspectives, including learning analytics, history, and systems engineering, all attempting to understand the breadth of human behavior using quantitative ethnographic approaches, which our colleague Ron Serlin has described as a "unified method" that integrates both qualitative and quantitative approaches to data and its analysis.

Although the papers collected here differ in their goals and domains of study, they represent the beginning of a rich conversation about research methods, tools, and the philosophical dilemmas of modeling human interaction. This is not coincidence. Fostering this dialog was the purpose of the conference and of these proceedings.

We believe the effort is both timely and essential, as we have seen a growing need for a professional and scientific discourse around the use of data to investigate human experience and culture with the advent of new kinds—and much larger quantities—of information about human activity in recent years.

Technological developments, from ubiquitous cell phones to cloud computing, have increased our ability to quantify aspects of human experience from health and culture to experience and consciousness—and this ability, in turn, has raised new possibilities and new concerns about how, whether, and why to use such tools.

The result has been opening of new branches and areas of research, such as learning analytics and digital humanities, but also changes in existing disciplines such as statistics, where the focus has shifted from answering simple questions by generalizing based on sparse amounts of data towards making sense of enormous amounts of data, thus allowing more complex inferences.

Our goal in starting a scientific dialogue around the concept of quantitative ethnography is to foster a community of researchers interested in sharing ideas, resources, and inspiration, hence we have also organized both the event and these proceedings so as to avoid creating intellectual silos by placing the different contributions into predefined categories of object or method. Our aim, rather, has been to maximize participants (and readers) opportunities to learn from each other—through the papers collected here, and at the conference through paper presentations, symposia, and keynotes aimed at presenting, comparing, contrasting, and synthesizing current work and setting future directions for the community of scholars interested in quantitative ethnography.

We hope that future ICQE meetings will be the place for scholars to present their research, to learn and share new methodological techniques, and to develop a robust methodological and philosophical discourse exploring new analytic approaches based on big data which create new and fruitful intersections between classical ethnographic methodologies and modern statistical methods.

Both approaches share the challenge of creating meaning and understanding in the face of rapid, technologically-driven expansion in the amount and kind of data available to researchers. By unifying tools, techniques, and epistemologies from both traditions—and by developing new methods at their intersection—quantitative ethnography is an approach to data analysis designed to allow researchers to tell textured stories at scale.

We believe that this conference and these proceedings are an important first step in that direction.

September 2019

Morten Misfeldt
Brendan Eagan

Organization

Program Committee Chairs

Morten Misfeldt   University of Copenhagen, Denmark
Brendan Eagan   University of Wisconsin-Madison, USA

Program Committee Members

Simon Buckingham Shum   University of Technology Sydney, Australia
Torben Elgaard Jensen   Aalborg University, Denmark
Aroutis Foster   Drexel University, USA
Karin Frey   University of Washington, USA
Eric Hamilton   Pepperdine University, USA
Srecko Joksimovic   University of South Australia, Australia
Ingo Kollar   Augsburg University, Germany
Simon Knight   University of Technology Sydney, Australia
Vitomir Kovanovic   University of South Australia, Australia
Peter Levine   Tufts University, USA
Toshio Mochizuki   Senshu University, Japan
David Williamson Shaffer   University of Wisconsin-Madison, USA
Amanda Siebert-Evenstone   University of Wisconsin-Madison, USA

Conference Committee Chair

Amanda Siebert-Evenstone   University of Wisconsin-Madison, USA

Conference Committee Members

Naomi C. Chesler   University of Wisconsin-Madison, USA
Brendan Eagan   University of Wisconsin-Madison, USA
Lew Friedland   University of Wisconsin-Madison, USA
César Hinojosa   University of Wisconsin-Madison, USA
Sarah Jung   University of Wisconsin-Madison, USA
Jeff Linderoth   University of Wisconsin-Madison, USA
Morten Misfeldt   University of Copenhagen, Denmark
Hyunju Park Johnson   University of Wisconsin-Madison, USA
Carla Pugh   Stanford University, USA
Andrew Ruis   University of Wisconsin-Madison, USA
Hanall Sung   University of Wisconsin-Madison, USA
Abby Wooldridge   University of Illinois, USA

Symposium Committee Chair

Simon Buckingham Shum   University of Technology Sydney, Australia

Symposium Panel

Karin Frey   University of Washington, USA
Adam Lefstein   Ben Gurion University of the Negev, Israel
Jun Oshima   Shizuoka University, Japan
David Williamson Shaffer   University of Wisconsin-Madison, USA

Contents

Short Papers

Full Papers

Examining Identity Exploration in a Video Game Participatory Culture

Amanda Barany$^{(\boxtimes)}$ ⓘ and Aroutis Foster ⓘ

Drexel University, Philadelphia, PA 19104, USA
amb595@drexel.edu

Abstract. To adapt to the needs of a 21st century context, educational researchers and practitioners could benefit from leveraging the potential of virtual learning environments such as games and the participatory cultures that surround them to support learning as a transformational and intentional process of identity exploration. This research offers a theoretically-comprehensive look into how a participant in an online game community forum engaged in identity exploration processes. Publicly-available longitudinal data was downloaded from *Kerbal Space Program (KSP)* players, which informed the development of an illustrative case study selected to elucidate how individual processes of identity exploration manifest. Lines of player data were deductively coded as representative of identity exploration and visualized using Epistemic Network Analysis to represent shifts in integration of identity constructs over time. Findings suggest that player participation in the community forum can support statistically significant identity change and highlight future areas of research in this field.

Keywords: Identity exploration · Video games · Informal learning · Epistemic Network Analysis · Case study

1 Problem and Significance

The rapid development of digital media tools in the 21st century has prompted changes in where, when, and how students of all ages learn. Computers and mobile devices have made foundational knowledge about most domains readily available to meet the customized needs of learners and their diverse learning environments [1]. Tools for media sharing and creation have allowed learners to actively explore their interests through participatory cultures that span many digital and non-digital venues, developing group-mediated expertise and eventually helping to enculturate novices [2, 3]. Most promisingly, video games and simulations offer valuable opportunities for learners to try new roles and develop skills in environments that can authentically mirror real-world professions and practices [4]. While digital media are a valuable tool for supporting such learning processes [5], however, their affordances often remain underutilized or poorly integrated in formal learning settings [6].

A vital step in harnessing the potential of gaming participatory cultures may involve a reconceptualization of 21st century learning as a self-directed process of *identity exploration*, which can support *identity change* over time in terms of "adaptive

B. Eagan et al. (Eds.): ICQE 2019, CCIS 1112, pp. 3–13, 2019.
https://doi.org/10.1007/978-3-030-33232-7_1

learning, motivation, and the development and achievement of educational goals" [7, p. 251]. Identity exploration involves deliberate self-reflection and self-assessment that helps learners set goals for personal growth and take steps to enact them [8]. There is growing evidence that video games have particular potential for supporting identity exploration and change [e.g. 9], and that shifts in academic learning, motivation, and interest can result [10]. As such, further research is needed to build understanding of how learners engage in identity exploration and change over time through games and the participatory cultures that develop around them [11]. This research addresses this need by examining the processes of identity exploration that a participant enacted in an online gaming community forum, which may offer insight to educators and institutions as they leverage or learn from such sources to support identity exploration. The research question asks: What processes of identity exploration manifest through engagement with the participatory cultures that develop around video games?

2 Projective Reflection as a Framework for Assessing Identity Exploration

Projective Reflection offers one conceptual and theoretical framework for understanding the way learners engage in self-transformation, or identity change, in immersive interactive environments such as games and virtual worlds (See Fig. 1, Table 1). This framework supports assessment of internal aspects of individual change as it is shaped by social and institutional features of a digital environment [12]. Projective Reflection uses four theoretical constructs to conceptualize identity in gaming contexts as participants first project out what they will need to do in the future to achieve a desired self, and then reflect on their current self as influenced by their contexts and self-knowledge. As such, identity change manifests over time as students

Fig. 1. The Projective Reflection framework and identity constructs.

Table 1. Projective reflection identity constructs.

PR Constructs	Definitions
1. Knowledge and game/technical literacy	• Changes in what a player knows about a topic (i.e. rocket engineering) from the beginning to the end of an experience
2. Interests and valuing	• Moving from recognition of the general relevance of the topic to awareness of personal awareness and meaning • Recognizing the value of the topic beyond the game context • Shifts in caring about the game or relevant participatory culture
3. Self-organization and self-control	• Changes in behavior, motivation, and cognition toward a goal: – Self-regulated learning: conducted independently – Co-regulated learning: conducted with real/virtual experts – Socially-shared regulation: conducted with peers
4. Self-perceptions and self-definitions	• Shifts in identification with roles (i.e. engineer, designer, CEO) • Changes in self-confidence and self-concept: – How a participant sees himself/herself as a member of the community – Changes in how they feel about game/community roles and what they would like to pursue

intentionally reflect on and then work toward changes they find valuable in terms of: (a) relevant knowledge and technical literacy [i.e. 13], (b) interest and valuing [i.e. 14], (c) patterns of self-organization and self-control [i.e. 15], and (d) self-perceptions and self-definitions related to a targeted domain [i.e. 16]. Identity exploration is characterized not by the manifestation of any one construct in isolation, but by a learner's ability to *integrate* reflection on and enactment of change in all four areas over time.

3   Methods

The site for this proposed research is the community page for the single-player space flight simulation video game *Kerbal Space Program (KSP)* (See Figs. 2, 3). Both the game and community forum are hosted on the Steam online gaming and community platform. *KSP* was selected for this research because it was designed to intentionally support a "vibrant, diverse community of space and explosion enthusiasts" [17, p. 1] who can encouraged to engage in a variety of roles through gameplay and community participation, such as designers, space engineers, corporation leaders, scientists, artists, and storytellers. Steam was selected as the data collection site based on accessibility of anonymized information on patterns of gameplay and community engagement through their Application Programming Interface.

Fig. 2. The Ship Design Interface in *Kerbal Space Program*.

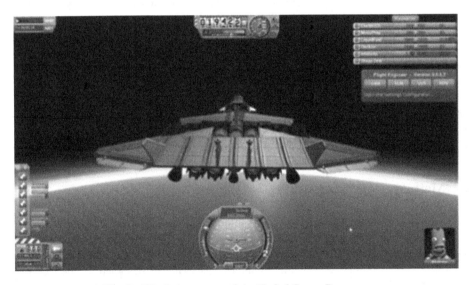

Fig. 3. Piloting a spacecraft in *Kerbal Space Program*.

3.1 Data Collection

To collect information on the patterns of participation enacted by users on the *KSP* community page, data was collected from the Steam Web API, which supports downloading of anonymized, public user data related to gameplay and participation. Public, player-to-player discussion data was scraped directly from the community page using a data mining script. From these two data sources, a network of approximately 42 highly-connected users were sampled from the population of over 230,000 *KSP* players on Steam. The sample consisted of the subset of participants who have replied to at least one discussion board post made by a player they added as a Steam friend, or who have received at least one reply from a friend. Sampling using this method identified users who have connected to peers in multiple ways, offering more insights into the connections between community participation and individual identity exploration processes. As the player in the network with the highest individual centrality measures in the network of users, ShipShape (pseudonym) was selected as a case study that would be illustrative of the identity exploration trajectory of a player that posted and engaged regularly with peers on the community forum discussion boards. From the chronological list of *KSP* forum data, posts made by ShipShape or referencing Ship-Shape were isolated and organized into a chronological account of his forum participation.

3.2　Data Analysis

This research leverages both qualitative case study [18] and quantitative ethnographic approaches [19] to understand identity exploration processes (conceptualized by PR) as manifested across five years of *KSP* forum posts made by ShipShape. The approaches used in tandem were valuable as a way to (a) develop a detailed narrative of the player's participation over time and (b) engage in systematic inquiry into *how* he changed, respectively. To prepare the data, each of ShipShape's 268 chronological posts were deductively coded by hand [20] as reflections on or manifestations of change across the four PR constructs (using a binary system of 1/0 for the presence/absence of the construct). For example, a post describing strategy for successful ship building would be coded as demonstrative of *KSP* design knowledge, while a post in which the player reflects on their enjoyment of design would be coded as reflection on *KSP* interest and valuing. Once a primary round of coding was completed by a graduate-level researcher, code revision and recoding continued as an inductive process of adding codes (filling in), examining existing codes based on new perspectives (extension), exploring new relationships between existing codes (bridging), and identifying new categories (surfacing) until all incidents can be readily classified and theoretical saturation is reached. Final codes were then reviewed by a faculty researcher with theoretical experience in games and identity, and verbal agreement was reached on all code discrepancies.

Epistemic Network Analysis (ENA) was then used to visualize the relationships between PR constructs across two time periods using ENA1.5.2 Web Tool [21]. Units of analysis were defined as all lines of data associated with the first half (Time 1, posts 1–134) and second half (Time 2, post 135–268) of ShipShape's chronological posts; these were further segmented by the day and hour posted. The ENA algorithm uses a moving window to construct a network model for each line in the data, showing how codes in the current line connect to codes within the recent temporal context [22], defined here as 4 lines (each line plus the 3 previous lines) in a given conversation. Conversations were segmented by any gap of more than a week between posts, as it was deemed unlikely based on a review of the data that ShipShape would refer back to his own reflections beyond this point.

The case study was developed using Stake's [18] process of searching for correspondence, or consistency across data sources to understand "behavior, issues, and contexts with regard to our particular case" (p. 78). Deductive codes were reviewed and subjected to inductive process of thematic expansion and refinement grounded in what emerged from the data with regards to identity exploration and change. This included iterative researcher memoing on trajectories of change that emerged, with final agreement reached by a content expert. The final illustrative case study, written in narrative form, offered a description of ShipShape's longitudinal identity exploration (change), contextualized by the Epistemic Network Analysis.

4 Quantitative Findings

Along the X axis (dimension 1 after means rotation), a two sample t test assuming unequal variance showed Time 1 (mean = 0.11, SD = 0.33, N = 118 was statistically significantly different at the alpha = 0.05 level from Time 2 (mean = −0.11, SD = 0.43, N = 116; t(216.65) = −4.50, p = 0, Cohen's d = 0.59). Along the Y axis, a two sample t test assuming unequal variance showed Time 1 (mean = 0.00, SD = 0.51, N = 118 was not statistically significantly different at the alpha = 0.05 level from Time 2 (mean = 0.00, SD = 0.49, N = 116; t(231.86) = 0, p = 1.00, Cohen's d = 0.00). These findings suggest that statistically significant differences exist between the manifestation of connections between constructs in Time 1 versus Time 2 (See Figs. 4, 5).

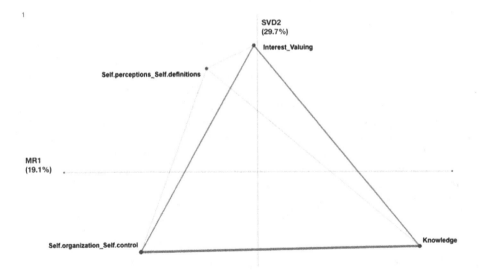

Fig. 4. Relationships between Projective Reflection constructs in Time 1 (posts 1–134).

A difference model subtracted the strength of connections between the four identity constructs in Time 1 from the strength of connections in Time 2 (See Fig. 6). In the difference model, the red line connecting Knowledge to Interest/valuing suggests that

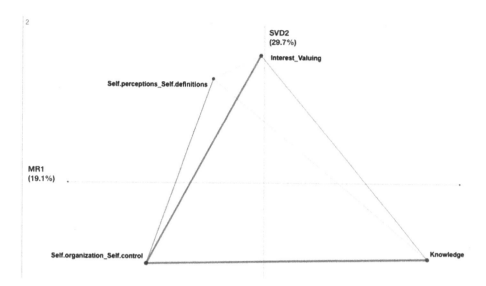

Fig. 5. Relationships between Projective Reflection constructs in Time 2 (posts 135–268).

ShipShape may have been more likely to connect his emerging knowledge to initial interests in Time 1, as compared to his later written forum posts. Blue lines in the difference model suggest that ShipShape was more likely to connect reflections on his Self-organization/self-control to both Interest/valuing and Self-perceptions/definitions in Time 2. Connections between Knowledge and Self-organization/self-control persist from Time 1 to Time 2 (See Figs. 4, 5), and therefore cancel out in the difference model (See Fig. 6).

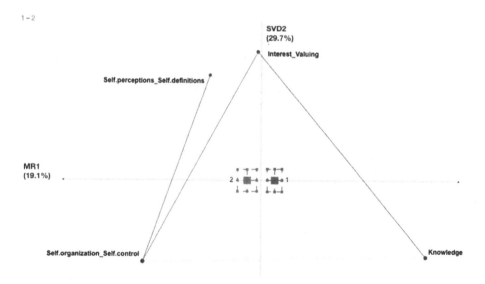

Fig. 6. The difference network illustrating connections between identity constructs made more frequently in Time 1 (red) versus Time 2 (blue). (Color figure online)

From an identity perspective, these findings suggest that ShipShape enacted statistically significant differences in the ways he connected identity constructs over time. Epistemic Networks for Time 1 and Time 2, as well as the difference model suggest that his trajectory of identity exploration may have become more *integrated*, as ShipShape made stronger connections across more identity constructs over time. Findings from the qualitative case study offer insights into how this trajectory of identity exploration manifested for ShipShape.

5 Qualitative Case Study

Examinations of ShipShape's (i)dentity characteristics in his posting to the *KSP* forum revealed that he primarily responded to a picture-heavy thread that invited players to write about what they had done in the game that day with supportive pictures and art. Typical reflections included reflections on his self-organization and self-control strategies such as goal-setting and chronological accounts of his trial and error in ship designs and launches, as well as his strategies of ship modifications over time. His narrative style also showcased his increasingly-detailed knowledge of ship parts and their functions, game goals and strategies, and the use of game modifications that could create design effects or serve different functions. For example, about 200 days into his posting tenure he described the ship changes he made to improve the life-expectancy of his pilots: "I have increased the consumption rates of life supplies to 4X without changing any of the other values, this at least makes it so if it says 20 units of whatever for one kerbal [pilot] each unit will last 6 h or one kerbal day and also makes some of the larger containers not so [overpowered]." This connection between reflections on Knowledge and Self-organization/self-control persisted across his entire tenure of forum participation (years 1–5), as illustrated by the strong, consistent line connecting these constructs in the epistemic networks from both Time 1 and Time 2 (See Figs. 4, 5).

During Time 1, ShipShape more frequently stated his emerging interest in and valuing of *KSP* and was more likely to share such reflections in relation to demonstrations of his growing knowledge of the game. For example, he shared in his first post that he had so far completed "346 h play time and still love it!" and detailed the parts used in the design of "my 100th space station." This not only indicated that ShipShape had likely mastered game literacy before he felt compelled to post, but also illustrates the stronger connection between interest and knowledge that manifested in Time 1 (See Fig. 4). In sum, ShipShape's initial reflections tended to focus on explicit descriptions of his emerging expertise and his enthusiasm for gameplay.

By Time 2, his posts had evolved to reflect increasingly personal interest in and valuing of the *aesthetics* of his designs, and even the artistic quality of his screenshots, which he connected to his self-organization and self-control strategies. Forum posts were more likely to consist of specific discussions of a complicated or difficult in-game goal that ShipShape had set for himself, and the outcome of his attempts to achieve the task. For example, on several occasions he described successfully landing a ship (a difficult feat), only to attempt to "hop" the ship into a more visually-pleasing position on land for the purpose of collecting screenshots, which risked tipping the ship and failing the mission. This exemplifies the increasingly robust relationship that emerged between Self-organization/self-control and Interests/valuing in Time 2 (See Fig. 6).

ShipShape only rarely referenced his own perceptions of self in Time 1 posts, preferring to focus on descriptions of his design processes. After a few weeks of regular posting, however, another user mentioned him by name as a designer of "advanced capitol ships," to which ShipShape replied "I do everything except SSTOs [single-stage to orbit ships] and VTOL [virtual takeoff and landing ships]." This interaction perhaps prompted later reflections on his emerging awareness of self as a ship designer and artist. During Time 2, ShipShape began to affirm how he saw himself as a designer by identifying *KSP* gameplay and the sharing of his designs as a creative outlet. Typically, ShipShape would describe self-perceptions and self-definitions when it connected to (a) his identification of a new design or artistic challenge he felt inspired to tackle (he later intentionally decided to design an SSTO, for example), or (b) his desire to turn capture his designs as artistic screenshots and design blueprints that he created using an external design software and posted to the forum. Over a year into his participation, ShipShape created a personal company logo with his username and imagery that he began to integrate into his photos of space landings and ship designs, solidifying his exploration of the artist/designer role. These activities may have contributed in part to the increasing connections between Self-organization/self-control and Self-perceptions/self-definitions that manifested in Time 2 (See Fig. 6).

6 Discussion and Implications

Overall, ShipShape demonstrated an increasingly-integrated process of identity exploration based on shifts in his enactment of PR constructs over time. He consistently demonstrated an awareness of the self-organization and self-control strategies he enacted as he played *KSP* and engaged with the community, and regularly leveraged his emerging knowledge of modding and design. Over time, connections between reflections on self-organization and self-control and his interest in tackling design challenges and creating beautiful products deepened. He also developed more intentional awareness and enactment of his preferred roles as designer and artist and connected them to the strategies he enacted.

Findings offer insights into how informal participatory cultures that emerge around game environments can supporting increasingly-explicit enactment of learning as an intentional process of self-transformation, or identity exploration that leads to change over time. Though some constructs of identity exploration (i.e. self-perceptions and self-definitions) were referenced less frequently and subsequently less integrated into his longitudinal trajectory, this suggests that modifications could be made to the design of the forum, the discussion prompts, or the platform itself to further scaffold these kinds of reflections. Findings also suggest that such venues have potential to support increasing integration of identity exploration processes over time. This work also points to areas in need of further research, such as the examination of identity exploration trajectories across students with different social situations (i.e. less-central members of the community). Exploring epistemic networks across different units of time could also reveal richer, more incremental information about the nature of individual change.

This research is the result of preliminary examinations of *Kerbal Space Program* forum data using case study and Epistemic Network Analysis methods as part of a dissertation study. As such, there are limitations that must be addressed in future work. The sampling method, which was implemented to identify a network of highly-connected users, simultaneously omits valid trajectories of identity exploration that may not involve friend connections or forum posts. Similarly, the case selection of a single, particularly-central member of a friend network on the *KSP* community forums may not be representative of broader patterns of identity exploration as enacted by members as a whole. As such, a broader examination of participants at different positions in the sample network, and across the larger community, are essential for more thoroughly understanding how players engage in identity exploration.

It is imperative that we as educational researchers and practitioners, particularly in the field of K-12 public education, continue striving to adapt learning practices to the needs of the 21st century. Collins and Halverson [1] advocate for radical and expansive educational reform that not only leverages the benefits of emerging technologies in classroom settings, but also reconceptualizes learning informed by practices that have emerged in online participatory cultures. Understanding and promoting identity exploration practices is one way to equip students with necessary tools for success in a 21st century life and career context [7]. This work contributes to this aim by examining identity exploration as it is enacted by participants in an online gaming community.

References

1. Collins, A., Halverson, R.: The second educational revolution: rethinking education in the age of technology. J. Comput. Assist. Learn. **26**(1), 18–27 (2010)
2. Jenkins, H., Purushotma, R., Weigel, M., Clinton, K., Robison, A.J.: Confronting the Challenges of Participatory Culture: Media Education for the 21st Century. MacArthur, Cambridge, MA (2009)
3. Gee, J.P.: Affinity spaces: how young people live and learn online and out of school. Phi Delta Kappan **99**(6), 8–13 (2018)
4. Shaffer, D.W.: How Computer Games Help Children Learn. Macmillan Publishers, New York (2006)
5. US Department of Education: Reimagining the role of technology in education: 2017 National Education Technology Plan update. Office of Educational Technology, Washington, DC (2017)
6. Adams Becker, S., Freeman, A., Giesinger Hall, C., Cummins, M., Yuhnke, B.: NMC/CoSN Horizon Report: 2016 K-12 Edition. The New Media Consortium, Austin (2016)
7. Kaplan, A., Sinai, M., Flum, H.: Design-based interventions for promoting students' identity exploration within the school curriculum. In: Karabenick, S.A., Urdan, T.C. (eds.) Motivational Interventions: Advances in Motivation and Achievement, vol. 18, pp. 243–291. Emerald Group Publishing Limited, Bingley, UK (2014)
8. Kaplan, A., Garner, J.K.: A complex dynamic systems perspective on identity and its development: The dynamic systems model of role identity. Dev. Psychol. **53**(11), 2036–2051 (2017)
9. Bagley, E.A., Shaffer, D.W.: Stop talking and type: comparing virtual and face-to-face mentoring in an epistemic game. J. Comput. Assist. Learn. **31**(6), 606–622 (2015)

10. Flum, H., Kaplan, A.: Identity formation in educational settings: a contextualized view of theory and research in practice. Contemp. Educ. Psychol. **37**(3), 240–245 (2012)
11. Shah, M., Foster, A., Barany, A.: Facilitating learning as identity change through game-based learning. In: Baek, Y. (ed.) Game-Based Learning: Theory. Strategies and Performance Outcomes. Nova Publishers, New York, NY (2017)
12. Foster, A.: CAREER: projective reflection: learning as identity exploration within games for science. National Science Foundation, Drexel University, Philadelphia, PA (2014)
13. Kereluik, K., Mishra, P., Fahnoe, C., Terry, L.: What knowledge is of most worth: Teacher knowledge for 21st century learning. J. Digit. Learn. Teacher Educ. **29**(4), 127–140 (2013)
14. Wigfield, A., Eccles, J.S.: Expectancy–value theory of achievement motivation. Contemp. Educ. Psychol. **25**(1), 68–81 (2000)
15. Vygotsky, L.S.: Thought and Language. The MIT Press, Cambridge (1934/1986)
16. Kaplan, A., Flum, H.: Identity formation in educational settings: a critical focus for education in the 21st century. Contemp. Educ. Psychol. **37**(3), 171–175 (2012)
17. About Kerbal Space Program. www.kerbalspaceprogram.com/en/page_id=7. Accessed 31 July 2018
18. Stake, R.E.: The Art of Case Study Research. Sage Publications, Thousand Oaks (1995)
19. Shaffer, D.W.: Quantitative Ethnography. Cathcart Press, Madison (2017)
20. Krippendorff, K.: Content Analysis: An Introduction to its Methodology. Sage Publications, Thousand Oaks (2004)
21. Marquart, C.L., Hinojosa, C., Swiecki, Z., Shaffer, D.W.: Epistemic Network Analysis, Version 0.1.0. http://app.epistemicnetwork.org, last accessed 2019/05/10
22. Siebert-Evenstone, A.L., Irgens, G.A., Collier, W., Swiecki, Z., Ruis, A.R., Shaffer, D.W.: In search of conversational grain size: modeling semantic structure using moving stanza windows. J. Learn. Analytics **4**(3), 123–139 (2017)

Using ENA to Analyze Pre-service Teachers' Diagnostic Argumentations: A Conceptual Framework and Initial Applications

Elisabeth Bauer[1]([⊠]), Michael Sailer[1], Jan Kiesewetter[2],
Claudia Schulz[3], Jonas Pfeiffer[3], Iryna Gurevych[3], Martin R. Fischer[2],
and Frank Fischer[1]

[1] Ludwig-Maximilians-University of Munich, Leopoldstr. 13, 80802 Munich,
Germany
`Elisabeth.Bauer@psy.lmu.de`
[2] University Hospital, Ludwig-Maximilians-University of Munich,
Pettenkoferstraße 8a, 80336 Munich, Germany
[3] Technical University of Darmstadt, Hochschulstraße 10, 64289 Darmstadt,
Germany

Abstract. Diagnostic argumentation can be decomposed referring to the dimensions of content (see Toulmin 2003) and explicated strategy use indicated by epistemic activities (see Fischer et al. 2014). We propose a conceptual framework to analyze these two dimensions within diagnostic argumentation and explore its use within initial applications using the method of Epistemic Network Analysis (Shaffer 2017). The results indicate that both approaches of solely analyzing the dimension of content and solely analyzing the dimension of epistemic activities offer less insights into diagnostic argumentations than an analysis that includes both dimensions.

Keywords: Diagnosing · Argumentation · Pre-service teachers

1 A Diagnostic Argumentation Framework

Diagnosing is considered as a specialized type of scientific reasoning: When being confronted with a problem, diagnosticians generate hypotheses, gather and evaluate evidence for and against these and draw diagnostic conclusions (see Fischer et al. 2014). In this regard, it can be decomposed in the application of a diagnostic strategy and relevant concepts (Coderre et al. 2003). One example from medicine would be to specifically generate and evaluate evidence required for the conclusion or exclusion of several hypothesized differential diagnoses. Likewise, this strategy can be applied to various other domains and problem sets as for example in the context of teaching: Teachers need to diagnose single students' learning progress, understanding and learning preconditions. They also need to be able to communicate their diagnoses professionally (Lawson, Daniel 2011), e.g. to colleagues or in interdisciplinary professional communication. By relating arguments for and (if applicable) against

B. Eagan et al. (Eds.): ICQE 2019, CCIS 1112, pp. 14–25, 2019.
https://doi.org/10.1007/978-3-030-33232-7_2

differential diagnoses and rebutting potential counterarguments, they formulate a diagnostic argumentation.

In this paper, we therefore propose that it is a relevant learning objective for university students within several domains to formulate good diagnostic argumentations as a matter of professionalizing their communication. Corresponding interventions like role-play or peer-feedback are very resource-intensive and hardly applicable on a large scale (see Gartmeier et al. 2015). To approach this issue, it might be feasible to automate the analysis of diagnostic argumentations, for example using methods of natural language processing (Schulz et al. 2018). Such automation might be a useful basis for large-scale interventions. However, an automated analysis firstly requires a basic definition of and detailed insights into diagnostic argumentative structures, particularly regarding the quality characteristics of diagnostic argumentation. The literature and previous research in diagnosing rather focus on diagnostic decision-making processes and suggest different sequences of activities like the hypothetico-deductive type of reasoning (see Coderre et al. 2003). Such in-process reasoning consists of inferences, which are based on diagnostic schemata (Charlin et al. 2007). These process-related models assume a chronological temporality that is not necessarily applicable to post-hoc argumentation. Moreover, argumentation serves the purpose of explicitly presenting previous inferences in a comprehensible and conclusive manner (see Berland, Reiser 2009). We argue that comprehensibility and conclusiveness express in the explicated application of diagnostic strategy and concepts. Hence, analyzing the quality of diagnostic argumentations requires a framework that captures both, the application of strategy and concepts.

Since to our knowledge there is no such framework, we suggest the literature of argumentation schemes as a potential starting point. More specifically, we refer to the Toulmin argumentation model (Toulmin 2003) to analyze the application of concepts within a case-specific content dimension of diagnostic argumentation. The basic assumption of the Toulmin model is that conclusions (C) need to be grounded (G) and logically warranted (W) to represent an argument. In a more complex form, the scheme suggests to include more categories as for example evidence (E) that supports the ground (G), or a backing (B) of the warrant (W). The scheme becomes applicable to discursive argumentation by adding the element of the rebuttal (R) that attacks an initial conclusion (C_i) and might obtain its replacement or limitation by adding a qualifier (Q) to the final conclusion (C_f). The model is well transferable to diagnostic concepts as conclusions and rebuttals represent (differential) diagnoses that require being grounded on accumulated evidences; diagnostic conclusions also require being warranted by the knowledge of relations between symptoms and disorders. This knowledge is ideally backed by diagnostic guidelines or epidemiological data. Referring to the Toulmin model, the comprehensibility and conclusiveness of single arguments can be judged by several quality criteria as the acceptability, relevancy and sufficiency of the grounds in supporting the conclusion or the applicability of the warrant to the case under discussion (Toulmin 2003).

However, the Toulmin model does only account for the application of concepts within the case-specific content dimension of diagnostic argumentation. Analyzing the explication of a diagnostic strategy requires the integration of a second, strategy-related dimension within diagnostic argumentation. The strategy dimension can be conceptualized referring to epistemic activities (Fischer et al. 2014) as hypothesis generation (HG), evidence generation (EG), evidence evaluation (EE) and drawing of conclusions (DC). These activities were found to be relevant across a wide range of reasoning and argumentation (Hetmanek et al. 2018), e.g. the analysis of pre-service teachers' problem solving (Csanadi et al. 2018). The study found differences in the activity use of individual vs. collaborative learners, which became only observable by analyzing co-occurrences of epistemic activities. These were depicted as activity patterns by applying the method of Epistemic Network Analysis (Shaffer 2017). Building on this research, we suggest analyzing co-occurrences of diagnostic concepts and epistemic activities to extract similar patterns from diagnostic argumentations.

An exemplary mapping of epistemic activities on the structure of the Toulmin model (2003) is shown in Fig. 1. Introducing this framework we suggest that the structure of diagnostic argumentation might be better captured by analyzing the application of case-specific content and epistemic activities in combination rather than separately. This might point to more implicit aspects within diagnostic argumentations in terms of the application of strategy and concepts. For example a diagnostic conclusion that is linked to the activity of drawing conclusions (DC) might rather be considered as a final and therefore rather certain conclusion (C_f) which is sufficiently grounded; a diagnostic conclusion that co-occurs with the activity of generating evidence (EG) might indicate that the diagnostician considers the evidence to finally support or exclude a diagnosis as being insufficiently grounded. Therefore, we want to explore which insights about diagnostic argumentation can be gained from (1) analyzing co-occurrences solely within the dimension of content as well as from (2) analyzing co-occurrences solely within the dimension of epistemic activities and finally (3) from analyzing co-occurrences across both dimensions. We expect that the combination of both dimensions provides not only the insights of both separate analyses but might reveal some additional information about more implicit structures within diagnostic argumentation.

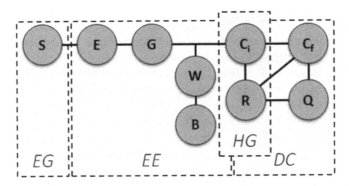

Fig. 1. Diagnostic Argumentation Framework. Abbreviations: S = Source of evidence. Adapted from Toulmin (2003): E = Evidence, G = Grounds, W = Warrant, B = Backing, C_i = initial conclusion, R = Rebuttal, Q = Qualifier, C_f = final conclusion. Adapted from Fischer et al. (2014): EG = generating evidence, EE = evaluating evidence, HG = generating hypotheses, DC = drawing conclusions.

2   Method

To explore the application of the framework we used data from a previous study. N = 118 pre-service teachers participated in a simulation-based training designed to foster their diagnostic reasoning and argumentation regarding school students' ADHD and dyslexia. The pre-service teachers were confronted with eight simulated cases about single school students displaying either behavioral problems or performance impairment specifically related to reading and writing. While working on the cases, they could generate pieces of evidence, for example by looking at reports of the school student's work and social behavior, reviewing exercises or tests, and by examining reports of conversations e.g. with the parents. The participants' final task was to indicate a diagnosis and formulate a diagnostic argumentation referring to their diagnostic strategy as well as evidences and diagnoses that they considered. We coded the resulting argumentations regarding two dimensions in two independent coding processes. In a first round, we coded the four diagnostic activities hypothesis generation (HG), evidence generation (EG), evidence evaluation (EE) and drawing conclusions (DC) (Fischer et al. 2014). For this purpose, fifteen percent of the argumentations were coded by four coders, resulting in an interrater reliability of Krippendorff's α_U = .65. In a second round, we coded the concept dimension, using a coding scheme that included 26 different diagnoses and 36 different categories of evidence. The coding manual offered specific coding recommendations for each of the eight cases on both sub-dimensions diagnoses and evidences. We coded fifteen percent of the argumentations per case with two coders each. Across the cases and sub-dimensions we reached interrater reliabilities ranging from Krippendorff's α_U = .73 to α_U = 1.00. Because of the case-specificity of the concept coding, we focus on the data of one training case for the current purpose. The third training case of the fictitious school student Anna displaying symptoms of ADD was considered as representative for the data because of its medium difficulty. The case-specific interrater reliability relating to the content dimension was Krippendorff's α_U = 1.00 for the diagnoses and Krippendorff's α_U = .81 for the evidences. For the current analysis, we accumulated the single categories by the correctness with respect to the principal case solution. This resulted in four sub-dimensions labeled with the codes "correct diagnoses" (diaC), "false diagnoses" (diaF), "correct evidences" (eviC) and "false evidences" (eviF).

To explore co-occurrences of codes within pre-service teachers' coded diagnostic argumentations of the training case Anna, we applied the method of Epistemic Network Analysis (Shaffer 2017). The ENA algorithm operationalizes sequences of codes by a moving window, which we defined as three lines, to construct a network model that shows the connections between codes. It further aggregates the resulting networks using a binary summation that reflects the presence or absence of co-occurrences of each pair of codes. The resulting networks are visualized using network graphs, where nodes correspond to codes and edges reflect the relative frequency of co-occurrence between each pair of codes. This results in two representations for each diagnostic argumentation: a plotted point in the low-dimensional projected space, plus a weighted network graph for every single argumentation. To make initial attempts in exploring the potential value of the combined dimensions concepts and epistemic activities suggested by the framework, we performed the ENA three times: (1) once only including the content dimension of the four codes correct and false evidences and diagnoses; (2) once only including the four epistemic activities hypothesis generation, evidence generation, evidence evaluation and drawing conclusions; (3) and once including all of these eight codes. To better examine the resulting networks, we created three plots for every of the tree analyses: The overall network across all participants' diagnostic argumentations for the respective case; one exemplary diagnostic argumentation formulated by the participant with the user ID 76N8058 who's argumentation quality was considered as high although the final diagnostic conclusion was false (see Table 1); and a comparison plot subtracting the overall network and the network of user 76N8058 indicating their discrepancies.

Table 1. Diagnostic argumentation formulated by the participant with the user ID 76N8058. Translated from German to English. Abbreviations: eviC = correct evidences, eviF = false evidences, diaC = correct diagnoses, diaF = false diagnoses, HG = hypothesis generation, EG = evidence generation, EE = evidence evalua-tion, DC = drawing conclusions.

	HG	EG	EE	DC	diaC	diaF	eviC	eviF
First, I observe Anna during class and initially I considered it to be only a temper, but over a longer period it became clear to me that she had a problem	0	1	1	0	0	0	0	0
For this reason, I had a look on her performance records, which corresponded to her general attention and her German grades	0	1	1	0	0	0	0	0
Then I talked to other subject teachers about Anna's problems	0	0	1	0	0	0	0	0
That made me reject my initial suspicion that she might have dyslexia	0	0	0	1	1	0	0	0
Eventually, she might have ADD, so I observe her during school recess, particularly her interaction with other kids, and talk to her mother	1	1	0	0	1	0	0	0
Her behavior during recess doesn't match the pattern of ADD, however, I took notice	0	0	1	0	0	1	0	0

(continued)

Table 1. (*continued*)

	HG	EG	EE	DC	diaC	diaF	eviC	eviF
of the problematic family environment								
Perhaps, the financial worries and absence of her mother are causal for her performance drop and her flight into her daydreaming and drawing	1	0	0	0	0	0	0	1
As a next step, I would consult a psychologist who might talk to Anna as well and check if my suspicion was correct	0	1	0	0	0	0	0	0
Besides, I make him aware of her seemingly depressive characteristics that he might keep an eye on	0	0	1	0	1	0	0	0

3   Results

Figure 2 shows the resulting plots of the first analysis solely including the content dimension. In Fig. 2a we see that the two strongest lines connect the codes correct and false evidences as well as correct evidences and correct diagnoses. Accordingly, in the context of three lines, correct evidences mostly co-occurred with false evidences and correct diagnoses. These co-occurrences represent firstly the selection and integration of different pieces of evidence in a grounding and secondly the grounding of correct diagnostic conclusions by acceptable pieces of evidence. The second strongest line is depicted between correct evidences and false diagnoses, which represents again the grounding of conclusions. In this case, the line can either indicate that false diagnostic conclusions were partially supported by correct evidences; however, it is more likely that the line represents the correct exclusion of false diagnoses grounded on correct evidences. The argumentation network of user 76N8058 shown in Fig. 2b presents a very different pattern of a triangle relating false diagnoses, false evidences and correct diagnoses. The network omits the code of correct evidences. Therefore, the user only related false evidences to correct and incorrect diagnoses. Integrating only the false evidences in a ground, the user most likely made a logically acceptable and yet false conclusion by accepting false diagnoses and excluding correct diagnoses.

The results of the second analysis solely considering the dimension of epistemic activities are shown in Fig. 3. In the overall network across all users (3a) we see that the thickest line relates the activities evidence evaluation and drawing conclusions. This relation shows that the evaluations performed were mostly considered as sufficient to draw rather certain conclusions. The second strongest relation is depicted between

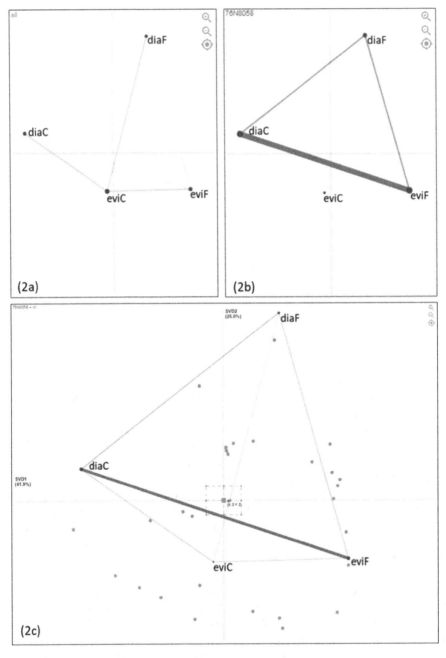

Fig. 2. Results of the first analysis including the content dimension only. Epistemic networks across all users (2a), one exemplary user (2b) and the difference between the networks (2c). Abbreviations: eviC = correct evidences, eviF = false evidences, diaC = correct diagnoses, diaF = false diagnoses.

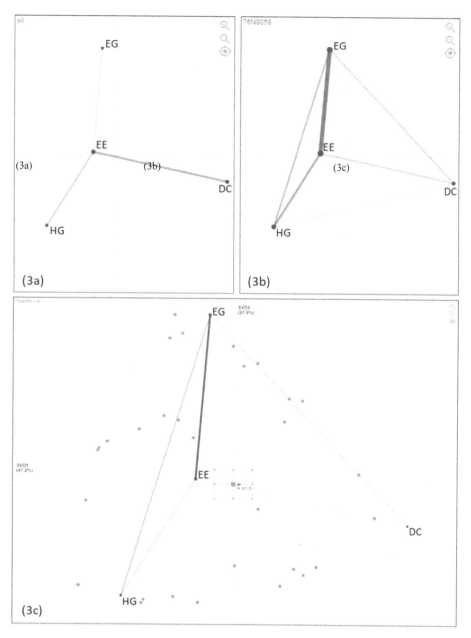

Fig. 3. Results of the second analysis including the epistemic activity dimension only. Epistemic networks across all users (3a), one exemplary user (3b) and the difference between the networks (3c). Abbreviations: HG = hypothesis generation, EG = evidence generation, EE = evidence evaluation, DC = drawing conclusions.

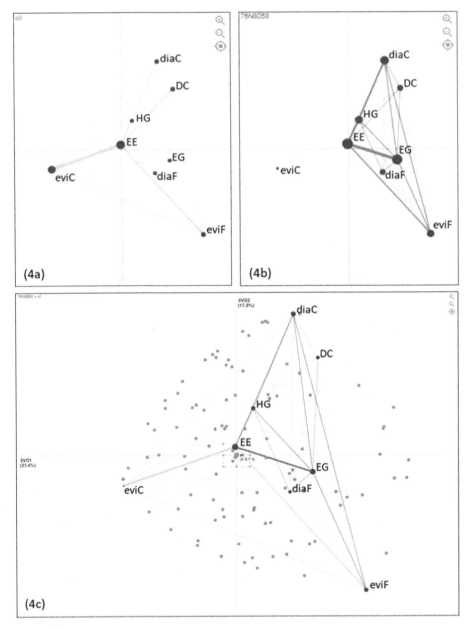

Fig. 4. Results of the third analysis including both dimensions. Epistemic networks across all users (4a), one exemplary user (4b) and the difference between the networks (4c). Abbreviations: eviC = correct evidences, eviF = false evidences, diaC = correct diagnoses, diaF = false diagnoses, HG = hypothesis generation, EG = evidence generation, EE = evidence evaluation, DC = drawing conclusions.

evidence evaluation and hypothesis generation. This line represents a higher degree of uncertainty in interpreting evaluated evidences. Looking at the argumentation network of user 76N8058 (3b) we see again some differences from the overall network. Above all, user 76N8058 rather refers to the activity of evidence generation compared to other users (3c). Hence, evidence generation and the other three epistemic activities are stronger related in the argumentation network of 76N8058. In addition, the link between hypothesis generation and evidence evaluation is slightly more apparent which indicates less certainty in concluding with respect to the evaluated evidences.

Figure 4 shows the resulting networks of the third analysis, which combines the two dimensions of content and epistemic activities. The resulting overall network (4a) shows the strongest line between evidence evaluation and correct evidences. This main line forms three triangles with false evidences, with drawing conclusions and with correct diagnoses. This shows that the majority of argumentations focus on the evaluation of the correct evidences but also evaluate false evidences partially with respect to correct evidences. These evaluations are used as grounding for the correct diagnoses, with more and less degrees of certainty. Comparing the overall network (4a) with the network of user 76N8058 (4b), one of the most obvious differences is the absence of the link between evidence evaluation and correct evidences; this is due to the general absence of the evaluation of the code for correct evidence which we also saw as a result from the first analysis including only the content dimension. All of the other codes co-occur within the diagnostic argumentation 76N8058. Apart from the absence of correct evidences, all lines from the overall network (4a) are also visible in the argumentation of user 76N8058 (4b). Nevertheless, the combination of both dimensions in this third analysis provides two additional insights in the user's diagnostic argumentation, particularly when considering the comparison plot of subtracted networks (4c). Firstly, we see that in argumentation 76N8058 (4b) both correct and false diagnoses rather co-occur with hypothesis generation as compared to other users (see 4c); at the same time the strength of the lines between diagnoses and the activity of drawing conclusions are comparable between network 76N8058 and the overall network (see 4c). Argumentation 76N8058 therefore expresses a comparably higher proportion of uncertainty, particularly with respect to the correct diagnoses but also regarding the false diagnoses. Secondly, we see a stronger link between all of the other codes (apart from correct evidences) with evidence generation. This represents a more transparent explanation of how the user generated the false evidences; moreover, the lines linking evidence generation with correct and false diagnoses show that user 76N8058 points to the necessity of generating further evidence as grounding for any final diagnostic conclusion.

4   Discussion

From the results of the first analysis solely including the content dimension we can draw conclusions about pre-service teachers' case-specific application of diagnostic concepts within diagnostic argumentation. Focusing on the content dimension allows comparing a diagnostician's performance to the correct answer or an expert solution. By applying the Toulmin model (2003) to the content dimension within diagnostic argumentation, we can progress beyond the correctness of a solution as a single indicator for diagnostic argumentative quality. In our first analysis we saw that the pre-service teacher with the user ID 76N8058 drew the false diagnostic conclusion probably because he or she missed central pieces of correct evidence. However, the false diagnostic conclusion was grounded on evidences. We argue to further focus on analyzing diagnostic argumentation in terms of quality criteria as for example the acceptability, relevancy and sufficiency of the grounds in supporting a conclusion (Toulmin 2003). These criteria may facilitate the analysis of single arguments' comprehensibility and conclusiveness, independently from the criteria of correctness.

The results of the second analysis solely considering the dimension of epistemic activities (see Fischer et al. 2014) in diagnostic argumentation presents insights into the explicated application of a diagnostic strategy. We see for example that pre-service teachers' single argumentations can differ largely in their report of evidence generation. However, reporting activities of evidence generation may indicate higher transparency regarding the quality of evidences for example in terms of appropriate application of methods or selection of sources of information (see Fischer et al. 2014). Moreover, analyzing epistemic activities in diagnostic argumentation provides information about different degrees of certainty in diagnosing. The exemplary analysis indicated that pre-service teachers differ in terms of the proportion to which they combine evaluative with hypothetical or concluding parts within their diagnostic argumentations. This result might even go beyond pointing to the sufficiency of the available evidences to draw a conclusion; the indication of uncertainty in solving a case might also partially reflect a subjective self-evaluation. We suggest to further research this question by manipulating variables like prior knowledge and complexity of the diagnostic task (see Charlin et al. 2007).

The results of the third analysis combining the two dimensions of content and epistemic activities capture the same insights as the previously discussed analyses referring to the two dimensions separately. Moreover, it combines the advantages of analyzing the two dimensions for example by indicating the certainty in referring to a correct or false diagnostic conclusion with respect to a sufficient grounding on correct or false evidences. Co-occurrences of certain epistemic activities with certain diagnostic concepts may also be able to indicate rather implicit assumptions within diagnostic argumentations. One example is the argumentative relation between diagnoses and the activity of evidence generation. This refers to the prior or further necessity of generating evidence as grounding for a final diagnostic conclusion. Consequently, the evaluated evidence was considered as insufficient for drawing a diagnostic conclusion. This reflects an implicit evaluation of certain evidences that might not necessarily be explicitly stated.

It has to be noted that the validity of the codes relating to the content dimension is very limited due to the simplifying reduction resulting in the four codes of case-specifically correct and false evidences and diagnoses. We precede our analyses by stronger differentiating codes within the content dimension. These could be additionally grouped in terms of other characteristics as for example their significance for diagnosing the respective case (see Charlin et al. 2007). It might also be interesting to add a code for the "source of evidence" (see Nicolaidou et al. 2011) as a corresponding content-related code for the activity of evidence generation.

Concluding from the previous considerations, we consider the presented initial applications of ENA to analyze diagnostic argumentations referring to the proposed framework (Fig. 1) as a promising approach to better understand diagnostic argumentations of pre-service teachers and eventually other diagnosticians.

References

Berland, L.K., Reiser, B.J.: Making sense of argumentation and explanation. Sci. Educ. **93**(1), 26–55 (2009)

Charlin, B., Boshuizen, H., Custers, E.J., Feltovich, P.J.: Scripts and clinical reasoning. Med. Educ. **41**(12), 1178–1184 (2007)

Coderre, S., Mandin, H., Harasym, P.H., Fick, G.H.: Diagnostic reasoning strategies and diagnostic success. Med. Educ. **37**(8), 695–703 (2003)

Csanadi, A., Eagan, B., Kollar, I., Shaffer, D.W., Fischer, F.: When coding-and-counting is not enough: using epistemic network analysis (ENA) to analyze verbal data in CSCL research. Int. J. Comput.-Support. Collaborative Learn. **13**(4), 419–438 (2018)

Fischer, F., et al.: Scientific reasoning and argumentation: advancing an interdisciplinary research agenda in education. Frontline Learn. Res. **2**(3), 28–45 (2014)

Gartmeier, M., et al.: Fostering professional communication skills of future physicians and teachers: effects of e-learning with video cases and role-play. Instr. Sci. **43**(4), 443–462 (2015)

Hetmanek, A., Engelmann, K., Opitz, A., Fischer, F.: Beyond intelligence and domain knowledge. Scientific reasoning and argumentation as a set of cross-domain skills. In: Fischer, F., Chinn, C.A., Engelmann, K., Osborne, J. (eds.) Scientific Reasoning and Argumentation: The Roles of Domain-Specific and Domain-General Knowledge, pp. 203–226. Routledge, New York (2018)

Lawson, A.E., Daniel, E.S.: Inferences of clinical diagnostic reasoning and diagnostic error. J. Biomed. Inform. **44**(3), 402–412 (2011)

Nicolaidou, I., Kyza, E.A., Terzian, F., Hadjichambis, A., Kafouris, D.: A framework for scaffolding students' assessment of the credibility of evidence. J. Res. Sci. Teach. **48**(7), 711–744 (2011)

Schulz, C., et al.: Challenges in the Automatic Analysis of Students' Diagnostic Reasoning. arXiv preprint arXiv:1811.10550 (2018)

Shaffer, D. W.: Quantitative ethnography. Cathcart Press, Madison (2017)

Toulmin, S.E.: The Uses of Argument. Cambridge University Press, Cambridge (2003)

The Multimodal Matrix as a Quantitative Ethnography Methodology

Simon Buckingham Shum[1(✉)], Vanessa Echeverria[1,2],
and Roberto Martinez-Maldonado[1]

[1] Connected Intelligence Centre, University of Technology Sydney,
PO Box 123, Broadway, Sydney, NSW 2007, Australia
{simon.buckinghamshum, vanessa.echeverria,
roberto.martinez-maldonado}@uts.edu.au
[2] Escuela Superior Politécnica del Litoral, ESPOL,
Campus Gustavo Galindo Km 30.5 Vía Perimetral, Guayaquil, Ecuador

Abstract. This paper seeks to contribute to the emerging field of Quantitative Ethnography (QE) by demonstrating its utility to solve a complex challenge in Learning Analytics: the provision of timely feedback to collocated teams and their coaches. We define two requirements that extend the QE concept in order to operationalise it such a design process, namely, the use of *co-design methodologies*, and the availability of *automated analytics workflow* to close the feedback loop. We introduce the *Multimodal Matrix* as a data modelling approach that can integrate theoretical concepts about teamwork with contextual insights about specific work practices, enabling the analyst to map between higher order codes and low-level sensor data, with the option add the results of manually performed analyses. This is implemented in software as a workflow for rapid data modelling, analysis and interactive visualisation, demonstrated in the context of nursing teamwork simulations. We propose that this exemplifies how a QE methodology can underpin collocated activity analytics, at scale, with in-principle applications to embodied, collocated activities beyond our case study.

Keywords: Multimodal · Learning analytics · Teamwork · CSCL · Sense making

1 Introduction and Background

Quantitative Ethnography (QE) is a methodological approach that respects the insights into specific cultural practices gained from the interpretive disciplines developed in ethnographic and other qualitative traditions, but seeks to apply the power of statistical and other data science techniques to qualitatively coded data, such as observational fieldnotes, interviews, or video analysis [36]. To date QE has been exemplified primarily by the development of the Epistemic Network Analysis (ENA) tool, to examine the relationships between coded data elements (e.g. [9, 16, 35]).

In this paper, we propose that a modelling methodology and analytics workflow, developed to process and visualise multimodal data from nursing team simulations,

© Springer Nature Switzerland AG 2019
B. Eagan et al. (Eds.): ICQE 2019, CCIS 1112, pp. 26–40, 2019.
https://doi.org/10.1007/978-3-030-33232-7_3

also exemplifies QE principles. The intended contribution is thus twofold: (i) an articulation of two key requirements in order to operationalise QE for a learning analytics application, and (ii) a demonstration of QE's value in implementing a multimodal learning analytics pipeline and end-user tool, which has been piloted with nursing academics and students. While Herder et al. [19] have described the full automation of ENA from *online* student teams' interaction data, generating a real-time dashboard for teachers, to our knowledge, this is the first time that QE has been applied to the analysis of multimodal sensor data, combined with human observational data, to generate a feedback visualisation about *collocated* teamwork.

In the next section we introduce the challenge of ascribing meaning to multimodal traces of learning activity. Section 3 introduces the specific educational challenge driving this work, namely, how to better inform the debriefings for collocated nursing teams on a simulated hospital ward. Section 4 provides a definition of our requirements in order to operationalise QE for such a learning analytics application. Section 5 introduces the *Multimodal Matrix*, inspired by QE principles in combination with a teamwork activity theory, illustrating its application to the nursing context, and briefly describes the visualisations it enables, informed by co-design. Section 6 summarises progress to date, and outlines some future work trajectories.

2   From Clicks and Data Streams, to Constructs

An important challenge in the fields of learning analytics (LA) and educational data mining (EDM) is to provide more timely, useful evidence to aid teachers in pedagogical decision-making, and students in understanding what actions they can take to maximise their opportunities of learning, e.g. [2, 15]. While there has been substantial interest in creating visualisations and dashboards to communicate data to different stakeholders, recent reviews question the impact of these learning analytics interfaces [4, 20, 25]. Clearly, at the user interface level, the representations need to be intelligible to users [1] but the problems in fact go deeper into the infrastructure [17]. On what basis can low-level system logs serve as proxies for higher order constructs? Once that relationship has been established, data can then be rendered in ways that are intelligible to people without a strong analytical background. Imbuing data with contextual meaning brings key stakeholders (such as teachers and students) into the sensemaking loop, whereas until quite recently, analytics for human activity remained the preserve of researchers.

This is fundamentally a modelling problem, long recognised in assessment science, and now manifesting in LA and EDM in various forms. In assessment science for technology-enabled learning such as educational gaming, we find techniques such as Stealth Assessment [37] and Evidence-Centred Design [3, 26]. In an LA context, this challenge has been dubbed as mapping "from clicks to constructs" [6, 39].

One particular strand of research within EDM focuses on adding meaning to student data *before* doing any data processing, through the use of *alphabets* to encode sequences of logged interactions. One example of this was presented by Perera et al. [30], who use alphabets as rules to encode low-level events from an online system into items representing a higher level of abstraction. This is performed as a pre-processing step before conducting sequence pattern mining to facilitate the interpretability of the results. A similar methodological approach was suggested by Martinez-Maldonado et al. [22] who used different alphabets to interrogate a multimodal educational dataset (speech acts and actions on a shared device by groups of students) to find sequential patterns that differentiated cohorts of students. The limitation reported by authors in these studies is that an item can only contain a certain amount of contextual information. The more that information is associated with a particular logged action, the more complex the task to mine useful partners becomes.

The challenge of developing multimodal learning analytics can be seen as an extreme case in which the aim is to capture rich contextual information in a learning situation [28]. These innovations offer exciting opportunities for educational research and practice through the analysis of multiple, intertwined streams of learner data (e.g. related to gestures, physical positioning, gaze, speech and physical manipulation of objects). Consequently, the user interfaces resulting from these multimodal innovations can be even more complex than regular LA systems [33] as they may pose serious challenges for teachers and students in terms of sense making of multiple sources of evidence. In fact, a recent review [10] identified that most current multimodal learning analytics tools are aimed at helping *researchers* to annotate multiple data streams to identify patterns of meaningful learning constructs. The current multimodal analytics prototypes, presented by Echeverria et al. [12, 13] and Ochoa et al. [29], are initial attempts to map low-level data with higher order constructs in the user interface. Worsley [40] proposed the use of the concept of epistemological framing [34] as a potential way for understanding human cognition through multimodal data. In this case, the epistemic frames can serve to typify of certain high-level activity (e.g. a person is discussing) based on a combination of low level behaviours that can potentially be detected via sensors (e.g. prolific gestures, an up straight posture, gaze at peers and animated talk and facial expressions).

In sum, there is a growing interest in mapping from low-level data to meaningful constructs, and related attempts have been made in the areas of EDM, data-driven assessment and multimodal LA. In the next section we introduce the applied challenge that has motivated the need to integrate theoretical concepts about teamwork with contextual insights about specific work practices, to enable the mapping of low-level sensor data with codes that could be used to operationalise meaningful constructs.

3 Timely Feedback on Nursing Teamwork Simulations

Nursing simulations play an important role in the development of teamwork, critical thinking and clinical skills and prepare nurses for real-world scenarios. Students from the UTS Bachelor of Nursing experience many hypothetical scenarios across different stages of their professional development. In these scenarios students, acting as Registered Nurses (RNs), provide care to a patient, who has been diagnosed with a specific condition. Manikins, ranging from newborn to adult, give students the opportunity to practise skills before implementing them in real life. Simulations are sometimes recorded and played back to students so that strengths and areas for improvement can be observed in facilitated debriefing sessions [18].

The manikin ("Mr. Lars") was programmed by the teacher to deteriorate over time, dividing the task into two phases. In phase one a group of four students assess and treat Mr Lars for chest pain. These RNs in different roles communicate with Mr. Lars, apply oxygen, assess his pain, perform vital sign observations, administer Anginine according to the six rights, connect him to an ECG, identify his cardiac rhythm, document appropriately and call for a clinical review. In phase two, the same group of students takes over Mr. Lars's care at which point he loses consciousness due to a fatal cardiac rhythm, and the team must perform basic life support. Each simulation lasted an average of 9.5 min. Fuller procedural details are provided in [12, 14].

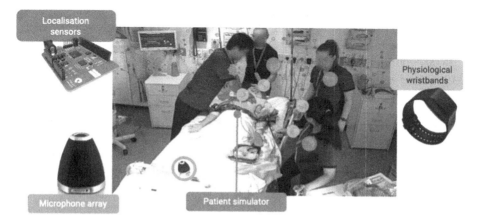

Fig. 1. Data collection from a nursing simulation in a lab scenario, using a range of sensors.

3.1 Instrumenting Simulations to Detect Teamwork

Several sensors and equipment were utilised to track interactions, summarised below:

Indoor Localisation: Students' movement around the manikin was captured auto-matically through ultra-wideband (UWB) wearable badges (Pozyx.io[1]). This system is composed of a set of anchors to sense the physical space, which are mounted on the walls, and several wearable tags or badges attached to people or objects (such as the resus trolley). Figure 1 illustrates the distribution of the anchors across the simulation room (blue squares).

Patient Simulator: Some student and patient actions were automatically logged by the high-fidelity Laerdal SimMan 3G[2] manikin including placing the oxygen mask, setting oxygen level, attaching blood pressure monitor, reading blood pressure, administering medicine, attaching the ECG device, starting CPR, and stopping CPR. Proprietary Laerdal Software exported the actions and their timestamp in a .txt file.

---

[1] Pozyx developer kit and a multitag-positioning system: https://www.pozyx.io.

[2] Laerdal simulation manikins: https://www.laerdal.com/nz/products/simulation-training/emergency-care-trauma/simman-3g.

Microphone Array: A six-channel high-quality USB microphone array (Microcone) was located at the base of the patient's bed to detect nurses' conversations. Microcone Recorder application for MacOS was used to automatically track multiple people speech. Six .wav files were saved at the end of each session, one per channel. In addition, the application generated a .csv file including the total duration of the session, start and end timestamps where speech was detected and the person who was speaking (previously configured in the application).

Physiological Wristbands: Empatica E4[3] wristbands included a photo plethysmography (PPG) sensor to measure Heart Rate continuously, an electrodermal activity (EDA) sensor to measure skin conductance, a 3-axis accelerometer to detect movement and activity, and an optical thermometer to sense physical activity. Each wristband exported an EDA.csv file containing the timestamp when the Empatica started to capture data and EDA values; and an ACC.csv file with x, y and z accelerometer values.

In addition to these sensors and equipment, all the sessions were recorded by the video camera system installed in the lab room, comprising three fixed cameras and several microphones in the ceiling.

Two researchers and a teacher were present in each session. Besides the data outlined above, other data gathering included observation notes and recordings of the group debriefing. These were transcribed for analysis. Data analysis involved two researchers independently screening the video recordings of the sessions looking for moments of interest that could serve to derive multimodal observations for further analysis. More details on the context and instrumentation are provided elsewhere [12].

4   Defining Requirements for QE Enabled Feedback

In his presentation of the QE concept, Shaffer [36] set the goal of designing ways to model and analyse data that harmonise qualitative and quantitative methodologies. Clearly there are many facets to the QE concept, but in our reading, of particular importance is the requirement that *all analysis techniques can read from, and write to, a common data representation.* This emphasis seems to us to be distinctive, clearly moving beyond mixed methods, and pivotal to enabling ethnography, and the social sciences more broadly, to move into data science and real time analytics. Building on this, we introduce two additional requirements (points 1 and 4 below) to specify the design process and enabling infrastructure required to deliver timely, analytics-driven feedback:

Using QE to inform the design of timely, analytics requires:

1. *Co-design with stakeholders in order to gain insights into current and envisioned work practices.*
2. These insights inform the modelling and analysis of qualitative and quantitative data.
3. Analysis techniques can read from, and write to, a common data representation.
4. *This is executed by an automated analytics workflow.*

---

[3] Empatica wristbands: https://www.empatica.com/en-int/research/e4.

We emphasise *co-design* because we are developing feedback tools for use by real users (educators and students), so simply from sound user-centred design principles, this is good practice: we need to understand what feedback will be of most value. Moreover, human-centred design goes deeper than good user interfaces: it shapes the data we gather, and how we model it. The analytics challenge requires us to devise a way to model the sensor data in ways that respect, and will enable, culturally meaningful interpretations of work practices when visualized. Co-design provides us with a way to understand work practices in great specificity.

Secondly, we emphasise *automated analytics workflow* because this is the only way to make sense of large data sets sufficiently quickly to serve our purposes, namely, to close the feedback loop to educators and students in a timely manner for post-simulation debriefings. We turn now to the question of how we bring the insights from co-design to the modelling of multimodal data, and automate visualisation generation to inform the team debriefing.

5   The Multimodal Matrix as a QE Modelling Methodology

In order to address the above requirements, we have developed a modelling approach and data representation named the *Multimodal Matrix* (Fig. 2), comprising the following conceptual elements: *dimensions of collaboration, multimodal observations, segments,* and *stanzas,* which are elaborated in subsequent sections.

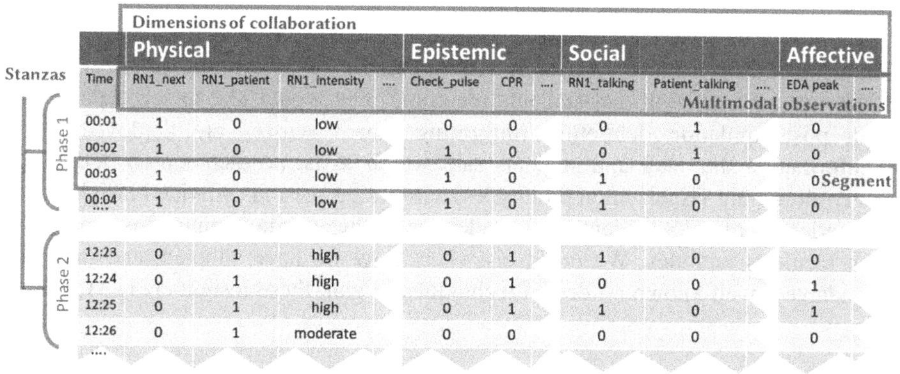

Stanzas	Time	Physical				Epistemic			Social			Affective	
		RN1_next	RN1_patient	RN1_intensity	....	Check_pulse	CPR	....	RN1_talking	Patient_talking	....	EDA peak	....
Phase 1	00:01	1	0	low		0	0		0	1		0	
	00:02	1	0	low		1	0		0	1		0	
	00:03	1	0	low		1	0		1	0		0Segment	
	00:04	1	0	low		1	0		1	0		0	
Phase 2	12:23	0	1	high		0	1		1	0		0	
	12:24	0	1	high		0	1		0	0		1	
	12:25	0	1	high		0	1		1	0		1	
	12:26	0	1	moderate		0	0		0	0		0	
	....												

Column header note: "Dimensions of collaboration"; "Multimodal observations"

Fig. 2. Schematic design of the Multimodal Matrix, from [12]

This matrix provides the common, integrating representation to hold data and analysis results from qualitative and quantitative methods: it can be populated with categorical data *automatically* from a full sensor/analytics pipeline, *semi-automatically* in which human input augments the analysis and/or workflow, or manually from conventional qualitative or quantitative data analysis. Qualitative codes are modelled by combining events from multiple sources (columns) into *segments*, and by combining multiple segments. Temporally dependent codes can be modelled into meaningful *stanzas* by combining *segments* (rows of events).

The matrix thus seeks to provide a representation to help make sense of low-level sensor data through the introduction of qualitative coding derived from top-down (theory) and bottom-up (context-specific phenomena) sources:

Theory: A framework for collaborative activity (ACAD) was used to define key constructs for combining lower level events into higher order codes (see below). Obviously, this could be replaced by any other theory/framework that served the analyst's interests and stakeholders' needs. The Multimodal Matrix enables us to introduce theoretical perspectives. We draw on the Activity-Centred Analysis & Design (ACAD) framework [24], which defines *physical, epistemic* and *social* dimensions as critical, as follows (p.2065):

- *"the set (physical) component* – which includes the place in which participants' activity unfolds, the physical and digital space and objects; the input devices, screens, software, material tools, awareness tools, artefacts, and other resources that need to be available
- *the social component* – which includes the variety of ways in which people might be grouped together (e.g. dyads, trios, groups); scripted roles, divisions of labour, etc.
- *the epistemic component* – which includes both implicit and explicit knowledge-oriented elements that shape the participants' tasks and working methods."

To these three dimensions we add *affective* states of engagement, worry or anticipation, this being particularly important in the healthcare professions. It will be seen below how these broad categories help to make sense of the data.

Insights into Work Practices: Multiple sources: (i) insights from nursing professionals about what makes a nurse's position meaningful when performing different tasks; (ii) information from staff and students regarding they would like to see captured to inform post-simulation debriefing (informing which sensors are deployed); and (iii) information that staff and students said would assist post-simulation debriefing (from co-designing visualization prototypes). Co-design sessions provided insights into the experiences of UTS students and academics in the specific simulations run in the Health faculty's facilities. We detail elsewhere how we have adopted, and in some cases adapted, well known co-design techniques to gain these insights [11, 31, 32].

5.1   Application of the Multimodal Matrix to Nursing Team Simulations

Each data stream captured by the sensors and devices listed above was encoded into columns in the multimodal matrix based on meaning elicited from subject matter experts, the learning design, or literature. The data streams were manually synchronised at 1 Hz, down-sampling data streams from sensors that had a higher frequency. The multimodal observations used in our studies, and their relationship with the dimensions of collaboration, are depicted in the edited excerpt from a nursing teamwork simulation in Table 1.

Table 1. Edited excerpt from a nursing teamwork simulation encoded in the Multimodal Matrix

time	Physical						Epistemic			Social						Affective	
	RN1.patient_bed	RN1.next_to_patient	RN1.around_patient	RN1.bed_head	RN1.trolley_area	RN1.physical_intensity	RN1.check_pulse	RN2.check_pulse	RN1.compressions	RN1.speaking	RN2.speaking	patient.speaking	RN1.listening	RN2.listening	patient.listening	RN1.EDA_peak	RN2.EDA_peak
3:22.0	0	1	0	0	0	L	0	0	0	0	0	1	1	1	0	0	0
3:22.1	0	1	0	0	0	L	0	0	0	0	0	1	1	1	0	1	0
3:22.2	0	1	0	0	0	L	0	0	0	0	0	1	1	1	0	0	0
3:22.3	0	1	0	0	0	L	0	0	0	0	0	1	1	1	0	0	0
3:22.4	0	1	0	0	0	L	0	0	0	0	0	1	1	1	0	0	0
3:22.5	0	1	0	0	0	L	0	1	0	0	0	0	0	0	0	0	0

5.2   Multimodal Observations

From the data collected, we were now in a position to associate multimodal observations, optionally in combination, with one or more dimensions of collaboration. Space precludes a very detailed description, but our goal in this paper is to convey the way in which the coding of data works.

Segments: Segments are considered the smallest unit of meaning. Thus, for this particular example in teamwork nursing simulations, we took a segment of one second. This small value was selected because we needed to analyse moment-to-moment critical reactions from nurses during the performance of the activity, this being a high-stakes activity.

Stanzas: Segments can be grouped according to criteria to show meaningful relationships. In the nursing simulations, stanzas were defined to capture key phases in the collaborative task (e.g. see rows grouped by phase in Fig. 2). For this particular example, two stanzas were defined, based on two critical actions in the learning design: (i) when the patient asks for help and (ii) when the patient loses consciousness.

The major column headings in Table 1, drawn from the ACAD framework, are described next.

ACAD Physical Dimension. Embodied strategies during high-stakes teamwork scenarios are critical in healthcare education [23]. This provides an example of how qualitative insights into the work practice shape the quantitative modelling: *what makes position meaningful,* for *these stakeholders,* in *this simulation?* In other simulations, position might take on other significances, or with more advanced students (for instance) there might be other learning outcomes, which will focus on other key behaviours. Based on interviews with four nursing teachers [14], we identified five meaningful zones which are associated with a range of actions nurses must perform: (i) *the patient's bed*, for cases in which nurses were located on top of or very close to the patient; (ii) *next to patient*, for cases in which nurses were at either side of the bed; (iii) *around the patient*, for cases in which nurses were further away from the bed, from 1.5 to 3 m away of the bed); (iv) *bed head*; which is an area where a nurse commonly stands to clear the airway during CPR; and (v) *trolley area*, for cases in which nurses were getting medication or equipment (a localisation badge was attached to the trolley). Indoor localisation data was automatically encoded into these meaningful zones (Fig. 3).

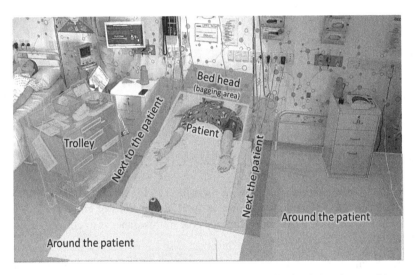

Fig. 3. Nurses' positions were classified into zones, reflecting insights from subject matter experts regarding what makes position significant in teamwork [12]

A Kalman filter was applied to remove noisy data points, and a cluster analysis was performed (k = 16) to assign one meaningful zone to each point. The first five columns in the Physical dimension group from Table 1 illustrate the meaningful zones for RN1 (e.g. RN1.patient_bed). Each cell has a value of "1" if that zone is occupied, or "0" otherwise. For instance, the row in the first second [0,1,0,0,0] means that RN1 was next to the patient. In addition to movement, nurses' *physical intensity* is studied in the literature [8], ranging from low (e.g. walking, talking, manipulating medical tools) to high (e.g. performing a CPR). We defined *low (L)*, *medium (M)* and *high (H)* levels, where high = performing CPR. The last *Physical* column shows that RN1's physical intensity at 03:22.0 was low (L).

ACAD Epistemic Dimension. In the matrix, each column represents who performed an action (e.g. RN1.check_pulse). For example, the first column in the Epistemic dimension group is RN1.check_pulse = 0, meaning that RN1 did not check the pulse at time 03:22.0, while RN2.check_pulse = 1 at 03:22.5.

ACAD Social Dimension. Verbal communication plays an important role in the management and coordination of patient care and teamwork strategies [41]. From the video recordings, we manually transcribed and synchronised the speech for each nurse and the patient using NVivo software. Start and ending points were annotated, along with the speaker and listener identification to further model the interactions. With this information, we created a sparse matrix, with 1's when a nurse was speaking/listening at a specific time, or 0 otherwise. The first column in the Social dimension group shows RN1.speaking, RN2.speaking and patient.speaking, and the following three columns are listening interactions. We can observe how the patient-nurse speaking-listening interaction is represented for the first four seconds: row [0,0,1]; [1,1,0] means that the patient is speaking while RN1 and RN2 are listening.

Affective Dimension. Physiological data can be effectively used to aid nurses in recalling confronting experiences in order to develop coping strategies [27]. An increase in EDA, specifically, is typically associated with changes in arousal states, commonly influenced by changes in emotions, stress, cognitive load or environmental stimuli. We automatically identified peaks in EDA data as a minimum increase of 0.03 μs [5], using EDA Explorer [38]. Each cell contains a value of "1" when a peak in that timeframe was detected. For example, the RN1.EDA_peak column shows that RN1 had an EDA peak at 03:22.0.

We should emphasise that the primary classification of an atomic event under one of the four column headings should not be read as a rigid constraint. For instance, common sense tells us that "social" actions may also have physical attributes. An arbitrary number of codes can be defined as combinations of any events, constrained only by the analyst's interests.

Judgements about whether an attribute is, for instance, "High/Medium/Low" are again, modelling decisions, which may be based on theory, evidence or intuition (see the above example rationale for setting activity peak thresholds for EDA data). Binary 0/1 variables are required, but apply only for the specified duration (row labels). For instance, if higher definition tracking of certain columns is required than in the above example, one could sample every 0.1 s to detect changes.

To summarise, the Multimodal Matrix provides a 'container' to make modelling decisions explicit. The merits of those decisions derive from the integrity of the underlying assumptions about the structure of the phenomena, and how they can be modelled. In the concluding section, we reflect on the commitments that this approach is itself requiring on behalf of the analyst.

5.3   Generation of Visual Feedback

We are now ready to return to the applied outcome, namely, to improve learning by providing better feedback to the nursing teams and teachers. Thus, the final step is to generate representations that offer insights. Figure 4 shows examples of the visualisations that are now being evaluated with the academics and students, informed by the

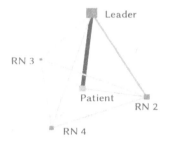

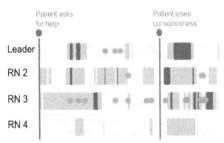

Patient-centred verbal communication within the nursing team as a sociogram

EDA peaks (orange) potentially signifying stress or other forms of cognitive arousal

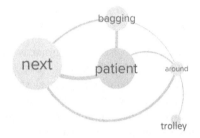

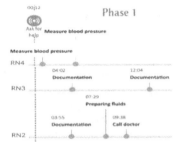

Patient-centred movement of nurses around the zones of the simulation ward

Clinical actions performed by nurses on the manikin patient and other devices

Fig. 4. Example visualisations generated from the multimodal teamwork activity data

initial co-design sessions. See [12] for a fuller design rationale behind these, and the results of user trials.

6 Summary and Future Work

In this paper, we have proposed a Quantitative Ethnographic methodology and analytics platform, to tackle the problem of making multimodal data streams meaningful for interpreting collocated teamwork. We accomplish this using a data representation called the Multimodal Matrix, which organises quantitative data in relation to codes derived from two sources: qualitative insights from stakeholders into their work practices (i.e. undertaking and coaching nursing simulations), plus theoretical insights into how collaborative teamwork can be analysed (ACAD framework). Rows in the matrix may be populated automatically or semi-automatically, and additionally with results from manual data analyses. This modelling approach thus enables us to build a principled bridge between multimodal logs generated from sensors, and higher order constructs such as a curriculum outcome (e.g. *'competence in patient-centred teamwork'*), and its constituent skills (e.g. *'patient-centred talk'; 'correct positioning during CPR'*). When those signals are combined meaningfully, they may serve as proxies for these competencies, and once visualized, may provoke deeper reflection and discussion in debriefing sessions.

To deliver real-time, scalable feedback to nursing teams, we have highlighted the need not only for an integrative representation, but also *co-design processes* that provide stakeholders with voices to shape the design, and an *automated workflow* capable of generating feedback representations sufficiently quickly. The analytics workflow is now sufficiently automated (there remain a few manual elements) to generate visualizations for student debriefs within a few minutes of completing a simulation, and the Multimodal Matrix has provided the database to generate visualizations to meet stakeholders' requests for debriefing support, which are now being evaluated with staff and students, with promising results [12].

Fuller workflow automation must overcome challenges such as identifying who is performing an action and connecting siloed data sources from different products (e.g. manikin and wristband data) before the scripts can integrate and visualize them. Although not the focus of our work to date, attempts to automate other forms of qualitative analysis—e.g. attending to what, how or when utterances are made—does of course present complex challenges, requiring different kinds of textual content analysis, sequence analysis [22], location analysis [21], speech content analysis [7], and so forth, which are the focus of multimodal learning analytics research. Analyses of how space is used can draw additionally on knowledge of team roles, and hence power. For instance, knowing that a nurse is the team leader leads one to expect different activity from that role at a critical moment. We are interested to learn of other examples in which positional data might be combined with other data classes in order to quantify phenomena that qualitative analyses have concluded are significant.

We believe that this is the first application of a QE approach for *collocated teamwork*, working from multimodal data streams rather than clickstreams. The principles underpinning Quantitative Ethnography, codified in the Multimodal Matrix, have assisted us in making significant advances in tackling an extremely complex challenge: delivering timely analyses and feedback on embodied groupwork.

To conclude, we would want to reflect critically on the Multimodal Matrix as a lens. It is agnostic both theoretically and empirically, in the sense that it can be used to structure data about any form of temporal activity, with columns labelled according to any constructs of interest, and myriad visualisations can be generated from this. On that basis, we envisage that it should be of service for analysing diverse forms of human activity. That being said, no symbol system is completely neutral, but always privileges certain information, and ignores others. All symbol systems require the structuring of thought and data in particular ways: writing has different affordances to oral communication, visual representations vary in their perceptual affordances from each other, multimedia adds new dimensions, and so forth. It is possible that the tabular representation used in the Multimodal Matrix, in which rows represent time windows, and columns discrete streams of data, may prove problematic for certain forms of ethnographic or other qualitative analysis. Since we have highlighted that a hallmark of QE is the ability for quantitative and qualitative methods to read and write a common representation, we welcome commentary on whether this representation is indeed sufficiently expressive to meet the needs of other QE approaches.

References

1. Alhadad, S.S.J.: Visualizing data to support judgement, inference, and decision making in learning analytics: insights from cognitive psychology and visualization science. J. Learn. Anal. **5**(2), 60–85 (2018)
2. Bakhshinategh, B., Zaiane, O.R., Elatia, S., Ipperciel, D.: Educational data mining applications and tasks: a survey of the last 10 years. Educ. Inf. Technol. **23**(1), 537–553 (2018)
3. Behrens, J.T., Dicerbo, K.E., Foltz, P.W.: Assessment of complex performances in digital environments. Ann. Am. Acad. Polit. Soc. Sci. **683**(1), 217–232 (2019)
4. Bodily, R., Verbert, K.: Trends and issues in student-facing learning analytics reporting systems research. In: Proceedings of the International Learning Analytics and Knowledge Conference, LAK 2017, pp. 309–318 (2017)
5. Braithwaite, J.J., Watson, D.G., Jones, R., Rowe, M.: A guide for analysing electrodermal activity (EDA) & skin conductance responses (SCRs) for psychological experiments. Psychophysiology **49**(1), 1017–1034 (2013)
6. Buckingham Shum, S., Crick, R.D.: Learning analytics for 21st century competencies. J. Learn. Anal. **3**(2), 6–21 (2016)
7. Chandrasegaran, S., Bryan, C., Shidara, H., Chuang, T.-Y., Ma, K.-L.: TalkTraces: real-time capture and visualization of verbal content in meetings. In: Proceedings of the SIGCHI Conference on Human Factors in Computing Systems, CHI 2019, pp. 577:571–577:514 (2019)
8. Chappel, S.E., Verswijveren, S.J.J.M., Aisbett, B., Considine, J., Ridgers, N.D.: Nurses' occupational physical activity levels: a systematic review. Int. J. Nurs. Stud. **73**(August), 52–62 (2017)
9. Csanadi, A., Eagan, B., Kollar, I., Shaffer, D.W., Fischer, F.: When coding-and-counting is not enough: using epistemic network analysis (ENA) to analyze verbal data in CSCL research. Int. J. Comput.-Support. Collab. Learn. **13**(4), 419–438 (2018)
10. Di Mitri, D., Schneider, J., Klemke, R., Specht, M., Drachsler, H.: Read between the lines: an annotation tool for multimodal data for learning. In: Proceedings of the International Conference on Learning Analytics and Knowledge, LAK 2019, pp. 51–60 (2019)
11. Echeverria, V.: Designing and validating automated feed-back for collocated teams using multimodal learning analytics. Ph.D. in Learning Analytics, University of Technology Sydney (UTS), Sydney, Australia (in preparation)
12. Echeverria, V., Martinez-Maldonado, R., Buckingham Shum, S.: Towards collaboration translucence: giving meaning to multimodal group data. In: Proceedings of the SIGCHI Conference on Human Factors in Computing Systems, CHI 2019, p. 39 (2019)
13. Echeverria, V., Martinez-Maldonado, R., Chiluiza, K., Buckingham Shum, S.: DBCollab: automated feedback for face-to-face group database design. In: Proceedings of the International Conference on Computers in Education, ICCE 2017, pp. 156–165 (2017)
14. Echeverria, V., Martinez-Maldonado, R., Power, T., Hayes, C., Buckingham Shum, S.: Where is the nurse? Towards automatically visualising meaningful team movement in healthcare education. In: Penstein Rosé, C., et al. (eds.) AIED 2018. LNCS (LNAI), vol. 10948, pp. 74–78. Springer, Cham (2018). https://doi.org/10.1007/978-3-319-93846-2_14
15. Gašević, D., Dawson, S., Siemens, G.: Let's not forget: learning analytics are about learning. TechTrends **59**(1), 64–71 (2015)
16. Gašević, D., Joksimović, S., Eagan, B.R., Shaffer, D.W.: SENS: network analytics to combine social and cognitive perspectives of collaborative learning. Comput. Hum. Behav. **92**, 562–577 (2018)

17. Gibson, A., Martinez-Maldonado, R.: That dashboard looks nice, but what does it mean?: towards making meaning explicit in learning analytics design. In: Proceedings of the Australian Conference on Computer-Human Interaction, OzCHI 2017, pp. 528–532 (2017)
18. Green, A., Stawicki, S.P., Firstenberg, M.S.: Medical error and associated harm-the the critical role of team communication and coordination. In: Vignettes in Patient Safety, pp. 1–13. IntechOpen, London (2018)
19. Herder, T., et al.: Supporting teachers' intervention in students' virtual collaboration using a network based model. In: Proceedings of the International Conference on Learning Analytics and Knowledge, LAK 2018, pp. 21–25 (2018)
20. Jivet, I., Scheffel, M., Specht, M., Drachsler, H.: License to evaluate: preparing learning analytics dashboards for educational practice. In: Proceedings of the International Learning Analytics and Knowledge Conference, LAK 2018, pp. 31–40 (2018)
21. Martinez-Maldonado, R.: I spent more time with that team: making spatial pedagogy visible using positioning sensors. In: Proceedings of the International Conference on Learning Analytics & Knowledge, LAK 2019, pp. 21–25 (2019)
22. Martinez-Maldonado, R., Dimitriadis, Y., Martinez-Monés, A., Kay, J., Yacef, K.: Capturing and analyzing verbal and physical collaborative learning interactions at an enriched interactive tabletop. Int. J. Comput.-Support. Collab. Learn. 8(4), 455–485 (2013)
23. Martinez-Maldonado, R., Echeverria, V., Santos, O.C., Dos Santos, A.D.P., Yacef, K.: Physical learning analytics: a multimodal perspective. In: Proceedings of the International Conference on Learning Analytics and Knowledge, LAK 2018, pp. 375–379 (2018)
24. Martinez-Maldonado, R., Goodyear, P., Kay, J., Thompson, K., Carvalho, L.: An actionable approach to understand group experience in complex, multi-surface spaces. In: Proceedings of the SIGCHI Conference on Human Factors in Computing Systems, CHI 2016, pp. 2062–2074 (2016)
25. Matcha, W., Gasevic, D., Pardo, A.: A systematic review of empirical studies on learning analytics dashboards: a self-regulated learning perspective. IEEE Trans. Learn. Technol. (2019, in press)
26. Mislevy, R.J., Behrens, J.T., Dicerbo, K.E., Levy, R.: Design and discovery in educational assessment: evidence-centered design, psychometrics, and educational data mining. J. Educ. Data Mining 4(1), 11–48 (2012)
27. Müller, L., Rivera-Pelayo, V., Kunzmann, C., Schmidt, A.: From stress awareness to coping strategies of medical staff: supporting reflection on physiological data. In: Salah, A.A., Lepri, B. (eds.) HBU 2011. LNCS, vol. 7065, pp. 93–103. Springer, Heidelberg (2011). https://doi.org/10.1007/978-3-642-25446-8_11
28. Ochoa, X.: Multimodal learning analytics. In: The Handbook of Learning Analytics, pp. 129–141. SOLAR, Alberta (2017)
29. Ochoa, X., Chiluiza, K., Granda, R., Falcones, G., Castells, J., Guamán, B.: Multimodal transcript of face-to-face group-work activity around interactive tabletops. In: Proceedings of the CROSS-MMLA Workshop on Multimodal Learning Analytics Across Spaces, pp. 1–6 (2018)
30. Perera, D., Kay, J., Koprinska, I., Yacef, K., Zaïane, O.R.: Clustering and sequential pattern mining of online collaborative learning data. IEEE Trans. Knowl. Data Eng. 21(6), 759–772 (2008)
31. Prieto-Alvarez, C., Martinez-Maldonado, R., Shum, S.B.: Mapping learner-data journeys: evolution of a visual co-design tool. In: Proceedings of the ACM Australian Computer-Human Interaction Conference, OzCHI 2018, pp. 205–214 (2018)
32. Prieto-Alvarez, C.G., Martinez-Maldonado, R., Anderson, T.D.: Co-designing learning analytics tools with learners. In: Learning Analytics in the Classroom: Translating Learning Analytics for Teachers, pp. 93–110. Routledge, London (2018)

33. Sarter, N.B.: Multimodal information presentation: design guidance and research challenges. Int. J. Ind. Ergon. **36**(5), 439–445 (2006)
34. Scherr, R.E., Hammer, D.: Student behavior and epistemological framing: examples from collaborative active-learning activities in physics. Cogn. Instr. **27**(2), 147–174 (2009)
35. Shaffer, D.W.: Epistemic frames for epistemic games. Comput. Educ. **46**(3), 223–234 (2006)
36. Shaffer, D.W.: Quantitative Ethnography. Cathcart Press, Madison (2017)
37. Shute, V.J., Ventura, M.: Stealth Assessment: Measuring and Supporting Learning in Video Games. MIT Press, Cambridge (2013)
38. Taylor, S., Jaques, N., Chen, W., Fedor, S., Sano, A., Picard, R.: Automatic identification of artifacts in electrodermal activity data. In: Proceedings of the Annual International Conference of the IEEE Engineering in Medicine and Biology Society, EMBC 2015, pp. 1934–1937 (2015)
39. Wise, A., Knight, S., Buckingham Shum, S.: Collaborative learning analytics. In: Cress, U., Rosé, C., Wise, A., Oshima, J. (eds.) International Handbook of Computer-Supported Collaborative Learning. Springer, Cham (in press)
40. Worsley, M., Blikstein, P.: A multimodal analysis of making. Int. J. Artif. Intell. Educ. **28**(3), 385–419 (2018)
41. Zhang, Z., Sarcevic, A.: Constructing awareness through speech, gesture, gaze and movement during a time-critical medical task. In: Boulus-Rødje, N., Ellingsen, G., Bratteteig, T., Aanestad, M., Bjørn, P. (eds.) ECSCW 2015: Proceedings of the 14th European Conference on Computer Supported Cooperative Work, 19-23 September 2015, Oslo, Norway, pp. 163–182. Springer, Cham (2015). https://doi.org/10.1007/978-3-319-20499-4_9

nCoder+: A Semantic Tool for Improving Recall of nCoder Coding

Zhiqiang Cai[1]([⊠]) [iD], Amanda Siebert-Evenstone[2] [iD],
Brendan Eagan[2] [iD], David Williamson Shaffer[2] [iD], Xiangen Hu[1,3] [iD],
and Arthur C. Graesser[1] [iD]

[1] The University of Memphis, Memphis, TN 38152, USA
zhiqiang.cai@gmail.com
[2] University of Wisconsin-Madison, Madison, WI 53706, USA
[3] China Central University, Wuhan, Hubei, China

Abstract. Coding is a process of assigning meaning to a given piece of evidence. Evidence may be found in a variety of data types, including documents, research interviews, posts from social media, conversations from learning platforms, or any source of data that may provide insights for the questions under qualitative study. In this study, we focus on text data and consider coding as a process of identifying words or phrases and categorizing them into codes to facilitate data analysis. There are a number of different approaches to generating qualitative codes, such as grounded coding, a priori coding, or using both in an iterative process. However, both qualitative and quantitative analysts face the same coding problem: when the data size is large, manually coding becomes impractical. nCoder is a tool that helps researchers to discover and code key concepts in text data with minimum human judgements. Once reliability and validity are established, nCoder automatically applies the coding scheme to the dataset. However, for concepts that occur infrequently, even with an acceptable reliability, the classifier may still result in too many false negatives. This paper explores these problems within the current nCoder and proposes adding a semantic component to the nCoder. A tool called "nCoder+" is presented with real data to demonstrate the usefulness of the semantic component. The possible ways of integrating this component and other natural language processing techniques into nCoder are discussed.

Keywords: Coding · Grounded coding · A priori coding · Automatic coding · Grounded theory · Qualitative analysis · Quantitative analysis · Latent Semantic Analysis · Topic modeling · Machine learning

1 Introduction

When researchers analyze text data, they often search for culturally relevant and meaningful aspects of a discourse. For example, people that are interested in understanding how students think during science curriculum might look for science content knowledge (e.g. nitrogen cycle knowledge) and for scientific practices (e.g. developing and using models). However, in order to make the claim that a student exhibits

B. Eagan et al. (Eds.): ICQE 2019, CCIS 1112, pp. 41–54, 2019.
https://doi.org/10.1007/978-3-030-33232-7_4

scientific knowledge or practices, researchers need to provide evidence for their interpretation. One way to find evidence in data—and eventually qualitative interpretations—is to develop a set of codes that allow researchers to systematically categorize phenomena in their data to help identify patterns [1, 2].

However, in the age of digital learning environments and ubiquitous data collection, we have more information than ever about what students are doing and how they are thinking. The sheer volume of this data can render traditional qualitative methods unfeasible. One way to address this issue is to create automated codes that apply some set of rules to a dataset to assign values to each piece of data.

To aid researchers, Shaffer and his colleagues created a system for scaffolding automated classifier development. The nCoder is a learning analytics platform used to develop, refine, validate, and implement automated coding schemes. The nCoder is designed specifically for working with large and small sets of text data, such as interviews, transcripts, logfiles, and other text data. Users can generate codes by defining a construct, identifying common words associated with the construct, testing their code, and then updating their construct definition or wordlist until the researcher achieves an acceptable level of agreement between their coding and the automated coding scheme.

In this paper, we first give an in-depth description and analysis of the nCoder coding process. Then we discuss a particular problem with *recall* during this process. Recall is the ratio of the number of items coded by a classifier to the total number of items that should be coded. In nCoder, it is possible to have a situation, when the frequency of a code is low, for the recall to be low, even if the kappa is high enough to achieve a statistically significant rho. That is, it is possible to achieve acceptable agreement and generalize that agreement to the dataset yet have a potentially high rate of false negatives. After identifying and explaining this problem, we describe a potential solution to this problem – a newly developed tool equipped with a semantic component that helps solving the low recall problem. This tool is called "nCoder+", which implies that this tool is simply an add-on to nCoder. A real data set will be used to illustrate the usefulness of this tool. We will end the paper with discussions on integrating this new component and other possible techniques into nCoder.

2   Approaches to Coding

Coding is a process of assigning meaning to a given piece of evidence. Evidence may be found in a variety of data types, including documents, research interviews, posts from social media, conversations from learning platforms, or any source of data that may provide insights for the questions under qualitative study. In this study, we focus on text data and consider coding as a process of identifying words or phrases and categorizing them into codes to facilitate data analysis.

There are a number of different approaches to generating qualitative codes, such as grounded coding, a priori coding, or using both in an iterative process [3]. Grounded coding, also referred to as inductive, emergent, or bottom-up coding, is an exploratory process which allows a researcher to discover new concepts and theories that emerge from the text data [4]. Researchers are encouraged to read through the data line by line and identify concepts that may construct in-depth understanding of the data [4, 5]. One major challenge in grounded coding is generating new concepts. Since it is exploratory, a researcher may iteratively refine the definition of the concepts, which implies data re-coding. When the data size becomes too large, grounded coding by hand can become time-consuming if not impractical.

A priori coding is another identification process, in which a coder identifies pre-defined concepts from an existing theoretical framework. A priori coding, also referred to as theoretical, deductive, or top-down coding, starts with a theory or set of constructs and then searches for the ideas in the data rather than using the ideas in the data. In our example above, someone interested in identifying science practices from the Next Generation Science Standards may search for qualitative evidence of students using one or multiple of the eight science and engineering practices.

Whether a researcher starts with the data and creates categories or starts with categories and identifies them in the data, researchers face challenges of coding reliably and consistently. First, researchers often work to validate codes in their data to ensure a common understanding of concepts between two raters[1]. Inter-rater reliability (IRR) measures assess whether two raters assign codes consistently in the same way. To determine whether two raters have identified the same properties in the data, researchers may use tests of agreement, such as Cohen's kappa, to quantify to what degree the two raters agree with each other. A common heuristic in studies of CSCL (Computer Supported Collaborative Learning) is for researchers to sample 10–20% of their data to assess IRR [6]. Again, when the data size is large, manually coding this much data can become difficult if not impossible, especially if multiple IRR analyses need to be performed to achieve acceptable reliability between raters.

Advances in computer and natural language processing technologies have made another coding approach more accessible to researchers: *automated coding*. An auto-mated coder or classifier is an algorithmic process that identifies whether a piece of data belongs in a certain category or class. For example, topic modeling is capable of automatically finding latent topics in a text data set and "code" the data by topic proportion scores [7, 8]. While this approach is automated and requires no human coding, researchers may find that not all topics can be easily interpreted and verified.

An ideal computer coding tool does not have to be fully automated. Instead, it should have enough flexibility to allow researchers to discover concepts they think important. Once some concepts are discovered, the tool should be able to "learn" from a relatively small sample of human coded data and code the data in a way close enough to the human coder. nCoder is such a tool. Before diving deep into nCoder, we first briefly introduce the concept of "codebooks", which plays an important role in the coding process.

---

[1] Not all researchers perform IRR tests. For example, researchers may use *social moderation*, where two or more raters code all of the data and resolve differences until they all agree on the code (Herrenkohl and Cornelius) [14].

3 Codebooks

During coding, researchers often create a *codebook* to organize and summarize information about codes. Code books are often used to communicate ideas for IRR and are often reported in the methods sections of resulting publications. Within a codebook, the name, the definition, and examples of the code are included. For example, in analyzing text data from presidential primary debates, environmental issues could be an interesting code. In a codebook, this code may look like the one shown in Table 1.

Table 1. Code book for environmental issues

Name	Definition	Example	Classifier	IRR
Environmental Issues	Referring to harmful effects of human activity on the biophysical environment	"Governor Pataki, you've indicated you believe climate change is real and caused at least in part by human activity"	\benvironment ,\bclimate ,\bpollut ,\bgreenhouse ,\bdegradation ,\bglobal warm ,\bhabitat ,\bextinction ,\bsuperfund ,\btoxic ,\bconservation ,\bsustainability ,\brunoff ,\bnatural resource	$\kappa = 1.00$ $rho = 0.01$ $n = 40$

If the researchers used automated classifiers, then the word lists, patterns, and/or rules that were applied to the data to assign the code may also be included in the codebook.

If the researchers choose to validate their codes, another section of the codebook is the results of their IRR analyses. In our case, we have added our classifier list and IRR between the human rater and classifier. In a full analysis, we would also validate our code between two human raters as well as between the second human rater and the classifier. In this analysis, we focus on this first iteration in the coding process.

The nCoder tool helps researchers build and validate their codebooks. A critical component of any statistical analysis of text data is some form of coding scheme that identifies key concepts and clusters of terms in the data. To generate valid insights, however, this coding process has to be compared to the work of human raters. nCoder minimizes the amount of data that needs to be hand-coded by using cutting-edge statistical techniques to establish the reliability and validity of codes. Once codes are validated, nCoder automatically applies a coding scheme to larger datasets quickly and efficiently, even coding new data as it becomes available. In the next section we describe how the nCoder scaffolds codebook creation and helps researchers perform code validation processes.

4   nCoder

nCoder is available as a free online tool (https://app.n-coder.org/) and an R package (nCodeR) [9] that helps researchers to code large amounts of text data by supporting the development, refinement, validation and application of automated coding schemes. nCoder also employs Shaffer's rho in reliability analyses, which again is available online (https://app.calcrho.org) as an R package (rhoR) [9, 10]. Both the R packages and webkits have been used to analyze large-scale datasets of many kinds, including chat, email, online actions, surgical performance, and brain scan data [9–14].

For this paper, we focus on code generation which is summarized in Fig. 1 as a flow chart for developing automated codes. To create a new coding project, users upload their data as either a CSV (Comma Separated Value) or Microsoft excel file. The data file may contain any number of columns, with the first row containing column names. After the file is loaded, the user is asked to specify the text column, which is used as the text data for coding. The detailed coding steps are described below.

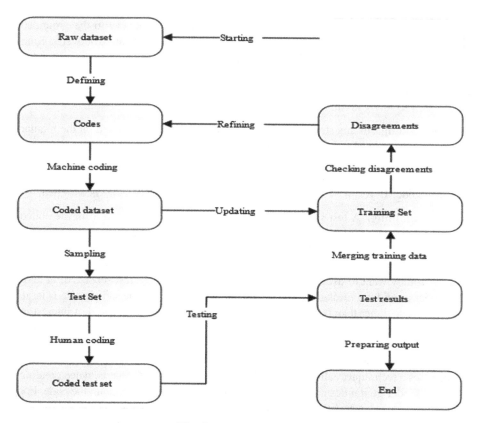

Fig. 1. nCoder flowchart

Defining Codes. The first step in coding is defining a code. In nCoder, a code definition contains three elements: a name, a description, and a list of regular expressions (RegExs). The name is the identifier of the code, which should therefore be unique. The name should also be short and meaningful so that when it is written in an article, people can easily understand and remember it. The description explains what the code is about. The description should also be clear, brief and accurate. The RegEx list specifies language patterns that can be used to code the data. For example, the RegEx "\benvironment" means that if there is a word starting with "environment" in a text cell, the code is considered present. Regular expressions are powerful in representing language patterns. nCoder provides an interface to designate word boundaries "\b". nCoder also helps users create complex combinations of words. For example, a user may want to find instances of "clean" but not "clean water". The user would search for a regular expression that begins with clean and does not include water, which is "^(?:(?!\bwater).)*\bclean(?!.*\bwater)". Such an expression may be easy for a more expert programmer and nCoder helps novice users create more nuance expressions. Users add words or expressions until they are ready to test their classifier. nCoder does not allow the user to freely go through the data because in a later step, the tool needs to sample "fresh" data for testing purpose. This restriction seems contradictory to the grounded theory method, which is built on the assumption that the concepts are discovered from navigating the data. The current version of nCoder does not allow users to do grounded coding within the tool and this feature may be built into subsequent releases. This may not be really a problem of nCoder, if we assume that the user has another offline tool (e.g., excel) to navigate through the data.

Machine Coding. In this step, the dataset is automatically coded based on the regular expressions provided by the user. nCoder uses a simple regular expression matching algorithm to automatically code the text data. It goes through each row of the specified text column, and checks if the text in the cell matches one of the regular expressions in a defined code. If yes, a "1" is assigned to the data. Otherwise, a "0" is assigned.

Test Set Sampling. A test set for each concept is randomly sampled from the coded dataset, where the minimum sample size is 10. The user may repeatedly increase sample size by 10 until the user is satisfied with the size of the test set. If a researcher wants to establish a kappa over a threshold of 0.65 they should use a test set size of at least 40. If they want to use a threshold of 0.9 they should use a test set size of at least 80. nCoder provides a measure "rho" to indicate whether or not the sample size is large enough. A rho higher than 0.05 when the kappa is higher than a threshold of interest in the test set, indicates that the test set size should be larger. When a targeted concept has a low occurrence in the dataset (e.g., <20%), a small test set may not be able to represent the targeted concept and can cause difficulties in achieving sufficient IRR. nCoder uses a technique, called "inflation" to solve this problem. That is, when creating test sets, nCoder guaranties at least 20% of the lines in the test set contain the code. For example, for a test set size of 10, nCoder would randomly pick 2 lines that the algorithm coded positively and then randomly select the final 8 lines from the rest of the data.

Human Coding. After the test set is generated, the sample texts are presented to the user for coding. The user selects a concept to code and the corresponding test items are displayed, together with a "Yes" button and a "No" button. The user codes the test items by pressing "Yes" or "No", with "Yes" indicating a "1", or "true positive" and "No" a "0" or "true negative". The test items can be added if the user is not satisfied.

Testing. A user sets a kappa threshold and presses the "Run Test" button to run an IRR test. nCoder shows three test numbers: a kappa for the test set, a kappa for the training set (the test items in earlier cycles), and a rho value. The kappa values measure the agreement between the human coding and the machine coding. The rho shows whether or not the kappa value generalizes to the untested items. When rho <0.05, acceptable reliability is established between the researcher and the machine coding. In this study, we present results that address this cycle of the process. However, for a complete analysis, we recommend three pairwise IRR checks. First, IRR should be checked between rater one and the automated code to ensure proper classifier rules. Next, IRR should still be established between two human raters as is typical in common IRR processes to achieve good conceptual validity. As a final check, IRR testing should check reliability between the second rater and the classifier to make sure all human and computer raters agree on a concept.

Merging Training Data. If the test result is not satisfactory (i.e., the rho is above the alpha level, typically 0.05), the test items are moved to the training data set. The training data set contains all tested items. The user could review each item in the training set and the coding from the human and the machine.

Checking Disagreement. nCoder automatically checks the disagreement between the human coding and machine coding. The disagreed items are displayed to the user for investigation.

Refining Codes. The user may remove the disagreements by removing, adding, or refining regular expressions, resulting in changes of the regular expression list.

Updating Training Data. Each time the regular expression list is changed, the concept is re-coded by the machine and the machine codes in the training data set is updated. At the same time, the disagreements are changed. The refining-updating cycle may continue until the minimum number of disagreements is reached. Zero disagreement is possible but it may sometimes involve complicated regular expressions.

Preparing Output. The testing-refining cycle may be continued until the rho values for all concepts are less than 0.05. The output of nCoder will be the original data with added new concept columns containing machine coded values.

5 Kappa, Shaffer's Rho, Sample Size and Recall

To ensure reliability, nCoder relies on kappa values and Shaffer's rho. In this section, we will review concepts related to kappa and take a mathematical look at Shaffer's rho. At the end of this section, we will address an unsolved problem in nCoder: when the *base rate* (i.e., the ratio of the occurrences to the data size) of a code is low, the coding recall of the concept could be low or unacceptable, even if the kappa has been shown to be statistically significantly above a threshold with rho.

Kappa is a statistical measure proposed by Cohen in 1960 [15], which has been used widely as an inter-rater reliability measurement [6]. It measures the degree of agreement between two or more coders controlling for chance agreement. In the case of nCoder, we may consider items coded by a human rater as "1"s as "human positives" and "0"s as "human negatives". The machine coded "1"s and "0"s are "machine positives" and "machine negatives", respectively. For each concept, an item may have a pair of human-machine coding as "1-1", "1-0", "0-1", or "0-0". If we take the human coding as actual truth and computer coding as prediction, kappa becomes a measure of the performance of the machine coding. Following the notions in literatures, we denote the proportion "1-1" by tp (true positive), "1-0" by fn (false negative), "0-1" by fp (false positive), and "0-0" by tn (true negative). Kappa is often written as

$$\kappa = \frac{p_0 - p_c}{1 - p_c},$$

where $p_0 = tp + tn$ is the total agreement between the human and the machine, and $p_c = (tp + fp)(tp + fn) + (tn + fp)(tn + fn)$ is the chance agreement between the human and the machine.

nCoder aims at minimizing the number of items a human rater has to code in a test set for the purpose of establishing reliability and providing a warrant for validity. That is, nCoder was designed to help researchers use the smallest sample size possible when establishing inter-rater reliability. The question is, how do we know that the machine coding is good enough? In other words, how much data does a trained human rater need to code in order to establish that the machine could reliably code untested data with a high enough level of agreement with a trained human rater? Shaffer and his colleagues provided the rho measure to answer this question. Roughly speaking, a rho <0.05 in nCoder means that we would be wrong less than five percent of the time if we concluded that if both the human rater and the machine were to code all the data, not just the sample or test set, that agreement between their ratings would be greater than the threshold of interest. Details about the computations of rho could be found in Shaffer [1] and Eagan et al. [6].

When a code has a low base rate, it is often practical to check all machine coded occurrences and get a low false positive rate by refining the regular expressions. However, checking false negative is often impractical. For example, if the base rate is about 5% and the data size is 10,000. Then the user may go through 500 items with machine coded "1"s and see if any item is a false positive. However, to check the false negative, the user would need to check 9,500 items. Thus, nCoder users may often end up with a coding which is of very few, if any, false positives but potentially a high false negative rate. To illustrate this issue, let's consider the situation when the false positive rate is zero. In this case, the false negative rate is a function of true positives and the kappa:

$$fn = \frac{2tp(1 - tp)(1 - \kappa)}{2tp + (1 - 2tp)\kappa}.$$

Since when there are zero false positives, $tp = 0$ implies $\kappa = 0$, the above equation holds only for $tp > 0$.

Figure 2 shows the curves of the false negative rate as a function of true positives for kappa = 0.65, 0.80 and 0.90, respectively. As an example, let's look at case when kappa = 0.65 (the top curve in blue). The curve shows that the maximum false negative rate is about 18% when the true positive rate is around 40%. When the true positive rate is very low, the false negative is also low. For example, for kappa = 0.65, when true positive rate is 5%, the false negative rate is also about 5%. Both rates are low. However, the false negative rate is about the same as true positive rate. In other words, the number of occurrences reported by the machine could be the same as those missed by the machine.

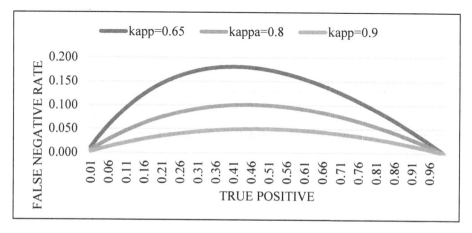

Fig. 2. False negative rate as a function of true positive and kappa when false positives = 0 (Color figure online)

Recall is a commonly used measure to indicate the capability of an algorithm or classifier in finding targeted items. Mathematically speaking, recall is the rate of true positives to the total truth, which can be written as:

$$recall = \frac{tp}{tp + fn}.$$

When the false positive rate is zero and $tp > 0$, for a given kappa, recall is a linear function of true positives:

$$recall = \frac{2(1 - \kappa)}{2 - k} tp + \frac{\kappa}{2 - \kappa}.$$

Figure 3 shows the recall lines as linear functions of true positives for kappa = 0.65, 0.80 and 0.90, respectively. When *tp* is small, the recall can be approximated by $\frac{\kappa}{2-\kappa}$. For example, when *tp* is small and kappa = 0.65, the recall is about 0.48. That means, when *tp* is small, a 0.65 kappa could not even guarantee a 50% recall.

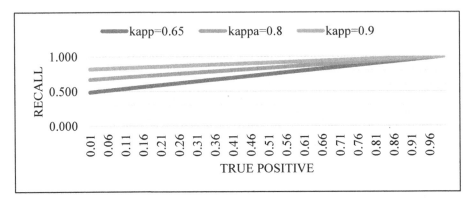

Fig. 3. Recall as linear a function of true positives for given kappa

6 nCoder+: Adding a Semantic Component to nCoder

To help improve recall when using nCoder, we built an add-on tool, called "nCoder+", that allows researchers to easily find false negatives, i.e., targets missed by nCoder. The tool is briefly described below and a real data example is used to demonstrate the tool's performance in the next section. In the final discussion section, we will talk about integrating this tool as a component into nCoder.

The core of nCoder+ is an LSA (Latent Semantic Analysis) component. LSA is a way to represent the "meaning" of words by vectors. Readers who are not familiar with LSA could refer to Landaure et al. [16]. The cosine between two vectors are often used to measure the similarity between two words. For a given word or phrase, the words with highest cosine values to the given word or phrase are called "nearest neighbors". Nearest neighbors make it possible to automatically find the most likely missing items.

Figure 4 shows a screenshot of nCoder+. On the left panel, there is a dropdown menu that allows user to select an LSA vector space. Below that is the "Keys" box that displays the regular expressions used to code the data. The "Words" box lists the neighbors to the keywords, together with the semantic cosine values as a measure of similarity of a neighbor to the keywords. The "neighbors" are sorted by similarity, so that the nearest neighbors are on the top of the list. The "Report" box below the neighbor list shows the result of each step. The table on the right panel shows the data under investigation. The box above the data table shows the content of any selected cell in the data table. The "Add Key" box is a place for users to enter new regular expressions. The "Test" button is used to check the validity of regular expressions. The "Update" button is used to add a new regular expression to the list.

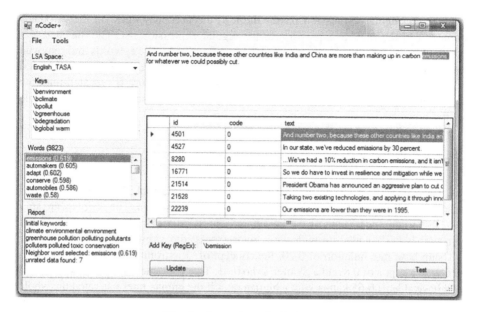

Fig. 4. Screenshot of nCoder+

nCoder+ starts from loading the regular expressions used in nCoder coding of a key concept and a csv data file with 3 columns: an id column, a text column and a binary code column. A "1" in the code column indicates that the text on the row contains the specific key concept.

After the regular expressions and the csv data file are loaded, nCoder+ extracts words from the text column and form a vocabulary. The words that match any one of the regular expressions are then identified as "keywords". For each vocabulary word, the LSA cosine values between this word and each of the keywords are computed. The maximum of these cosine values is taken as the similarity of the word to the keywords. Of course, if the word itself is a keyword, the similarity will be 1. The vocabulary words are then sorted by the similarity so that the nearest neighbors are on the top of the sorted word list. The word list is then displayed in the "Words" box with keywords excluded.

A user reviews the nearest neighbors from the top to see if there is any word that is highly associated with the key concept. If a word is identified, the user may click on it. All text items in the data with code "0" that contain the selected word will appear in the data table. The user may check the text items in the table and see if there are any missing items that should be coded.

When missing items are identified, the user may compose a new regular expression based on the selected word. To check whether the regular expression is well composed, the user may press the "Test" button. Then text items that match the new regular expression will be displayed in the data table. The user may revise the regular expression until it accurately flags the missing items.

Once the regular expression passes the test, the user may click on the "Update" button to add the regular expression to the regular expression list. The matched text items will be coded as 1. The selected word will be added into keywords and removed from the displayed word list. The similarity of the words will be re-computed, re-sorted and displayed as neighbors to the new keywords. This procedure is repeated until the user cannot find any new words in the word list that could be used to find possible missing items.

7 nCoder+ Validation

The "Primary Debates 2016" dataset was used to test the tool. The dataset contains transcripts of every debates held during the 2016 primary season. The scripts were split into 23,714 utterances. One of the co-authors ran nCoder to code the concept "Environmental Issues", ended up with the regular expressions shown in Table 1.

Using base rate inflation of 0.20, the concept of "Environmental Issues" was validated with a kappa of 0.8 and a Shaffer's rho 0.01. That guarantees that the coded values could have at least 0.65 kappa with a human rater if the human rater had rated the whole data set. The final data set flagged 120 utterances that contain the key concept. Another co-author checked all 120 utterances and found that the false positive rate was 0. However, the recall was not checked because that would mean to go through the rest of the 23,596 utterances and see if there are any more utterances that contain the key concept. With zero false positive, 0.65 kappa, and true positive = 120/23, 596 = 0.5%, the corresponding recall is about 0.5, which means that there could be as many as 120 items containing the "Environment Issues" that were not coded as "1".

The data and the regular expressions were loaded to nCoder+. A set of keywords were extracted from the regular expressions and the rest of the words were displayed in the order of semantic similarity to the keywords. The first new keyword we identified was "emissions". The tool showed up 7 items containing "emissions" that should be but were not coded as "1". The regular expression "\bemissions" was added to the regular expression list and 7 new items were added to true positives. Repeating this process, we identified 8 new keywords as shown in Table 2. nCoder+ showed 175

Table 2. Keywords, added regular expressions and additionally identified items

Neighbor word	Items involved	RegEx	Items coded
Emissions	7	\bemissions	7
Clean	39	\bclean.*(water|power|electric|energy)	23
Gases	1	\bgases	1
Gas	34	(\bgas.*energy)|(\benergy.*gas)	4
Waste	21	\bwaste.*water	2
Coal	60	(\buse of coal)|(\bcoal\b.*(energy|clean))	2
Carbon	4	\bcarbon	4
Solar	9	\bsolar	9
Total	**175**		**52**

items containing these 8 new keywords. With 8 carefully generated regular expressions, 52 new items were added to the true positives. Thus, assuming there were a total of 240 positive items, the recall is increased from 50% to 172/240 = 72%.

8 Discussions

nCoder is a tool that helps both qualitative and quantitative researchers in coding text data. The regular expression based automatic coding is simple and effective. When the data size is large, it is usually hard to define a complete set of regular expressions that represent a certain concept, which can result in low coding recall. This paper provided an effective tool, nCoder+, to find missing items and thus improve recall. The current nCoder+ tool is just a prototype for validating the idea. There are multiple ways to integrate this tool into nCoder. One way is to add it as a post process component and use it in the way we used in this paper. The other way is to integrate it as a component in the nCoder's testing process (see Fig. 1). At the nCoder testing step, a test set is sampled from data that has not been exposed to the researcher. A proportion (e.g., 20%) of the test items are sampled from the data with machine coded "1"s and the rest are randomly sampled from the remaining data. To make use of the semantic neighbors, a proportion of the "0" items could be those that contains nearest neighbors of the keywords. Thus, the test set contains three types of items: "1" items, "near neighbor" items and "remote neighbor" items. The "1" items will help determine the true positive; the "near neighbors" will help improve recall; and the "remote neighbors" will help determine the true negative.

This paper only considered integrating the idea of semantic neighbors into nCoder. Other NLP methods may also be useful to further improve nCoder. For example, topic modeling could help researchers to find meaningful concepts that researchers may otherwise miss. Neural network algorithms may also help in inferring concepts from given keywords. Our future work will be continuously incorporating latest advances in natural language processing and providing researchers with better coding tools.

Acknowledgements. The research was supported by the National Science Foundation (SBR 9720314, REC 0106965, REC 0126265, ITR 0325428, REESE 0633918, ALT-0834847, DRK-12-0918409, 1108845; DRL-1661036, 1713110; ACI-1443068), the Institute of Education Sciences (R305H050169, R305B070349, R305A080589, R305A080594, R305G020018, R305C120001), the Army Research Lab (W911INF-12-2-0030), and the Office of Naval Research (N00014-00-1-0600, N00014-12-C-0643; N00014-16-C-3027), the Wisconsin Alumni Research Foundation, and the Office of the Vice Chancellor for Research and Graduate Education at the University of Wisconsin-Madison. The opinions, findings, and conclusions do not reflect the views of the funding agencies, cooperating institutions, or other individuals.

References

1. Shaffer, D.W.: Quantitative Ethnography. Cathcart Press, Madison (2017)
2. Chi, M.T.H.: Quantifying qualitative analyses of verbal data: a practical guide. J. Learn. Sci. **6**, 271–315 (1997)

3. Saldaña, J.: The Coding Manual for Qualitative Researchers (2014). https://doi.org/10.1007/s13398-014-0173-7.2

4. Glaser, B.G., Strauss, A.L.: The Discovery of Grounded Theory: Strategies for Qualitative Research. Aldine Transaction, New Brunswick (1967)

5. Charmaz, K.: Constructing Grounded Theory. SAGE, London (2006)

6. Eagan, B.R., Rogers, B., Serlin, R., Ruis, A.R., Irgens, G.A., Shaffer, D.W.: Can we rely on IRR? testing the assumptions of inter-rater reliability. In: CSCL 2017 Proceedings, pp. 529–532 (2017)

7. Blei, D.M., Edu, B.B., Ng, A.Y., Edu, A.S., Jordan, M.I., Edu, J.B.: Latent Dirichlet allocation. J. Mach. Learn. Res. 3, 993–1022 (2003). https://doi.org/10.1162/jmlr.2003.3.4-5.993

8. Hu, Y., Boyd-Graber, J., Satinoff, B.: Interactive topic modeling. In: Proceedings of the 49th Annual Meeting Association for Computational Linguistics Human Language Technologies, pp. 248–257 (2011)

9. Marquart, C.L., Swiecki, Z., Eagan, B., Shaffer, D.W.: ncodeR (Version 0.1.2) (2018)

10. Eagan, B.R., Rogers, B., Pozen, R., Marquart, C., Shaffer, D.W.: rhoR: Rho for inter rater reliability (Version 1.1.0) (2016). https://cran.r-project.org/web/packages/rhoR/index.html

11. Gašević, D., Joksimović, S., Eagan, B., Shaffer, D.W.: SENS: network analytics to combine social and cognitive perspectives of collaborative learning. Comput. Hum. Behav. 92, 562–577 (2019)

12. Cai, Z., Pennebaker, J.W., Eagan, B., Shaffer, D.W., Dowell, N.M., Graesser, A.C.: Epistemic network analysis and topic modeling for chat data from collaborative learning environment. In: Proceedings of the 10th International Conference on Educational Data Mining, pp. 104–111 (2017)

13. Sullivan, S., et al.: Using epistemic network analysis to identify targets for educational interventions in trauma team communication. Surg. (United States) 163, 938–943 (2018). https://doi.org/10.1016/j.surg.2017.11.009

14. Shaffer, D.W., Ruis, A.R.: Epistemic network analysis: a worked example of theory-based learning analytics. In: Handbook of Learning Analytics Data Mining, in press (2017)

15. Cohen, J., Cohen, J.: A coefficient of agreement for nomial scales. Educ. Psychol. Meas. 20 (1), 37–46 (1960). https://doi.org/10.1177/001316446002000104a coefficient of agreement for nomial scales. Educ. Psychol. Meas. 20, 37–46 (1960). https://doi.org/10.1177/001316446002000104

16. Landauer, T., McNamara, D., Dennis, S., Kintsch, W.: Handbook of Latent Semantic Analysis (2007)

Examining the Dynamic of Participation Level on Group Contribution in a Global, STEM-Focused Digital Makerspace Community

Danielle P. Espino[(✉)], Seung B. Lee, Lauren Van Tress,
and Eric R. Hamilton

Pepperdine University, Malibu, CA 90263, USA
danielle.espino@pepperdine.edu

Abstract. Passive behavior in collaborative group settings is often associated with negative or no contributions to the group (social loafing). This paper examines low and high participation levels of students in a virtual collaborative group setting within a global, STEM-focused digital makerspace community. The results of using epistemic network analysis show that both high and low participation levels contributed to the overall balance of the group discourse, overcoming social loafing behavior. High participation level students provided social aspects that contributed to the development of a safe social space for sharing, while low level participation provided content focused dialogue for the group.

Keywords: STEM education · Global · Collaboration · CSCL · Online · Digital makerspace · Informal learning · Participation · Social loafing

1 Introduction

Technology affords the opportunity for students to connect and collaborate across boundaries more than ever before. However, whenever students come together to interact in any setting, a group dynamic emerges based on the convergence of individual student behaviors. One dynamic is social loafing, which is the tendency to do less work in a group setting, therefore relying on a few motivated students to carry the work for the group [7, 8]. Latané, Williams and Harkins [10] associate negative consequences to social loafing, describing it as a social disease that results in reduced efficiency. This pattern of behavior also carries into virtual settings, where only certain participants are actively involved in discussions with the group, while others are prone to passive, observer behavior [1, 15]. Passive behavior has also been linked to low performance, as higher participation in discourse is positively associated with achievement [2].

To address the social aspects in computer supported collaborative learning environments, Kreijns, Kirschner and Vermeulen [9] developed a framework comprised of three elements: sociability, social space, and social presence. Sociability is described as the "potential to encourage socioemotional interaction" and in turn, establish social space that ties participants together [9, p. 217]. A higher sense of sociability is positively associated with socioemotional interactions that impact the group's development in the environment. Integrating established work by Preece [14] focused on sociability as part

© Springer Nature Switzerland AG 2019
B. Eagan et al. (Eds.): ICQE 2019, CCIS 1112, pp. 55–65, 2019.
https://doi.org/10.1007/978-3-030-33232-7_5

of the social system, Kreijns et al. [9] expand on three components of sociability: purpose, ensuring the goal of the collaborative learning activity is clear to all to avoid frustration; people, where each participant serves in different, defined roles for the group; and policies, which are the protocols and rules established within the group. The presence of these three components contribute to positively fostering sociability.

The establishment of a sound social space and social presence relies strongly on the development of sociability. Of sociability's three components, the people component is the most challenging to manage. While purpose and policies can be structured and pre-determined in an expected way, the behavior of people in roles is unpredictable. In an informal learning setting where participant roles are naturally fluid and less defined, overcoming the possibility of social loafing and the negative association with passive behavior becomes a challenge. However, should a group achieve a sense of cohesiveness, then individuals are less likely to engage in behavior such as social loafing [11]. Through achieving sociability, a group is more likely to overcome the inherent challenges associated with group activity.

This paper examines the dynamic of group participation that emerges from low versus high participation levels of students involved in a global, STEM-focused collaborative digital makerspace environment. Students range from 10–19 years in age to participate in informal, after school clubs across four continents, developing digital media artifacts on STEM topics. The students share their digital media artifacts, which include videos and Microsoft Sway presentations, with students from different club sites during online global meet-ups. These meet-ups take place synchronously in a video conference platform and provide students with an opportunity to interact with students from other countries while focused on collaboratively building knowledge around STEM topics and media making.

2 Methodology

The data used to perform this analysis came from two separate meet-ups from 2017 and 2018 between students from Kenya, Finland and the United States. One line of data represented an utterance, or a turn of talk. Each line was coded individually by two raters for the following seven constructs, which were identified as the most salient through a grounded analysis of the data: Content Focus, Media Production, Curiosity, Feedback, Information Sharing, Social Disposition, and Participatory Teaching. A description of the codebook can be seen in Table 1. The raters then came to an agreement on the final coding for each line through social moderation [4, 5].

The participants of the online meet-ups comprised not only of student members, but also facilitators and expert observers; however, only student members were included in this analysis. The students were categorized into two groups, those with higher and lower levels of participation. For the purposes of this paper, participation level was determined by how frequently the participant spoke during the meet-up. The total number of lines spoken by each student was tabulated along with the average number of lines spoken by students during each meet-up. Students whose total line count was above the average were classified into the high participation group while those below the average were placed in the low participation group. This meta-data was added to the

Table 1. Codebook of constructs included in the analysis

Code	Description	Example
Content Focus	Dialogue focused on the meet-up's STEM-related educational content	*"I was working on one of the Sustainable Development Goals which was clean water and sanitation"*
Curiosity	Seeking clarification or further information for better understanding of STEM-related content or project	*"What would you be able to move stuff over this one kilometer, hypothetically, or do you need more research on that?"*
Feedback	Communicating one's opinions/ideas or sharing suggestions on projects	*"I think that could be something you could talk about as well, uh with your research... is how cost effective is the UV radiation for purifying water and how can you apply UV radiation in that scale..."*
Information Sharing	Sharing of personal experiences or contextual information relevant to the discussion (not explicit STEM facts)	*"...our local water source is from Lake Mead and we have to treat that water before we can use it. So, we have a wastewater treatment plant here outside of town..."*
Media Production	Dialogue related to the production of media artifacts	*"...we filmed last week so now I'm in the process of editing that video right now"*
Participatory Teaching	Helping others to learn STEM subject matter by providing factual information/content in explanation	*"...we have several ways in which you can use to purify water. We can boil, we can use chlorine and other chemicals...You can also use the UV rays to purify water, and uh according to research, it is one of the best way of purifying water because according to statistics, it is stated that it kills almost ninety-nine point ninety-nine percent of the micro-organisms..."*
Social Disposition	Demonstrating pro-social tendencies, especially in expressing appreciation, acknowledgement or validation	*"Yeah it was really great meeting you all and I think this is a really good opportunity for kids like us to connect with other people who share the same interests..."*

coded dataset for analysis and visualization using the Epistemic Network Analysis (ENA) Web Tool (version 1.5.2) [12].

For the ENA network models, an individual participant was defined as the unit of analysis. A moving stanza window of 5 lines was used to model the connections that the speaker made between the constructs within the recent temporal context. Each meet-up constituted a conversation to which the connections were limited. Within the ENA environment, the network for each unit of analysis was normalized to account for the differences in the number of lines spoken by the participant. This was critical to assessing the overall discourse patterns associated with each individual's participation, regardless of how frequently they had spoken during the meet-ups.

3 Results

The total and average number of lines spoken by students in the high and low participation groups are presented in Table 2. In the April 2017 meet-up, the average number of lines spoken by a student was 11.71, based on which 3 and 4 students were placed in the high and low participation groups, respectively. The same procedure was undertaken for the November 2018 session, with 2 and 4 students each categorized respectively into the high and low participation groups.

Table 2. Number of lines spoken by high and low participation groups during the two meet-ups.

	April 2017 meet-up			Nov 2019 meet-up		
	All	High	Low	All	High	Low
Student participants	7	3	4	6	2	4
Total lines spoken	82	64	18	118	117	71
Avg. lines spoken per student	11.71	21.33	4.50	31.33	58.5	17.75

Figure 1 displays the projected points (circle) and their means locations (square) for the students in the high participation group (red) and those in the low participation group (blue) across both meet-ups. The boxes around the mean locations represent the 95% confidence intervals. It can be seen that the students in the two groups are distinguished along the x-axis, with the high participation group concentrated on the

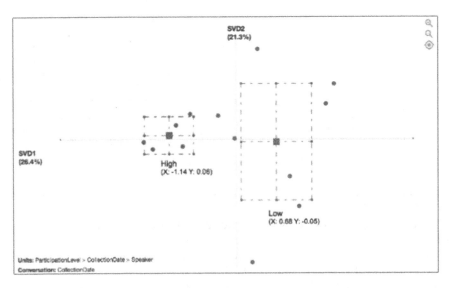

Fig. 1. Projected points and their mean locations shown for the high participation (red) and low participation (blue) groups. (Color figure online)

left side and the low participation group on the right. A two-sample Mann-Whitney U test found statistically significant difference between the distributions of the projected points of the high participation group (Mdn = −1.02, N = 5) and the low participation group (Mdn = 0.64, N = 8) along the x-axis (U = 0, p < 0.01, r = 1.00).

Figure 2 presents the ENA network models for the two groups as well as a subtracted graph depicting the differences between the two networks across both meet-ups. The nodes of the network represent the constructs (codes) and the weights of the

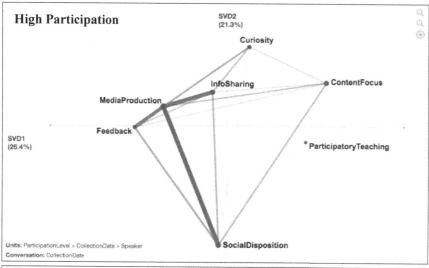

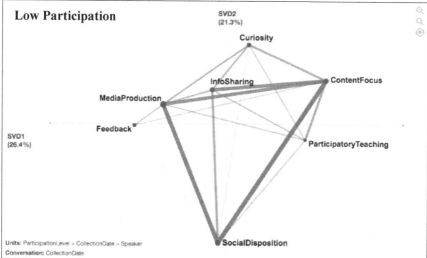

Fig. 2. ENA network models for the high and low participation groups.

thickness of the edges indicate the strength of connection between them. Along the x-axis, the ENA space is distinguished by Feedback and Media Production to the left and Content Focus and Participatory Teaching to the right. For the high participation group, the strongest connections can be seen on the left side of the ENA space between Media Production, Feedback, Info Sharing, and Social Disposition. In contrast, the network for the low participation group displays prominent connections on the right side of the ENA space, with strong linkages between Content Focus, Media Production, Social Disposition and Information Sharing. In addition, Participatory Teaching is moderately associated with several other constructs such as Content Focus and Information Sharing in the low participant group's network, whereas such connections were virtually nonexistent for the high participation group. The differences between the two networks can be seen more clearly in the subtracted network graph shown in Fig. 3.

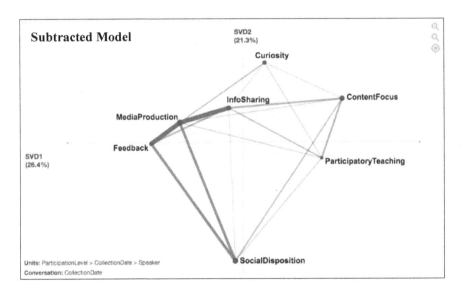

Fig. 3. Subtracted network showing the differences between the two groups.

Furthermore, the ENA network models of the two groups were examined separately for each meetup session to confirm the differences observed in the aggregate models. Figure 4 displays the subtracted models depicting the differences in the discourse patterns between high and low participation groups during each meet-up. Similar to the combined models, the disaggregated networks also demonstrate the tendency for students in the high participation group to connect more strongly to the constructs located on the left side of the ENA space, such as Feedback, and Media Production. The

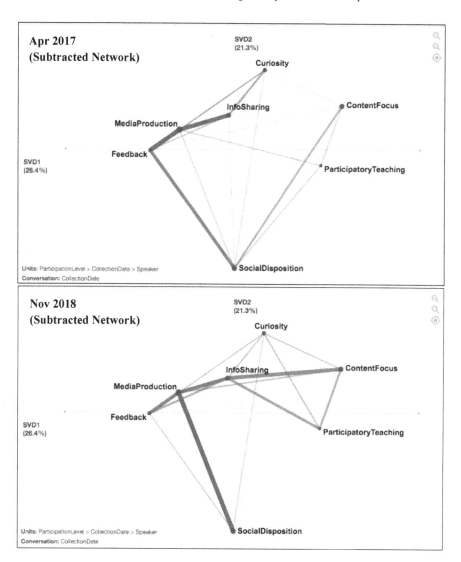

Fig. 4. Subtracted networks showing the differences between the high and low participation groups for each meet-up session.

discourse patterns for the low participation students in each meet-up are focused on the right side around Content Focus and Participatory Teaching, mirroring the associations made in the aggregate model.

Table 3. Excerpt from transcript of dialogue exchange between participants with low and high participation levels

Participation level	Speaker	Utterance
	Facilitator	*How do you guys feel about video? Does it make you like want to go and build a Rube Goldberg machine, because you know that's the purpose of the video, right?*
High	Student A	*Yeah*
	Facilitator	*Okay*
Low	Student B	*As for me, I enjoyed the video but I have a question. So, why did you...work on that machine?*
High	Student A	*What is the what? I didn't hear the middle part, sorry*
	Facilitator	*She's asking...like what's the function of the machine?*
High	Student A	*Um, basically, you sort of have an object at the beginning and it goes through a long obstacle course sort of. And at the end, the purpose of the one in the video was to hit the bell, um but a lot of other people make it to do maybe close, close a door or something. There's like a bunch of purposes that it can fulfill. You should try one*
	Facilitator	*Okay. Do you understand more about the machine? Or do you like need a little more elaboration? Great. Okay, yes. How about you, [Student C]?*
Low	Student C	*Yeah, I'd like to say because you're inviting other kids and other people to come to [a museum] to partake in this, correct?*
	Facilitator	*Yeah, I mean that was the idea in the video, right?*
High	Student A	*Hmmm*
Low	Student C	*So, I'd say in the video you might want to explain what is a Rube Goldberg machine*
High	Student A	*Yeah*
Low	Student C	*Because correct me if I'm wrong, but it's a machine that's overcomplicated to do a simple task, right?*
High	Student A	*Hmm-mm*
Low	Student C	*So, I think maybe just added at the beginning the video so it's not you have to do this one task you can do several things anyway you want. I think that would make it more broad and help like little kids understand who might not know what this is. Other than that, it was good editing, good lighting, overall good video*
High	Student A	*Thank you... Cool lizard*
	Facilitator	*Yeah. I wanted to mention that too. Yes, it's casually just chilling on his shoulder*
High	Student D	*Yeah, I was like... haha*
Low	Student C	*Yeah. This is Aries. He's a bearded dragon*
High	Student A	*Okay. Cool, cool. Cool little lizard*
Low	Student C	*It just chills*

4 Discussion

In examining these two meet-ups, both the high and low participation students had contributions that balanced the group's overall epistemic network. The high participation students had strong connections to Social Disposition, Feedback, Media Production and Info Sharing while low level participation students provided connections to Content Focus and Participatory Teaching. While high participation was associated with more social constructs, low participation balanced the dynamic with an association to content-oriented constructs.

In contrast to the lack of contribution that passive participation is commonly associated with, in this global, collaborative digital makerspace setting, those with low participation still provide substantive contribution to the overall group dynamic. The focus on content would not have occurred as strongly without their participation. With all students contributing to the group in a setting with defined purpose and policies set by the meet-up facilitator, sociability within these meet-ups was achieved. Those with high participation, associated with social constructs, were able to build on the group's established sociability to encourage the development of a social space and social presence through the promotion of socioemotional interaction conveyed through their dialogue. This helped to develop an environment that welcomed and valued contributions from those with low level participation.

The transcript excerpt in Table 3 provides an example of this, where participants with low participation provide substantive contributions to the conversation, followed by social encouragement from high participation level participants. Student B (low participation) asks a question answered by Student A (high participation), which then sparks substantive feedback from Student C (low participation) on suggestions for Student A's project. Student A and Student D (high participation) seem to demonstrate appreciation and social inclusion by commenting on Student C's pet lizard visible in the frame for the duration of the meet-up.

This strong sense of cohesion from the development of sociability was able to overcome the negative behavior of social loafing in a group. Less assertive participants are usually associated with either feeling intimidated, not welcome to contribute, or not knowledgeable enough to offer something useful to the group [13]. However, the online environment seems to contribute to mitigating passive behaviors, like social loafing, in a learning context. A study by Hudson and Bruckman examines participation patterns in an online learning environment, where they observe students to demonstrate less inhibition online versus face to face [6]. In particular, the study examines the case studies of two university students that demonstrated extreme ends of different personalities in the classroom that seemed to neutralize in the online setting. In the face to face environment, one student was extremely confident and seemed to dominate in verbal contributions and frequently directly addressed the professor. In contrast, the other student was shy and did not contribute as much to the conversation, nor addressed anything to the professor directly. However, when class shifted to the online space, this dynamic changed. The same shy student in the classroom became more willing to contribute dialogue to the conversation, even directing comments and questions to the professor.

While the online environment likely plays a factor in the dynamics of the behaviors observed in this study, the aspect of boundary-crossing distributed group collaboration is also likely to play a role in overcoming negative group behaviors such as social loafing. For example, the positive development of social aspects associated with sociability can also be attributed to the sense of community that can be developed over time by the culture of online synchronous interactions within a global, STEM-focused digital makerspace network [3]. Further examination of participation in distributed collaboration environments can provide additional insight into how sociability is established in such settings and create strengthened peer learning opportunities.

4.1   Future Study

In this examination, the achievement of sociability and group cohesion led to over-coming of the inclination of social loafing, and the collective sense to not let down the other group members [11]. This was observed in the context of similarly sized groups within meet-ups. Researchers have theorized that perceived equity of task can a role in a participant's likelihood to exert less effort [7, 10]. If a group is large, a participant is less likely to think their contribution matters or in contrast, the perception that the division of labor is not equitable as others are less skilled or motivated. In both instances, this leads to an individual becoming more inclined to exude social loafing behavior. However, the motivation involved with collaboration across boundaries may foster a sense of virtual migration where participants achieve a social space that overcomes anticipated negative group behavior.

More examples from this setting should be examined to further support and provide greater insight into how sociability is established in the group and its impact on the contributions of those with low level participation. Future work on this examination might consider group size to see if the cohesion and sociability carries on in larger numbers of participants within this online, global collaborative context. The impact of the cultural diversity of participants in this context should also be more thoroughly considered. While this study did not go into depth on consideration for the cultural norms of its participants, this bears further exploration to see if low and high partici-pation levels correlate to the anticipated behavior based on a participant's country of origin, or to what extent this differs.

Analysis of group behavior, like in this study, is usually limited to an examination of transcript dialogue. This does not take into consideration the possibility of nonverbal cues that might contribute to establishing sociability. Those with low participation levels may also be positively contributing to the development of social space and social presence through facial expressions that is not captured in transcript data. Examining video data and taking emotional affect into account in distributed collaboration settings might provide additional insights on how such settings create a positive environment for collaboration and contribution that overcomes the downfalls normally associated with group settings.

The continued work of computer-supported collaborative learning in a global context is still growing. This examination contributes insight to the potential of how such boundary crossing collaboration can overcome typical challenges of group dynamics to positively enhance the learning experience.

Acknowledgements. The authors gratefully acknowledge funding support from the US National Science Foundation for the work this paper reports. Views appearing in this paper do not reflect those of the funding agency.

References

1. Caspi, A., Gorsky, P., Chajut, E.: The influence of group size on nonmandatory asynchronous instructional discussion groups. Internet High. Educ. **6**(3), 227–240 (2003)
2. Cohen, E.G.: Restructuring the classroom: conditions for productive small groups. Rev. Educ. Res. **64**(1), 1–35 (1994)
3. Espino, D.P., Lee, S.B., Eagan, B., Hamilton, E.R.: An initial look at the developing culture of online global meet-ups in establishing a collaborative, STEM media-making community. In: Proceedings of the International Conference for Computer Supported Collaborative Learning (CSCL2019). Lyon, France, International Society for Learning Sciences (2019)
4. Frederiksen, J.R., Sipusic, M., Sherin, M., Wolfe, E.W.: Video portfolio assessment: creating a framework for viewing the functions of teaching. Educ. Assess. **5**(4), 225–297 (1998)
5. Herrenkohl, L.R., Cornelius, L.: Investigating elementary students' scientific and historical argumentation. J. Learn. Sci. **22**(3), 413–461 (2013). https://doi.org/10.1080/10508406.2013.799475
6. Hudson, J.M., Bruckman, A.S.: The bystander effect: a lens for understanding patterns of participation. J. Learn. Sci. **13**(2), 165–195 (2004). https://doi.org/10.1207/s15327809jls1302_2
7. Karau, S., Williams, K.: Social loafing: a meta-analytic review and theoretical integration. J. Pers. Soc. Psychol. **65**(4), 681–706 (1993)
8. Kwon, K., Hong, R.-Y., Laffey, J.M.: The educational impact of metacognitive group coordination in computer-supported collaborative learning. Comput. Hum. Behav. **29**(4), 1271–1281 (2013)
9. Kreijns, K., Kirschner, P.A., Vermeulen, M.: Social aspects of CSCL environments: a research framework. Educ. Psychol. **48**(4), 229–242 (2013)
10. Latané, B., Williams, K., Harkins, S.: Many hands make light the work: the causes and consequences of social loafing. J. Pers. Soc. Psychol. **37**(6), 822–832 (1979). https://doi.org/10.1037/0022-3514.37.6.822
11. Liden, R., Wayne, S., Jaworski, R., Bennett, N.: Social loafing: a field investigation. J. Manage. **30**(2), 285–304 (2004). https://doi.org/10.1016/j.jm.2003.02.002
12. Marquart, C.L., Hinojosa, C., Swiecki, Z., Eagan, B., Shaffer, D.W.: Epistemic Network Analysis (Version 1.5.2) (2018). http://app.epistemicnetwork.org
13. Piezon, S.L., Donaldson, R.L.: Online groups and social loafing: understanding student-group interactions. Online J. Distance Learn. Adm. **8**(4), 1–11 (2005)
14. Preece, J.: Online Communities: Designing Usability, Supporting Sociability. Wiley, New York (2000)
15. Rehm, M., Gijselaers, W., Segers, M.: The impact of hierarchical positions on communities of learning. Int. J. Comput.-Support. Collaborative Learn. **10**(2), 117–138 (2015)

What is the Effect of a Dominant Code in an Epistemic Network Analysis?

Rafael Ferreira Mello[1]([✉]) [iD] and Dragan Gašević[2] [iD]

[1] Universidade Federal Rural de Pernambuco, Recife, Brazil
rafael.mello@ufrpe.br
[2] Monash University, Clayton, Australia
dragan.gasevic@monash.edu

Abstract. This paper investigates how different configuration of epistemic network analysis parameters influence the examination of student interactions in asynchronous discussions in online learning environments. Specifically, the paper investigates strategies for dealing by unintended consequences of a dominant node in epistemic network analysis (ENA). In particular, the paper reports on a study that explored the effects of two different strategies including (i) the use of different dimensions calculated with singular value decomposition (SVD), and (ii) exclusion of a dominant code. Our results showed that the use of different SVDs did not change the influence of a dominant code in the graph. On the other hand, the exclusion of the dominant code led to an entirely different configuration in ENA. The practical implications of the results are further discussed.

Keywords: Epistemic network analysis · Dominant code · Graph analysis

1 Introduction

The technological advancements experienced over the past decade led to the increased adoption of digital learning environments that support online and blended learning [21]. These environments provide several tools that enable interactions between instructors and students. Among these, asynchronous online discussions represent one of the most commonly adopted tools for supporting social interactions and social-constructivist pedagogies [1]. While online discussions are widely used to support student learning, there are many challenges associated with their effective use by teachers and students. As with any learning tool, the simple provision of the tool does not mean that students will know how to use it, and more importantly, how to use it effectively [8].

In this regard, the Community of Inquiry (CoI) model [14] represents a pedagogical model that seeks to understand how asynchronous online communication shapes student learning and cognitive development. The CoI model defines three dimensions (called presences) that together provide a holistic overview of online learning experience: (1) Cognitive presence captures the development of desirable learning outcomes such as critical thinking and knowledge (co-)construction, (2) Social presence outlines the essential role of humanizing relationships among online course participants, and

B. Eagan et al. (Eds.): ICQE 2019, CCIS 1112, pp. 66–76, 2019.
https://doi.org/10.1007/978-3-030-33232-7_6

(3) Teaching presence describes the instructors' role before (i.e., course design) and during (i.e., facilitation and direct instruction) the course.

Several studies have attempted to analyze online discussions under the CoI perspective using manual content analysis [6], natural language processing techniques [11], social network analysis [7], among others. Although they provide valuable results, the existing research still lacks a more qualitative understanding of the discussions. More recently, several works [4, 13] proposed the adoption of Epistemic Network Analysis (ENA) [18] to produce in-depth insights in communities of inquiry. For instance, Rolim et al. [13] demonstrated how ENA could be used to identify links between indicators of cognitive and social presence across different groups of learners. Rolim et al. [13] also showed how the development of the relationship between social and cognitive presence changes over time in an online discussion.

ENA offers a promising direction to evaluate social interactions, with increasing adoption numbers [2, 4, 10]. Nevertheless, it is necessary to de ne several configurations in order to apply ENA adequately. An ENA configuration includes parameters related to the domain of the application, which the user needs to configure depending on the application (e.g., unit of analysis and codes), and the method, which can be changed in order to improve the interpretability of the results (e.g., dimensions identified through singular value decomposition). However, there have been fewer known examples of how changes in the ENA parameters could affect results and the interpretation of the results.

This paper presents the results of an exploratory study about the impact of different parameter configurations, related to the method, on the analysis of student interactions analyzed in terms of the CoI model. Specifically, the study analyzed the approaches that can be used to mitigate the impact of the dominance of a node in an epistemic network. The approaches included the use of different dimensions obtained with singular value decomposition (SVD) and the removal of the dominant code.

2   Background

Epistemic Network Analysis (ENA) [18] is a graph-based analysis technique, built on the theory of epistemic frames [17] and used to investigate associations between concepts. It was created to analyze problems with a relatively small set of concepts characterized by highly dynamic and dense interactions. The first application was to evaluate the student progress on epistemic games [19], which was an environment that simulates novices training to be professionals, further called virtual internships. Within this environment, the students were requested to interact with peers and mentors to develop prototypes for fictitious companies producing content, which was automatically analyzed with ENA.

ENA investigates the relationships among different concepts (called codes) for each analysis unit (e.g. individual students). Two codes are considered related if they appear in the same chunk of text, called stanza, which in [19], is the message of each student within the virtual internships environment. One or more stanzas can be collapsed into a single conversation (stanza window) in order to be analyzed together. Unlike other network analysis techniques, ENA was primarily designed for problems with a relatively small set of concepts characterized by highly dynamic and dense interactions.

To visualize epistemic networks, a two-dimensional representation of the analytics space, called projection space, is derived through a dimensionality reduction algorithm called singular value decomposition (SVD). Moreover, the networks of code relationships for a particular analysis unit (or group of analysis units) are also visualized as undirected graphs. It can also be used to compare the differences between different groups of analysis units in a visualization called subtraction network.

ENA has been largely used in educational settings [3, 4, 7, 10]. For instance, it can be used to examine students' cognitive connections during problem-solving [10], or dynamics and interactions in students' group discourse [3]. In general, ENA is used to investigate the associations between codes of a coding scheme (e.g., the topics extracted from students interactions in virtual internships [19]) where the coding scheme is applied to analyze transcripts of online discussions [3, 7].

Recently, Ferreira et al. [4] and Rolim et al. [12] proposed the adoption of ENA to assess the relationship between cognitive and social presence with the course topics within communities of inquiry. These papers qualitatively investigated the difference between two different groups of students under the perspective of ENA graphs. With this analysis, it was possible to investigate not only the statistical differences between groups of students but also the main characteristics of the students that led to these differences. Moreover, Rolim et al. [13] used ENA to investigate the connections between cognitive and social presence and how those developed over time.

Besides the application of ENA to several domains, the literature reports works related to the configuration of the analysis. For instance, Siebert et al. [20] highlighted the importance of adopting moving windows to established links between codes in ENA. The study showed that for groups interactions (e.g., student teams), the moving windows captured more information than the use of a single utterance as a stanza. Furthermore, Ruis et al. [15] reported the results of an experimental study which revealed that more relevant connections were captured when a moving window with a length of seven was adopted.

However, as ENA is a relatively new analytic technique, several aspects could be further explored in order to improve the readability and effectiveness of its results. In this paper, we address the issue of having a dominant code in an ENA model.

3   Problem Statement and Research Questions

Although [20] and [15] studied the influence of the moving window in the outcome of ENA, other parameters should be taken into consideration. For instance, there is a lacuna in the literature that investigates the impact of the usage of different codes and dimensions calculated through SVD on ENA. Specifically, in this work, we studied different parameter configurations in ENA when there are one or more dominant codes.

In this study, we consider a dominant code when it possesses one or more of the following characteristics:

- It has many more instances, at least the double, than the other codes in dataset used in ENA;
- It is the most connected node in the network according to the results of ENA;
- It is the main factor that explains one axis of the final graph generated by ENA.

This paper reports on a study that reproduced the same analysis performed by Rolim et al. [13], but with different input parameters to avoid dominant codes. The first strategy to reach this goal was the evaluation of different dimensions identified with SVD. As a hypothesis, we intend to evaluate if a different combination of dimensions of SVD could eliminate the code dominance in terms of explanation of one axis. As such, our first research question was:

Research Question 1:
What is the effect of using different dimensions of SVD on the final network when dealing with the problem of a dominant code?

In addition to examining the different SVD combinations, we were interested in exploring whether the exclusion of a dominant code (with more instances and connections) could provide additional insights into the outcome of an ENA. Thus, our second research question was:

Research Question 2:
Does the exclusion of a dominant code impact the information provided by the ENA?

4 Method

4.1 Data

The data used consisted of six offerings of a graduate level research-intensive online course in software at a Canadian public university between 2008 and 2011. In those six offerings, a total of 81 students posted 1,747 messages. As part of the assessment, the students were required to select one research paper on a course topic, record a video presentation, and post a URL to a new course online discussion, in which the other students would engage in the debate around their presentation.

During the first two offerings of the course, student participation was primarily driven by the extrinsic motivational factors (i.e., course grade), with limited scaffolding support. These students composed the control group in this study, which consisted of 37 students who produced 845 messages. After the first two course offers, the scaffolding of discussion participation through role assignments and clear instructions were adopted. In total, 44 students (treatment group) were exposed to this instructional intervention and produced a total of 902 messages [6].

The dataset was coded according to the indicators of social presence and the phases of cognitive presence. Initially, the coders annotated 1 and 0 for the presence and absence of an indicator of social presence following the scheme defined by Rourke et al. [14]. The coders achieved a high level of agreement, with all of the indicators reaching a percentage of agreement of at least 84% [9]. It is important to highlight that each message could have more than one indicator. Thus, the final number of annotations was 3,770, instead of 1,747 (the total number of messages). Besides, some of the indicators (i.e., Continuing a thread, Complementing, and Vocatives) were excluded from the analysis due to a disproportionately large number of messages with such codes [9]. Table 1 details the distribution of messages per indicator.

Table 1. Distribution of social presence indicators [9]

Category	Indicator	Messages	Percentages
Affective	Expression of emotions	288	16.5%
	Use of humor	44	2.52%
	Self-disclosure	322	18.4%
Interactive	Continuing a thread	1,664	95.2%
	Quoting from others messages	65	3.72%
	Referring explicitly to other's messages	91	5.21%
	Asking questions	800	45.8%
	Complementing, expressing appreciation	1391	79.6%
	Expressing agreement	243	13.9%
Cohesive	Vocatives	1,433	82%
	Addresses or refers to the group using inclusive pronouns	144	8.24%
	Phatics, salutations	1,281	73.3%

Cognitive presence was coded in one of the four phases of cognitive presence and the absence of any was coded as other. Initially, the messages were coded by two expert coders according to the phases of cognitive presence as suggested by Garrison et al. [5]. The coders achieved an excellent level of agreement for both presences reaching a (percentage of agreement = 98.1%, Cohens k = 0.974) with a total of only 32 disagreements which were resolved through discussion. Table 2 presents the number of messages coded into the cognitive presence's phases.

Table 2. Distribution of cognitive presence.

ID	Phase	Messages	Percentages
0	Other	140	8.01%
1	Triggering event	308	17.63%
2	Exploration	684	39.15%
3	Integration	508	29.08%
4	Resolution	107	6.13%
Total		1,747	100.00%

4.2 Epistemic Network Analysis

This paper addresses the problem of a dominant code using the work proposed by Rolim and colleagues [13]. Rolim et al. propose the adoption of ENA to analyze the relationship between social and cognitive presences in asynchronous online discussions.

The ENA initial configuration used binary codes representing the presence and the absence of the social presence indicators and the 5 categories (4 phases + others) of cognitive presence as ENA codes. The units of analysis and stanzas were individual students and students' discussion messages, respectively.

This configuration was also used to address the research questions studied in this work. Initially, we explored different configurations of SVD dimensions of for the mean network of all students together. We decided to investigate all the combinations of SVD1, SVD2 and SVD3 dimensions because they each explained a variance higher than 10%. Then, we evaluated a different configuration removing the most dominant codes. Finally, we analyzed the students in the treatment/control groups before and after the removal of the most dominant codes. The differences between the student groups on different configurations of SVDs were then compared by using a series of the Mann-Whitney [16] with $\alpha = 0.05$. The subtraction network was used to explain the qualitative differences between the student groups.

5 Results

As described before, for the purpose of this analysis, the dominant code has more instances and is highly connected in comparisons to other codes, or it is the main factor that explains one axis of the ENA.

Figure 1(a) presents the mean network produced by original ENA configuration for all students in our dataset using SVD1 and SVD2. It reveals the predominance of the codes salutation (higher number of instances and connections) and asking question (which explains the right side of SVD1). The rest of the indicators of social presence were plotted at the center of the graph. This configuration could bias the final analysis as these codes are dominating the analysis regarding the social presence.

Figures 1(b) and (c) show the networks for SVD1 x SVD3 and SVD2 x SVD3, respectively. In these alternative networks, all social presence indicators continue to be plotted in the middle of the graph while asking question and salutation have a predominant position. Finally, Fig. 1(d) presents the same network without the dominant codes. In contrast to the previous graphs, this one shows the indicators of social presence well divided into the network. On the other hand, cognitive presence changed from a well-distributed arrangement to a concentrated plot where four out of five categories were in the same quadrant.

We evaluated the differences between control and treatment groups in all configuration. For the initial three set-ups, different SVDs configuration, the statistical analysis reached the same result $U = 2165.50$; $p = 0.001$; $r = 0.37$ using Mann-Whitney test. On the other side, there was a slightly change in the results after the removal of the dominant code ($U = 2327.00$; $p = 0.001$; $r = 0.32$), but the difference

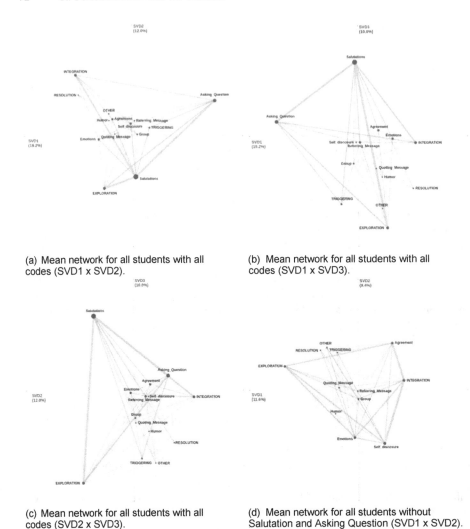

(a) Mean network for all students with all codes (SVD1 x SVD2).

(b) Mean network for all students with all codes (SVD1 x SVD3).

(c) Mean network for all students with all codes (SVD2 x SVD3).

(d) Mean network for all students without Salutation and Asking Question (SVD1 x SVD2).

Fig. 1. ENA network with different parameters configuration.

between the groups continued to be statistically significant. Figures 2(a) and (b) present the main difference, between the ENA original version (with SVD1 x SVD2) and the version without the dominant codes, in more details using subtraction networks between control (red) and treatment (blue) groups.

The analysis of the network graphs considering only the connections showed that in Fig. 2(a) the treatment group had stronger connections, while Fig. 2(b) presents that the control group had more dominant links. Moreover, these figures presented different behaviors related to the ENA codes, while the Fig. 2(a) has a distribution of cognitive presence over different quadrants, in Fig. 2(b) the social presence is more spread over the graph.

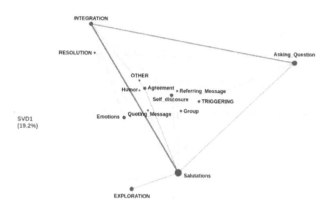

(a) Subtraction network between control (red) and treatment (blue) group with all codes.

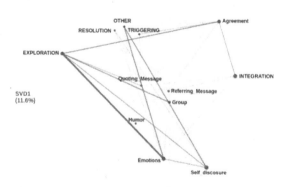

(b) Subtraction network between control (red) and treatment (blue) group without Salutation and Asking Question.

Fig. 2. ENA subtraction network with different parameters configuration. (Color figure online)

6 Discussion

The initial set-up of the ENA (Fig. 1(a)) led to the interpretation of the graphs only based on the categories of cognitive presence, as they were well-distributed across the different quadrants of the final graph. Thus, the evaluation of different configurations of the ENA could increase the readability of the final graph.

The study showed that even by changing the SVDs, the codes asking question and salutation remained to be dominant. The rest of the indicators of social presence continue to be plotted at the center of the graph. This result indicated that the replacement of the SVD dimensions did not change significantly the interpretation of the original results which were drawn by plotting SVD1 and SVD2. In addition to not reducing the effect of the dominant code, this approach of using other SVD dimensions (SVD1 x SVD3 and SVD2 x SVD3) decreased the variance explained by each axis used in the diagrams. On the other hand, after excluding the dominant codes, the ENA graph improved the presentation of the distribution of the codes related to social presence, as they are plotted in different quadrants (Fig. 1(d)) of the graph instead of just the middle as presented in Fig. 1(a). This new graph allows a better analysis about what is happening to social interactions on the online discussion.

The comparison of the control and treatment groups showed the impact on the final ENA network after the of removing dominant codes. Although the differences between the groups continued to be significant, the subtraction network revealed a shift in the most dominant relationships among codes from the treatment group to the control group. This happened because the students in the treatment group increased the numbers of the integration and resolution messages with the asking question and salutation indicators. Thus, the removal of these two codes (asking question and salutation) decreased the impact of the cognitive presence phases on the results. Figures 1(d) and 2 (b) groups exploration, resolution, triggering and other categories in the top-left corner of the graph. This grouping reduced the impact of the codes of cognitive presence and emphasized the role of the codes social presence in the ENA graphs.

Therefore, the results reveal that the best approach to deal with the problem posed in this study is the removal of the dominant code. However, these findings need to be further validated on different datasets.

7 Final Remarks

This paper analyzed the behavior of student interaction in online discussions under the perspective of the community of inquiry framework using different configurations of ENA parameters. The results showed that the variation of SVD dimensions had low influence on the final results. The removal of dominant codes presented a high impact on the analysis of different groups performance. The key implication of our results is that the removal of the codes can lead to an improved understanding of one of the two key constructs in the CoI model (social presence) but reduced insights into the other one (cognitive presence).

Based on these results, we can suggest for studies that aim to look at the holistic associations of different constructs (in our study - social and cognitive presences), the approach that removes dominant codes is undesirable. However, when the insight into the construct that had dominant codes in epistemic networks (i.e., social presence) is the goal, the removal approach could lead to plausible results.

There are some limitations of the present study which should be acknowledged. First, the data is from a single course at a single institution, although from six-course offerings. This can negatively affect the degree of potential generalization of the analysis. Besides, the method required many methodological decisions, such as deciding on the coding used for cognitive and social presences and the parameters that we variate. It might be the case that different decision would lead to different findings. To address these limitations, we intend to perform the same analysis using data from another scenario and applying different parameters as input to the ENA.

As future work, we intend to: (i) reproduce the same analysis using a different context of the application, such as data from other courses, and analyze other phenomena different from the model of community of inquiry; (ii) explore what is the impact of change of other ENA parameters such as the length of the sliding window for stanza, and the order which the instances were presented in the input file; and (iii) apply statistical tools to determine whether two graphs, for instance graphs generated with different SVDs, are significantly different, following for instance the methodology used by Swiecki and colleagues [22].

References

1. Anderson, T., Dron, J.: Three generations of distance education pedagogy. Int. Rev. Res. Open Distrib. Learn. **12**(3), 80–97 (2011)
2. Arastoopour, G., Shaffer, D.W., Swiecki, Z., Ruis, A., Chesler, N.C.: Teaching and assessing engineering design thinking with virtual internships and epistemic network analysis. Int. J. Eng. Educ. **32**(2), 1492–1501 (2016)
3. Cai, Z., Eagan, B., Dowell, N., Pennebaker, J., Shaffer, D., Graesser, A.: Epistemic network analysis and topic modeling for chat data from collaborative learning environment. In: Proceedings of the 10th International Conference on Educational Data Mining (2017)
4. Ferreira, R., Kovanović, V., Gašević, D., Rolim, V.: Towards combined network and text analytics of student discourse in online discussions. In: Penstein Rosé, C., et al. (eds.) AIED 2018. LNCS (LNAI), vol. 10947, pp. 111–126. Springer, Cham (2018). https://doi.org/10.1007/978-3-319-93843-1_9
5. Garrison, D.R., Anderson, T., Archer, W.: Critical thinking, cognitive presence, and computer conferencing in distance education. Am. J. Distance Educ. **15**(1), 7–23 (2001). https://doi.org/10.1080/08923640109527071
6. Gašević, D., Adesope, O., Joksimović, S., Kovanović, V.: Externally-facilitated regulation scaffolding and role assignment to develop cognitive presence in asynchronous online discussions. Internet High. Educ. **24**, 53–65 (2015)
7. Gašević, D., Joksimović, S., Eagan, B.R., Shaffer, D.W.: SENS: network analytics to combine social and cognitive perspectives of collaborative learning. Comput. Hum. Behav. **92**, 562–577 (2019)
8. Gašević, D., Mirriahi, N., Dawson, S., Joksimović, S.: Effects of instructional conditions and experience on the adoption of a learning tool. Comput. Hum. Behav. **67**, 207–220 (2017)

9. Kovanovic, V., Joksimovic, S., Gasevic, D., Hatala, M.: What is the source of social capital? the association between social network position and social presence in communities of inquiry (2014)

10. Nash, P., Shaffer, D.W.: Mentor modeling: the internalization of modeled professional thinking in an epistemic game. J. Comput. Assist. Learn. **27**(2), 173–189 (2011)

11. Neto, V., et al.: Automated analysis of cognitive presence in online discussions written in Portuguese. In: Pammer-Schindler, V., Pérez-Sanagustín, M., Drachsler, H., Elferink, R., Scheffel, M. (eds.) EC-TEL 2018. LNCS, vol. 11082, pp. 245–261. Springer, Cham (2018). https://doi.org/10.1007/978-3-319-98572-5_19

12. Rolim, V., Ferreira, R., Kovanović, V., Gašević, D.: Analysing social presence in online discussions through network and text analytics. In: 2019 IEEE 19th International Conference on Advanced Learning Technologies (ICALT), vol. 2161, pp. 163–167. IEEE (2019)

13. Rolim, V., Ferreira, R., Lins, R.D., Găsević, D.: A network-based analytic approach to uncovering the relationship between social and cognitive presences in communities of inquiry. Internet High. Educ. **42**, 53–65 (2019)

14. Rourke, L., Anderson, T., Garrison, D.R., Archer, W.: Assessing social presence in asynchronous text-based computer conferencing (1999)

15. Ruis, A., Siebert-Evenstone, A., Pozen, R., Eagan, B.R., Shaffer, D.W.: A method for determining the extent of recent temporal context in analyses of complex, collaborative thinking. In: International Society of the Learning Sciences, Inc. [ISLS] (2018)

16. Ruxton, G.D.: The unequal variance t-test is an underused alternative to student's t-test and the mann–whitney u test. Behav. Ecol. **17**(4), 688–690 (2006)

17. Shaffer, D.W.: Epistemic frames for epistemic games. Comput. Educ. **46**(3), 223–234 (2006)

18. Shaffer, D.W., Collier, W., Ruis, A.: A tutorial on epistemic network analysis: analyzing the structure of connections in cognitive, social, and interaction data. J. Learn. Anal. **3**(3), 9–45 (2016)

19. Shaffer, D.W., et al.: Epistemic network analysis: a prototype for 21st-century assessment of learning. Int. J. Learn. Media **1**(2), 1–21 (2009)

20. Siebert-Evenstone, A.L., Irgens, G.A., Collier, W., Swiecki, Z., Ruis, A.R., Shaffer, D.W.: In search of conversational grain size: modeling semantic structure using moving stanza windows. J. Learn. Anal. **4**(3), 123–139 (2017)

21. Siemens, G., Gašević, D., Dawson, S.: Preparing for the digital university: A review of the history and current state of distance, blended, and online learning (2015)

22. Swiecki, Z., Ruis, R., Shaffer, D.: Does order matter? investigating sequential and cotemporal models of collaboration. In: 13th International Conference on Computer-Supported Collaborative Learning (2019)

Tracing Identity Exploration Trajectories with Quantitative Ethnographic Techniques: A Case Study

Aroutis Foster[1](✉) , Mamta Shah[2], Amanda Barany[1] ,
and Hamideh Talafian[1]

[1] Drexel University, Philadelphia, PA 19104, USA
anf37@drexel.edu
[2] Elsevier Inc., Philadelphia, PA 19103, USA

Abstract. This paper is situated in a 5-year NSF CAREER project awarded to test and refine Projective Reflection (PR) as a theoretical and methodological framework for facilitating learning as identity exploration in play-based environments. 54 high school students engaged in *Virtual City Planning*, an iteratively refined course that provided systematic and personally-relevant opportunities for play, curricular, reflection and discussion activities in *Philadelphia Land Science*, a virtual learning environment, and in an associated curriculum enacted in a STEM museum-classroom. In-game logged data and in-class student data were examined using Epistemic Network Analysis. An illustrative case study revealed visual and interpretive patterns in students' identity exploration. The change was reflected in their knowledge, interest and valuing, self-organization and self-control, and self-perception and self-definition (KIVSSSS) in relation to the roles explored from the start of the intervention (Starting Self), during (Exploring role-specific Possible Selves), and the end (New Self).

Keywords: Identity exploration · Epistemic Network Analysis · Case study · Projective Reflection · Virtual learning environments · Game-based learning

1 Introduction

Virtual learning environments (VLEs) such as games are forms of play-based environments that can present players with opportunities for self-transformation [1] and enculturation [2]. Specifically, VLEs can be conducive for (a) catalyzing learner interest in science and exploring career roles as future possible selves, and (b) using these role-possible selves as anchors for cognitive, social and affective skill development that can prepare learners to explore and reconstruct their identities [3–5]. The theoretical promise of VLEs and advancements in game studies are well documented [6]; however, there is a scarcity of evidence-based practices to support researchers in designing and leveraging the affordances of gaming technologies. Additional work is needed to comprehend the theoretical and methodological processes by which VLEs affect educational outcomes [7]. With this emerging area of research comes the need for

© Springer Nature Switzerland AG 2019
B. Eagan et al. (Eds.): ICQE 2019, CCIS 1112, pp. 77–88, 2019.
https://doi.org/10.1007/978-3-030-33232-7_7

research methods that complement emerging theories to elucidate how learners evolve in data-rich and dynamic play-based environments such as VLEs [8].

To address these gaps, we introduce Projective Reflection [1] as a theoretical and methodological approach to frame and facilitate learning as identity exploration in digital and non-digital play-based environments. Projective Reflection (PR) refers to the processes by which a person engages in intentional exploration of role-possible selves (i.e. roles connected to future selves a learner may want to be) in play-based environments, while projecting forward and reflecting on who they are in relation to specific domains and careers such as a STEM professional [1].

In this study, Projective Reflection structured the design and implementation of *Virtual City Planning (VCP)*, a play-based course that included identity exploration experiences mediated by a VLE (*Philadelphia Land Science*), and classroom experiences based on *PLS*. Quantitative Ethnography (QE) techniques [9] were applied to visualize and interpret changes in identity exploration trajectories at the cohort and individual levels as a result of exploring the role-possible selves of an environmental scientist and urban planner in *VCP*. In doing so, the following research question was answered: *"What is the nature of a high school student's trajectory of identity exploration over time in a play-based course as defined by Projective Reflection?"*

2 Theoretical Framework

Projective Reflection (PR) is a theory and methodology of learning that integrates a focus on content ((I)dentity anchored in a specific community of practice and enacted locally) and on the self ((i)dentity engaged in role-possible selves inspired by the community of practice reflecting an individual goal) in an integrated manner (i/Identity). Play-based experiences informed by Projective Reflection are designed to promote intentional exploration of specific and targeted roles in an authentic socially-situated environment. *Virtual City Planning*, for example, immersed students in the role of urban planning interns in Philadelphia collaborating on the research and development of design proposals that address environmental and economic issues facing their communities.

Four theoretical constructs support exploration of identities through role-possible selves in PR to enable an integrated change in learners over time: (1) **K**nowledge (foundational, meta, humanistic) [10], (2) **I**nterests (situated/perceptual, epistemic/personal) and **V**aluing [11], (3) patterns of **S**elf-organization and **S**elf-control (co-regulation, socially-shared regulation, and self-regulation) [12], and (4) **S**elf-perceptions and **S**elf-definitions (self-concept, self-efficacy) [13] (**KIVSSSS**). Table 1 provides a more in-depth explanation of the constructs and sub-constructs as manifested by students in *Virtual City Planning*.

The PR constructs (KIVSSSS) are mapped along six questions: (1) what the learner knows – current knowledge, (2) what the learner cares about – self and interest/valuing, (3) what/who the learner expects to be throughout the virtual experience and their long term-future self, (4) what the learner wants to be – possible self, (5) how the learner thinks – self and interest, and (6) how the learner sees him/herself – self-perceptions and self-definitions. The six questions are used to scaffold and track the change in

Table 1. Projective Reflection construct definitions.

PR constructs	Definitions	Sample citations
Knowledge and game/technical literacy	Shifts in what a player knows about environmental science, urban planning, and urban planning systems from the beginning to the end of an intervention: • *Foundational knowledge*: awareness of complex and domain-specific content and processes that includes the ability to access information using digital technologies • *Meta-knowledge*: awareness of how to use foundational knowledge in relevant socially-situated contexts • *Humanistic knowledge*: awareness of the self and one's situation in a broader social and global context	[10]
Interest and valuing	• Caring about environmental science and urban planning issues and viewing them as personally relevant or meaningful • Shifts in identification with environmental science • Viewing environmental science and urban planning as being relevant to the community or the world • Seeing the need for environmental science for self and for use beyond school contexts	[11, 14, 15]
Self-organization and self-control	Shifts in behavior, motivation, and cognition toward a goal: • *Self-regulated learning:* goal-setting and goal-achievement conducted independently • *Co-regulated learning*: self-regulation processes supported by more knowledgeable real/virtual mentors • *Socially-shared learning*: self-regulation is socially-shared and defined in collaboration with peers	[12, 16, 17]
Self-perceptions and self-definitions	Shifts in how a participant sees himself/herself in relation to (environmental) science: • *Self-efficacy*: confidence in one's own ability to achieve goals and future roles • *Self-concept*: awareness of current aspects of self (i.e. skills, preferences, characteristics, abilities, etc.) • Specific roles one wants or expects to become in future	[13]

(a) learners' initial current self or starting self (SS) that is established at the start of an intervention, (b) their exploration of multiple role-possible selves (EPS measured repeatedly across an intervention), and the new self (NS) at the end of the intervention/experience [1] (See Fig. 1).

The Play, Curricular activity, Reflection, Discussion (PCaRD) pedagogical model for play-based learning (See [18] for more information) can be used to enact Projective Reflection in a game and game-based curricula through Play, and outside the game through Curricular activities that include opportunities for Reflection and Discussion. This process can provide students with opportunities to intentionally construct knowledge, cultivate interest and valuing for the academic domain, develop

PROJECTIVE REFLECTION

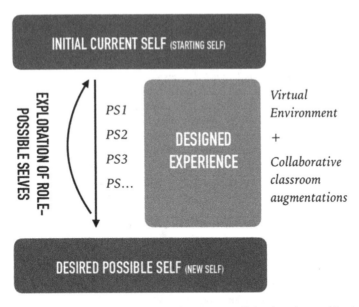

Fig. 1. The Projective Reflection framework for conceptualizing learning as identity change.

competence with the learning content and context, and explore possible selves in relation to the domain in an explicit manner [3]. While CaRD activities are anchored in the game and build on students' play experiences, they are also designed to align with target knowledge and attitudes within chosen academic domains. Learners are encouraged to engage in targeted and intentional reflection on aspects of self at repeated points throughout the situated designed-experience (SS, EPS, NS). For each student, the process of tracing change is carried out chronologically, guided by the six questions that elucidate how learners explored a possible role self as intentional changes in KIVSSSS - indicating the extent to which the identity exploration process was comprehensive or integrated.

3   Methods

This research was conducted as part of a 5-year (2014–2019) NSF CAREER project awarded to advance theory and research on promoting identity exploration and change in science using VLEs through Projective Reflection [1]. Building on this broader agenda, *Virtual City Planning (VCP)*, a play-based course, was designed, developed, implemented, and refined using design-based research [19] from 2016–2017 to help freshman high school students (a) construct foundational-meta-humanistic knowledge for urban planning which were aligned with the Next Generation Science Standards (NGSS) for high school environmental science, (b) generate and sustain interest in environmental science, (c) enable global and personal valuing in environmental science, and (d) explore multiple identities (role-possible selves) related to environmental science in an urban context.

3.1 Data Collection and Procedures

VCP1 (the first of the three iterations) was offered in Fall for 9 weeks with 20 students. It featured weekly use of both the virtual learning environment *Philadelphia Land Science (PLS)* and supportive real-world augmentation in the classroom. *VCP1* consisted of (a) internal aspects - VLE experiences informed by the mechanics of *PLS*, and (b) external augmentations - activities that occur in designed spaces outside of the *PLS*, but as a result of VLE roleplay. The PCaRD model informed the design of both internal aspects and external augmentations to facilitate identity change. Data sources included logged in-game data, pre-post assessments, classroom artifacts, written reflections, and researcher observations. Participating students role-played as urban planning and environmental science interns in a fictional firm called Regional Design Associates. Over nine weeks, students learned about the process of proposing a rezoning proposal for the city of Philadelphia that addresses the competing and complementary needs of four stakeholders. *PLS* and external curricular activities were designed to facilitate intentional shifts in what learners know, how they think, what they care about, how they see themselves, and what they want and expect to be in relation to ES and urban planning. In week 1, students learned about their teams and the expected workflow of *PLS*. Each student's *starting self* was documented through an intake interview in *PLS*, background survey, 5-point Likert-scale survey, and a focus group discussion. In weeks 2 and 3, students reviewed the 'Request for a Rezoning Proposal,' and researched stakeholders' concerns related to commercial, residential, environmental, civic, and social issues. This helped them understand what was expected of them as urban planners. In Weeks 4 and 5, students edited interactive models of Philadelphia using the 'iPlan' mapping tool to test the extent to which their rezoning proposal met stakeholder needs. They received feedback from the stakeholders and their mentors, leading to iterative refinements in weeks 6–8. Collected data included in-game logged data such as chats, notebook entries, and iPlan maps, as well as researcher observation memos. In week 9, students worked towards finalizing a written document explaining the rezoning plan and representing their plan on a map. Data was collected to examine each student's *new self* included an in-game exit interview and a focus group discussion. *Starting self* changes were documented through weeks 1–2 data, *exploring role-possible selves* in weeks 3–7 data, and *new self* in weeks 8–9 data.

4   Data Analysis

Student data was coded inductively and deductively to answer the research question. Quantitative Ethnography [9] was used to guide the data analysis procedures; researchers engaged in a deductive coding process for each student/case [20] using the qualitative data analysis software MAXQDA 2018. Lines of student data were coded as self-reflection on or demonstration of the four PR constructs and their sub-constructs, with agreement reached by two graduate-level coders. The qualitatively-coded data was then quantified. Each line was coded for the occurrence (1) or non-occurrence (0) of the four constructs of PR that defined identity exploration (e.g. knowledge) and sub-constructs (e.g. foundational knowledge) to prepare the data for Epistemic Network Analysis (ENA). Coding was completed by two graduate-level coders, who then discussed all coding inconsistencies until agreement was reached. While ENA and qualitative analyses were applied to all 54 students and the aggregated class data, this proposal focuses on a single student case: Zola (pseudonym).

Processes of identity exploration as defined by Projective Reflection are most valuable when students can enact them in an integrated fashion; that is, when students can regularly connect Knowledge gains, emerging personal Interests and Values, the enactment of Self-organization and Self-control strategies, and specific Self-perceptions and Self-definitions in a domain (KIVSSSS). Given that identity exploration is conceptualized as a developmental process of change over time, it is vital to examine identity integration as not only co-occurrences in a single piece of data (integration in that moment), but also as a longitudinal relationship between the codes manifested from one moment to the next across a meaningful unit of time (a conversation). To achieve this, ENA generates network visualizations of the co-occurrence of codes within a moving stanza window: codes applied to one line of student data are connected to each other and to codes applied to the previous 3 lines of chronological student data (as recommended by [21]). In this way, epistemic networks of construct relationships were leveraged to visualize chronological changes in Zola's process of identity exploration as it shifted from her Starting Self (weeks 1–2) to Exploring Possible Selves (weeks 3–7) to New Self (weeks 8–9). Mann-Whitney nonparametric tests were used to test whether change across each time period was statistically significant.

We referred back to interactions and activities coded in the data to close the interpretive loop and thus fully understand the phenomenon illustrated in Zola's chronological epistemic networks. This last step was relevant for both individual and group findings as instrumental case studies [22], enabling the researchers to highlight the dominant issue for this paper; that is, nature of participants' identity exploration.

4.1   Case Study: Zola

Zola was a 14-year old Caucasian female student. At the start of *VCP*1, in the pre-survey, Zola indicated averaging spending 2–3 h each week playing a variety of tabletop/board games and digital games on consoles and mobile computing devices to socialize with friends, to learn about a topic, and to practice new skills. Her responses were indicative of her technical literacy and media practices. Additionally, she strongly agreed in her confidence and ability to learn with VLEs.

Zola began the intervention (See Fig. 2) with confidence in her understanding of environmental science and the role of urban planners. More interestingly, Zola's responses demonstrated her attempts to grapple with the complexity of urban planning terms and concepts. For instance, when asked "What is a stakeholder? How do stakeholders affect Philadelphia?" Zola remembered:

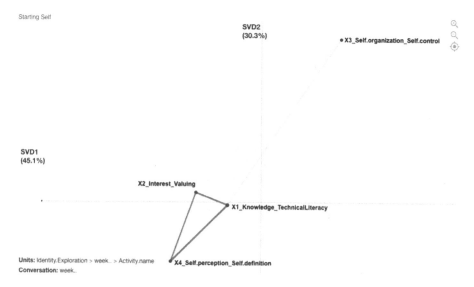

Fig. 2. Epistemic network for Zola's Starting Self in weeks 1–2 of participation in *VCP*.

A stakeholder is a company, group or a person who is expressing a want for the city and plans to act on it...If a stakeholder requests something and some people are against it, it may affect how people view the government/officials and people like that. On the positive side, stakeholders start the process of getting things the city wants happen, and if those things happen, they affect Philadelphia.

Zola's interest in learning about cities and the environment were scaffolded by her mother, who worked with urban planners. Zola's patterns of participation in *VCP*1 portrayed her as detailed and goal-oriented, with a desire to complete each task fully. Mentors commented in observation memos that Zola "took her time to complete the tasks even if it meant writing during break time." When asked if she saw herself in a future career that contributes to urban planning, Zola replied, "It's possible. I'm not exactly sure what it [an urban planning career] entails, but helping to change things about the city to make it better for the community sounds like a career I might want to pursue." Zola also described being able to explore "a lot of [STEM] careers that seem interesting" at school might further her awareness and shape her ability to pursue them.

As the course progressed, Zola demonstrated her emergent foundational, meta and humanistic knowledge as she role played as an urban planner. This knowledge increase led her to confidently justify her recommended changes. She provided detailed justifications for each zoning change she made to the Philadelphia map, and connected them to her stakeholder's needs and feedback (e.g. "Arnold suggested developing local businesses along the Schuylkill river, so I added a few commercial areas"). Concurrently, Zola described the zoning changes she would make to the city map to reflect her own values (e.g. an increase in affordable, medium-density housing "for struggling families, and if they have homes, they have a higher chance of getting jobs.").

Through weeks 3–7 (See Fig. 3), Zola actively participated in group meetings with peers, discussing the needs of their assigned stakeholder group, the numerical economic and environmental levels they needed to strive for to meet the stakeholder needs, and strategies for balancing various needs and numbers when creating their maps. On one occasion, Zola was observed brainstorming with a doctoral student on the technical aspects of *Philadelphia Land Science* (e.g. how to view her customer input map, notebook entry and land use codes at the same time). During this time, Zola also articulated how she reconciled what she personally valued with the values of assigned stakeholder groups. For instance, in a preference survey reflection, Zola expressed that she did not agree with certain changes that resulted from the stakeholder needs because they conflicted with her personal values (e.g. removing land for bluebirds). She reasoned that environment was important for her and as a result, "[I]t was hard to see how adding housing and businesses in place of green space could be good. However, the way the stakeholders talk about it, I really see the benefits of expansion." As the weeks progressed, Zola began using statements that indicated her continued embodiment of an urban planner role by suggesting zoning changes and using evidence to support her claims.

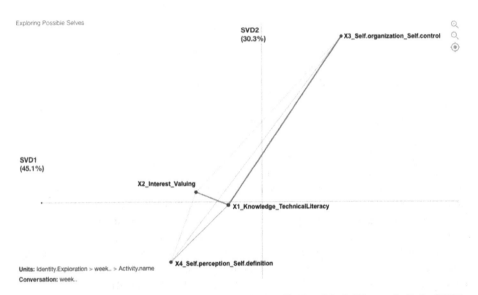

Fig. 3. Epistemic network for Zola Exploring a role-specific Possible Self in weeks 3–7 of *VCP*.

Towards the end of *VCP*1 (See Fig. 4) Zola's understanding of what urban planners do deepened, although her understanding of environmental science remained relatively unchanged. She wrote "[an urban planner] studies the housing and the business to know what need to be fixed and how to design artifacts" and "[environmental science] is something that you study outside. Such as the environment, animals, and other things in the environment." Whereas Zola's initial interest in learning about cities and the environment stemmed primarily from her mother's profession in urban planning, Zola completed the course expressing fascination, and confidence in her ability at intentionally making way for communities to flourish while also protecting the environment. She was curious to know what decisions make some cities establish this balance and some fail at it. When asked what she wanted to be in the future, she said that she was 'totally sure at this moment' that she wanted to be able to "evoke emotion from people. May it be through script writing for movies, illustrating boards, writing books to creating 3D models for special effects."

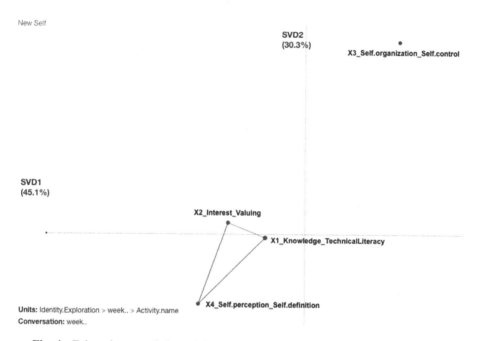

Fig. 4. Epistemic network for Zola's New Self in weeks 8–9 of participation in *VCP*.

Nonparametric Tests of Significance. Mann-Whitney tests were run to test whether there were statistically significant differences along the X and Y axes between Zola's Starting Self (weeks 1–2), Exploring Possible Selves (weeks 3–7), and New Self data (weeks 8–9).

No statistically significant differences were found between Starting Self and Exploring Possible Selves data. Along the X axis (SVD1), a Mann-Whitney test showed that Starting Self (Median = –0.44, N = 6) was not statistically significantly different at the alpha = 0.05 level from Exploring Possible Selves (Median = 0.53, N = 19 U = 50.00, p = 0.66, r = 0.12). Along the Y axis (SVD2), a Mann-Whitney test showed that Starting Self (Median = 0.29, N = 6) was not statistically significantly different at the alpha = 0.05 level from Exploring Possible Selves (Median = 0.42, N = 19 U = 51.00, p = 0.71, r = 0.11).

Similarly, no statistically significant differences were found between Starting Self and New Self data. Along the X axis (SVD1), a Mann-Whitney test showed that Starting Self (Median = –0.44, N = 6) was not statistically significantly different at the alpha = 0.05 level from New Self (Median = 0.53, N = 6 U = 14.00, p = 0.53, r = 0.22). Along the Y axis (SVD2), a Mann-Whitney test showed that Starting Self (Median = 0.29, N = 6) was not statistically significantly different at the alpha = 0.05 level from New Self (Median = 0.42, N = 6 U = 7.00, p = 0.06, r = 0.61).

While no statistically significant differences were found between Exploring Possible Selves and New Self data along the X axis, differences were identified across the Y axis. Along the X axis (SVD1), a Mann-Whitney test showed that Exploring Possible Selves (Median = 0.53, N = 19) was not statistically significantly different at the alpha = 0.05 level from New Self (Median = 0.53, N = 6 U = 47.00, p = 0.49, r = 0.18). Along the Y axis (SVD2), a Mann-Whitney test showed that Exploring Possible Selves (Median = 0.42, N = 19) was statistically significantly different at the alpha = 0.05 level from New Self (Median = 0.42, N = 6 U = 35.00, p = 0.12, r = 0.39).

Examination of epistemic networks across the three time periods for Zola suggest that her integration of Projective Reflection constructs shifted to a statistically significant degree from Exploring Possible Selves to New Self. However, a visual examination of the network for Exploring Possible Selves (See Fig. 3) as compared to the network for New Self (See Fig. 4) suggests that her connections between constructs became less pronounced over time and that her identity exploration in the final weeks may have been less integrated. Review of the Zola's data in the final weeks revealed that while Zola's reflections were qualitatively rich, a classroom absence limited the number of reflections she offered on her emerging New Self. Zola's epistemic networks proved valuable for illustrating some emerging connections she made across PR constructs during *Virtual City Planning*, however, these findings also illustrate the potential limitations of epistemic network analysis for illustrating identity exploration processes if implemented in isolation. Detailed knowledge of both the dataset and theoretical framework are essential for leveraging the benefits of epistemic networks. In-depth qualitative analysis of data can also serve as a valuable triangulation tool when analyzing processes of identity exploration.

5   Discussion

Scholars argue that opportunities for intentional and repeated examination of who a learner is and who he/she wants to become are valuable for supporting learner agency and participation in a constantly changing society [5, 13]. In this study, the Projective Reflection theoretical and methodological framework [1] was applied to design a play-based course and facilitate high school students' exploration of role possible selves in STEM; namely, environmental science and urban planning. As was reflected in an illustrative case study, Zola, a 14-year old female freshmen high school student was guided through systematic, targeted and intentional opportunities for exploring these identities in a personally relevant context (i.e. Philadelphia). Epistemic Network Analysis was applied (a) to visualize the associations she made between what she knew, how she thought, what she cared about, what she wanted to be, what she expected to be, and how she saw herself in relation to urban planning and environmental science (See Figs. 2, 3 and 4), and (b) to ascertain if the changes in her starting self-exploring, role possible selves, and new self was statistically significant. Future reports will involve an exploration of the modeled patterns of identity exploration in all 54 students across the three iterations of project and interpretation of how those changes reflected the distinct features of the three iterations of *Virtual City Planning* (e.g. 9 weeks, 8-weeks, and 4-weeks). Future studies will test and refine the design of virtual learning environments (VLEs) that can not only facilitate Projective Reflection more effectively, but also incorporate methods like Social-Epistemic Network Analysis in the assessment capacities of the VLEs such that learners have access to their patterns of identity exploration at distinct points (starting-self, exploration of role-possible selves, and new self), thus allowing a greater ownership of their learning processes.

References

1. Foster, A.: CAREER: Projective reflection: Learning as identity exploration within games for science. National Science Foundation, Drexel University, Philadelphia, PA (2014)
2. Shaffer, D.W.: Epistemic frames for epistemic games. Comput. Educ. **46**(3), 223–234 (2006)
3. Foster, A., Shah, M.: Knew me and new me: facilitating student identity exploration and learning through game integration. Int. J. Gaming Comput. Mediated Simul. **8**(3), 39–58 (2016)
4. Shah, M., Foster, A., Barany, A.: Facilitating learning as identity change through game-based learning. In: Baek, Y. (ed.) Game-Based Learning: Theory, Strategies and Performance Outcomes. Nova Publishers, New York (2017)
5. Thomas, D., Brown, J.S.: A New Culture of Learning: Cultivating the Imagination for a World of Constant Change. CreateSpace, Lexington (2011)
6. Clark, D.B., Tanner-Smith, E.E., Killingsworth, S.S.: Digital games, design, and learning: a systematic review and meta-analysis. Rev. Educ. Res. **86**(1), 79–122 (2016)
7. Foster, A.: Assessing learning games for school content: the TPACK-PCaRD framework and methodology. In: Ifenthaler, D., Eseryel, D., Ge, X. (eds.) Assessment in Game-Based Learning: Foundations, Innovations, and Perspectives. Springer, New York. https://doi.org/10.1007/978-1-4614-3546-4_11
8. Wise, A.F., Shaffer, D.W.: Why theory matters more than ever in the age of big data. J. Learn. Anal. **2**(2), 5–13 (2015)
9. Shaffer, D.W.: Quantitative Ethnography. Cathcart Press, Madison (2017)
10. Kereluik, K., Mishra, P., Fahnoe, C., Terry, L.: What knowledge is of most worth: teacher knowledge for 21st century learning. J. Digital Learn. Teach. Educ. **29**(4), 127–140 (2013)
11. Wigfield, A., Eccles, J.S.: Expectancy-value theory of achievement motivation. Contemp. Educ. Psychol. **25**(1), 68–81 (2000)
12. Vygotsky, L.S.: Thought and Language. The MIT Press, Cambridge (1986). (1934)
13. Kaplan, A., Sinai, M., Flum, H.: Design-based interventions for promoting students' identity exploration within the school curriculum. In: Karabenick, S.A., Urdan, T.C. Motivational Interventions: Advances in Motivation and Achievement, vol. 18, pp. 243–291. Emerald Group Publishing Limited, Bingley (2014)
14. Foster, A.: Games and motivation to learn science: personal identity, applicability, relevance and meaningfulness. J. Interact. Learn. Res. **19**(4), 597–614 (2008)
15. Hidi, S., Renninger, K.A.: The four-phase model of interest development. Educ. Psychol. **41**(2), 111–127 (2006)

16. Hadwin, A., Oshige, M.: Self-regulation, coregulation, and socially shared regulation: exploring perspectives of social in self-regulated learning theory. Teach. Coll. Rec. **113**(2), 240–264 (2011)
17. Zimmerman, B.J.: A social cognitive view of self-regulated academic learning. J. Educ. Psychol. **81**(3), 329–339 (1989)
18. Foster, A., Shah, M.: The play curricular activity reflection and discussion model for game-based learning. J. Res. Technol. Educ. **47**(2), 71–88 (2015)
19. Cobb, P., Confrey, J., DiSessa, A., Lehrer, R., Schauble, L.: Design experiments in educational research. Educ. Res. **32**(1), 9–13 (2003)
20. Krippendorff, K.: Content Analysis: An Introduction to its Methodology. Sage Publications, Thousand Oaks (2004)
21. Siebert-Evenstone, A.L., Irgens, G.A., Collier, W., Swiecki, Z., Ruis, A.R., Shaffer, D.W.: In search of conversational grain size: modeling semantic structure using moving stanza windows. J. Learn. Anal. **4**(3), 123–139 (2017)
22. Stake, R.E.: The Art of Case Study Research. Sage Publications, Thousand Oaks (1995)

Adolescents' Views of Third-Party Vengeful and Reparative Actions

Karin S. Frey[1]([✉]), Saejin Kwak-Tanquay[1], Hannah A. Nguyen[1],
Ada C. Onyewuenyi[2], Zoe Higheagle Strong[3], and Ian A. Waller[4]

[1] University of Washington, Seattle, WA, USA
`karinf@uw.edu`
[2] The College of New Jersey, Lawrenceville, NJ, USA
[3] Washington State University, Pullman, WA, USA
[4] University of California at Santa Barbara, Santa Barbara, CA, USA

Abstract. African-, European-, Mexican-, and Native-American adolescents ($N = 270$) described times they had responded to a peer's victimization with efforts to repair relationships or avenge the aggression. They provided ratings of self-evaluative emotions and judgements (e.g.., pride, shame, helpfulness) and explanations for each rating. Explanations were coded as exemplifying one of eight goals, and whether the action described in each condition promoted or threatened the desired goal. Complementary analyses utilizing the general linear model and epistemic network examined the rates of each type of goal and the connections between goals, and between goals and outcomes. Benevolence was the most frequently cited goal, and third-party reparative efforts were viewed as promoting benevolence and competence. Benevolence goals were both promoted and threatened by third-party revenge. Self-directed growth was cited most often following revenge, as an example of goal threat and often in conjunction with shame. Relationships between revenge and reparative efforts are explored.

Keywords: Third-party revenge · Reparative actions · Goals · Adolescents

1 Introduction

1.1 Third-Party Revenge and Repair

Lack of Specificity in Categorizing Third-Party Actions. Second-by-second observations indicate that young adolescents are targeted for an average of 1.1 aggressive events per hour on school grounds [1]. Thus, classmates often witness friends and fellow students being targeted and harassed [2]. Youth report feeling confused and anxious after witnessing aggressive incidents [3]. They may try to avoid involvement, side with the aggressor or victim, or attempt to stop the aggression [4, 5]. The actions of witnesses have important implications for themselves and others. Victims of aggression are less anxious and experience less peer rejection and victimization when classmates defend them [5–7]. Although there is evidence that defenders tend to be more assertive [8], empathetic, and morally engaged than their classmates [9], conclusions may be

© Springer Nature Switzerland AG 2019
B. Eagan et al. (Eds.): ICQE 2019, CCIS 1112, pp. 89–105, 2019.
https://doi.org/10.1007/978-3-030-33232-7_8

premature given the breadth of behaviors defined as *defending* [10]. In prior research, behaviors as diverse as consoling the victim, reconciling adversaries, and confronting the aggressor might be considered defense. Further, confrontation might involve physical aggression, abusive language and/or assertive reasoning that invokes moral constructs of harm and justice. Playground observations estimate that nearly half of direct confrontations include aggressive elements, sometimes alone or in combination with assertive behaviors—that is, behaviors that are strong requests or demands to desist [11]. A common pattern in these observations was that youth would intervene with assertive behaviors, then switch to aggression if the initial intervention failed.

Self-evaluation of Actions Reflects Complex Thinking. In addition to the lack of specificity, research on third-party intervention suffers from inattention to the diversity of outcomes experienced by all parties. Targets of aggression whose friends retaliate on their behalf, for example, might feel both anxious that the aggression may escalate and pleased that an ally would take on such a risk. Equally important is the experience of third-parties. Although much research has focused on defining specific roles that individuals *typically* occupy [4], observations indicate that over time, roles are reasonably fluid, such that those who are aggressors on some occasions are witnesses on other occasions [12]. Similar fluidity is to be expected in the ways that youth intervene, sometimes by retaliating, sometimes by attempting to repair relationships and resolve differences peacefully. As youth experience consequences of their actions, some may be perceived as more beneficial and be preferred overtime. Therefore, it is important to understand how youth construe their own actions after they have intervened as third-parties to aggressive events. Prior work indicates that youth feel proud, helpful and consistent with peer social norms when they had attempted to repair and resolve aggression that targets a peer [13]. Efforts to exact revenge on behalf of a peer, however, can arouse ambivalence. Youth who sought vengeance have reported being moderately proud, helpful and consistent with social norms. However, they also experienced relatively high levels of shame and guilt, and did not feel like their actions were as consistent with being a good friend as when they had made reparative efforts. The goal of the current study was to examine how third-party actors interpret and explain their self-evaluative emotions and judgements following their efforts to (a) avenge aggression to a peer and (b) to repair the situation peacefully.

1.2   Goals Structure Self-evaluation

How youth evaluate their actions will depend in part on the goals they had prior to acting. Goals influence emotional responses to outcomes by defining which outcomes are valued and which ones are disappointing [14]. Influential models of social decision-making and behavior [15, 16] posit people's responses in-the-moment depend on a rapid processing sequence that appraises the situation and constructs action goals (e.g., punish the aggressor; repair the relationship), and evaluates possible actions based on beliefs and values. Beliefs include self-perceived competence (e.g., I can prevail; I'm influential), social norms (e.g., retaliation is justified), and outcome expectancies (e.g., the situation may escalate; risky displays of loyalty may solidify this friendship).

Values are defined as trans-situational goals that serve as guiding principles in people's lives (e.g., acquire power; protect others), motivating actions consistent with values [17]. Unlike abstract values, action goals require that individuals coordinate multiple, possibly conflicting values in order to pursue concrete action. Thus, multiple and inconsistent values may be invoked when youth struggle to evaluate and derive meaning from their actions.

Unlike the sequence described above, impulsive responding typically proceeds directly from appraisal to action without attempts to coordinate or even articulate an action goal [18]. Impulsive actions, like goal-directed ones may nevertheless stimulate self-reflection for months afterward [19] as youth evaluate their actions vis à vis personal values, relationships and self-identity. Vengeful acts are often impulsive, stimulated by emotionally-arousing conditions that pre-empt goal-directed processing. Because witnessed aggression is not as emotionally arousing as experienced aggression, goal-directed actions may represent a larger percentage of third-party responses.

1.3   The Current Study

What considerations may motivate third-party responses to aggression? Both revenge and repair efforts are risky endeavors to undertake on behalf of someone else [20]. Motivations may include social norms that blur the distinctions between self and close others [20], efforts to deter aggression against the self, desires to enhance one's reputation and relationships [21], and a morally-engaged sense of social justice [22]. When the chosen action is resolution, actors have the potential to benefit all parties. Third-party revenge, however, is morally ambiguous. It combines antisocial elements (harming someone) with prosocial ones (protecting or obtaining justice for another). This mixed-methods study examined adolescents' self-evaluative emotions and judgements after they responded to aggression directed towards a peer or themselves. Specifically, we asked youth to describe times they had tried to (1) avenge harm to a peer and (2) seek a peaceful resolution. They rated and explained the emotions they felt and judgements they made after those actions.

Research Questions and Hypotheses. Compared to children, adolescents are more likely to endorse revenge and are less likely to endorse peaceful responses to aggression [23]. In some contexts, revenge may be considered the only socially acceptable response [20, 24]. Nevertheless, adolescents report feeling most authentic, most like their actions are consistent with their 'true self' when they act prosocially [25]. How do young people resolve these conflicting values? Using a diverse sample of US youth, the current study examined adolescent values in the context of actions taken following aggression directed at a peer.

Overall Levels of Perceived Goal Threat and Goal Promotion. Given that third-party revenge elicited high levels of shame and guilt and low levels of pride, perceived helpfulness, peer approval and feeling like a good friend relative to reparative efforts, we made the following predictions.

- Third-party repair would be viewed as *promoting* goals more than third-party revenge.

- Third-party revenge would be viewed as *threatening* goals more than third-party repair.

Levels and Links Between Specific Goals and Outcomes. Although benevolent intentions may motivate both vengeful and reparative responses to a peer's victimization, the high rates of shame and guilt that accompany revenge [13] suggest that revenge will be viewed as threatening benevolence goals more than promoting them, while reparative actions would be more closely linked to goal promotion. We also predicted that revenge would be associated with threats to security, competence and self-directed growth more than reparative actions. Finally, we predicted higher levels of power goals would be cited more in the revenge condition than in the reparative condition.

- Connections between goal promotion, and power and security goals would be stronger in the third-party revenge network than in the repair network; whereas connections between those goals and goal threat would be stronger in the third-party revenge network.
- Connections between goal threat and self-directed growth, competence and benevolence goals would be stronger in the third-party revenge network than in the repair network; whereas connections between those goals and goal promotion would be stronger in the third-party repair network.

2 Methods

2.1 Participants

Participants were 270 adolescents (53.7% female) ranging from 14 to 18 years of age. They were African Americans (25.9%), European Americans (24.4%), Mexican Americans (24.4%) and Columbian Plateau Native Americans (25.2%) living in urban, suburban and rural areas in Washington State and Idaho. Participants were paid $20 for interviews lasting 1.25 h. Interviews were completed during the summer with respect to the previous school year.

2.2 Procedures

Participant recruitment and data collection were conducted in tribal and public schools, community centers, and summer programs during the summers. In accordance with institutional review board guidelines and, when applicable, tribal authority research approval, researchers conducting the study obtained participant assent and parent or guardian consent via permission forms available at the sites used in the study.

Prior to conducting the interviews, research assistants completed a training and practice sequence in interview protocol and ethical guidelines. To maintain confidentiality and reduce potential interviewer effects, participants were interviewed in empty community center rooms, libraries, or classrooms by interviewers of the same race or ethnicity. The adolescents in the study were compensated $20 for participating.

In the interviews, participants were asked to describe their experiences with four types of actions following the victimization of a peer. The two actions investigated in the currents study were times in which the interviewee intervened after a peer had been victimized by attempting to (1) reconcile the involved parties or (2) avenge the victim. Any act of aggression was accepted as a perceived victimization event, including evenly matched conflicts and bullying that involved a power differential. After describing events, participants rated how they felt after the event with respect to three self-evaluative emotions and seven other emotions. They were also asked to rate the helpfulness of their interventions and how much peers would approve. Following each emotion, participants explained their ratings. Those explanations were subsequently transcribed and coded in order to understand the meaning and motivational implications of the participants' actions and self-evaluation.

2.3   Theoretical Foundation of Explanation Codes

We used *a priori* frameworks to code youths' explanations for the emotions they felt after each action and their evaluative judgements of the helpfulness and consistency with being a good friend. The first of two exhaustive and mutually exclusive systems was used to designate whether outcomes were described as consistent with the specified value (goal promotion), inconsistent with the value (goal threat), or not specified. The second coding system was derived from Schwartz' circumplex model of values (see Fig. 1)—a model previously tested with more than 80,000 participants around the globe [17, 26]. This model structure has been extensively validated with both adults and adolescents [27]. Among adolescents, endorsement of these values predicts adolescents' aggressive and prosocial behavior [28, 29] and self-evaluative emotions [30]. The work of Schwartz and colleagues has relied on questionnaires to assess values by

Fig. 1. Schematic of Schwartz' circumplex model of values and goals.

having respondents indicate the degree of similarity between themselves and hypothetical persons espousing each value. Since we were examining open-ended explanations for self-evaluation, we used Schwartz's definitions of each value in our code book.

To better fit our data and differentiate between similar codes, we slightly changed Schwartz's nomenclature [17, 26]. Achievement became Competence Display, Self-direction became Self-directed Growth, and Universalism became Justice and Welfare for All. We also did not attempt to code the two sets of sub-categories. Thus, Hedonism and Stimulation were coded as Seek Stimulation—the equivalent category in Schwartz's scheme. Conformity and Tradition are sub-categories that were coded as Uphold Norms. To create an exhaustive coding system, we added a category for explanations that were circular or provided insufficient evidence to code.

Codes and Reliability. Trained research assistants coded participants' event descriptions and associated rationales for emotions and judgements by event type. Augmented base rates were used to assess inter-rater reliability [31]. Coding began when Cohen's *kappa* reached 0.60 for each category, and reliability tests were subsequently administered at three-week intervals in order to prevent coder drift. The mean *kappa* obtained for all codes was .71. *Kappas* calculated separately for goal promotion and goal threat were $k = .66$ and $k = .79$, respectively.

Three goals were rarely cited and will not be discussed further (uphold norms, seek stimulation, and justice & welfare). For all values that were cited at a minimum rate of 0.20 in at least one condition, we provide operational definitions, examples and individual Cohen's *kappa* values below.

Benevolence. These goals cite caring for and showing concern for the welfare of close associates, or being loyal, helpful, honest, forgiving, or responsible. Examples of responses coded for benevolence included helping a friend to stay out of trouble (promotion) and expressing disappointment in oneself for causing harm to a peer (threat), $k = .69$.

Security. These indicate a desire for physical, emotional, and social safety; security, harmony, and stability in relationships; or a sense of belonging. Examples of responses coded for security goals included feeling relief over reconciling with a peer (goal promotion) and expressing concern that a victim's actions could lead to a fight (goal threat), $k = .75$.

Power. These indicate a wish to acquire social status and prestige by gaining and demonstrating control and dominance over people and resources. This includes enforcing personal norms through force or punishment, or comments that put oneself in a superior position compared to the other person. Examples include citing the influence one had over a victim's response to aggression (goal promotion) and expressing disappointment that one's efforts were insufficient to get an aggressor to back down (goal threat), $k = .71$.

Competence. These goals aim to gain a sense of personal success through demonstrating competence according to social standards or having an image of being successful, intelligent, ambitious, or influential. Examples of responses coded for competence included advising a friend in effective conflict negotiation (promotion) and losing control of one's emotions during conflict with others (threat), $k = .75$.

Self-directed Growth. These statements indicate the desire for independent or authentic thought and action; curiosity; self-respect and authenticity; choosing one's own goals or pursuing improvements in cognitive, social-emotional and physical skills for their intrinsic value. Examples of responses included learning how to become a better friend (promotion) and failing to uphold personal standards of goodness (threat), $k = .68$.

2.4   Analytic Plan

Two sets of complementary analyses were used to examine the types of explanations given for self-evaluative emotions and judgements in each condition. First, the number of references to each type were analyzed using repeated measures analyses of variance (SPSS 19) with action condition as the repeated variable. Given the large number of tests that we were performing, we elected to reduce our probability of making a Type I error by using a p-value of .01 (two-tailed) as our significance level.

We next performed Epistemic Network Analyses (ENA) using the ENA1.5.2. Web Tool [32]. The units of analyses were all lines of explanation associated with a single value of the action condition (e.g., repair) with subsets by participant. The resulting networks were aggregated across all participants within each action. Aggregations were created from unweighted summations in which the networks reflect the log of the product for each pair of codes. The ENA model normalized the networks for all units of analysis before they were subject to dimensional reductions, which accounts for the fact that different units of analysis may have different rates of goal codes. For the dimensional reduction, we used a singular value decomposition, which produced orthogonal dimensions that maximize the variance explained by each dimension.

3   Results

3.1   Repeated Measures Analyses of Explanations

The mean number of times a participant cited a type of goal within each action condition are provided in Table 1, along with F-values and effect sizes. The table also shows results for the rates at which the actions were judged to promote valued goals and to threaten the goals in question.

The ambivalence that we observed in self-evaluative emotions and judgements following third-party revenge is partly explained by the rates at which actions were viewed as threatening versus promoting valued goals. Third-party revenge was viewed as a greater threat than a facilitator of valued goals, whereas third-party repair showed the opposite pattern. Neither goals of benevolence nor security differed significantly across the two conditions. As predicted, power was cited more frequently in the revenge than in the repair condition. That was also true of competence and self-directed growth goals. Although we made no specific predictions for rates of the latter two

Table 1. Rates of goal citation during explanations of self-evaluation by condition

	Mean citations per action		F-value	p	Partial η^2
	Revenge	Repair			
Threatened goals	1.73	0.54	109.41	<.001	.30
Promoted goals	2.81	2.06	42.59	<.001	.14
Benevolence	1.88	1.82	<1	ns	.00
Security	0.68	0.51	6.27	ns	.02
Power	0.29	0.04	32.45	<.001	.11
Competence	0.49	0.91	32.78	<.001	.12
Self-directed growth	0.79	0.39	34.43	<.001	.12

goals, we did expect differential linkages between the goals and goal outcomes. These are explored next.

3.2   Epistemic Analyses of Goals Cited in Self-evaluative Rationales

The units for the ENA were all lines of explanation associated with a single value of the action condition (e.g., repair) with subsets by participant. The resulting networks are aggregated across all participants within each action. Aggregations were created from weighted summations in which the networks reflect the log of the product for each pair of codes. The ENA model normalized the networks for all units of analysis before they were subject to dimensional reductions, which accounts for the fact that different units of analysis may have different amounts of explanation codes. For the dimensional reduction, we used a singular value decomposition, which produced orthogonal dimensions that maximize the variance explained by each dimension.

In the graphs of the networks, lines connecting each type of goal reflect the relative frequency of co-occurrence between two codes, represented as plotted points. The position of the graph nodes for each goal are determined by an optimization routine that minimizes the differences between the plotted nodes and their corresponding network centroids. For a more detailed explanation of the quantitative methodology, see Shaffer, Collier and Ruis [33]. Because of this co-registration of network graphs and projected space, the positions of the plotted nodes – and the connections they define between goals – can assist in interpreting the dimensions of the projected space.

Networks for Revenge and Repair Efforts. As shown in Fig. 2, benevolence goals were both promoted and threatened by third-party revenge, as were security goals. This reflects the dual foci of the action. Benevolence towards the victim may increase intimacy and sense of belonging reflected in relationship security, whereas lack of benevolence towards the aggressor might heighten insecurity. Reparative actions strongly promoted both benevolence and competence goals, and to a lesser extent, security goals. Competence goals were promoted, and to a lesser extent, threatened by revenge, although they were cited less often in the context of revenge than of repair. Self-directed growth was cited most frequently as an explanation for self-evaluative feelings following revenge. It had strong links to goal threat, which was almost never true of reparative action. Comparisons between the two goal networks are highlighted by a difference network which is produced by subtracting the weights of corresponding connections between pairs of networks [33].

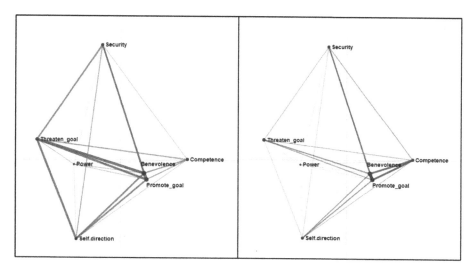

Fig. 2. Goal networks associated with self-evaluative emotions and judgements following vengeful actions on the left (red) and reparative actions on the right (blue). (Color figure online)

Centroids for the Two Networks Differ Significantly. The centroids presented in Fig. 3 summarize the dimensions of each network. Centroids statistically compare networks in their entirety, with non-overlapping confidence intervals indicating statistically significant differences. These can be compared statistically as well as visually. On the x-axis, a t-test assuming unequal variance showed that the centroid for third-party repair ($M = 0.33$, $SD = 0.61$) was significantly different from the centroid for third-party revenge ($M = -0.33$, $SD = 0.61$; t (583) = -12.90 $p < .001$, Cohen's $d = 1.07$). The corresponding values for the y-axis (repair, $M = 0.00$, $SD = 0.67$; third-party revenge, $M = 0.00$, $SD = 0.60$) was not significantly different, $t < 1$, ns. Comparing the networks suggests that the x-axis represents personal agency or efficacy, as it accounted for 17.2% of the variance. The y-axis appears to reflect Schwartz' [26] dimensions of protection (security) versus growth (self-direction) and accounts for 14.2% of the variance. The circumplex model places benevolence and competence in intermediate positions with regard to security and growth, as we see in the epistemic networks. Power, however, is generally conceptualized as occupying a position closer to security than is found in this graph of actions on behalf of a victimized peer.

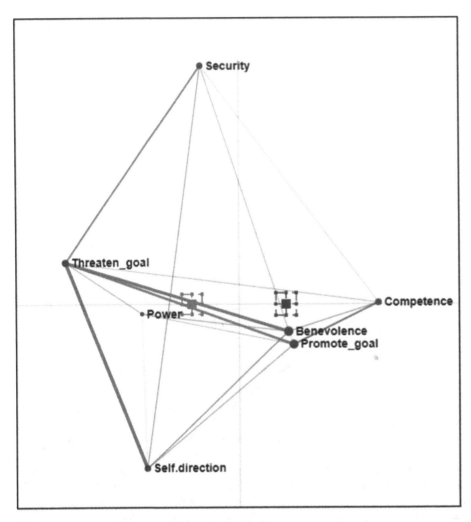

Fig. 3. Subtractive network with centroids. The solid boxes indicate the centroids calculated for revenge (red) and reparative actions (blue). Dotted boxes surrounding the centroids indicate 95% confidence intervals in each dimension. The connecting lines shown between nodes indicate the magnitudes of differences in the strength of connections between the two action conditions. Red indicates stronger connections in the goals cited for revenge. Blue indicates stronger connections in the goals cited for repair. (Color figure online)

4   Discussion

While age-related changes in endorsement of retaliation are concerning [23], they need to be situated within the larger context of how youth evaluate themselves when they retaliate or repair relationships. Adolescents' self-evaluative emotions and judgements documented in an earlier study [13] show unequivocally that adolescents experience pride, feel that they have performed well, and expect peer approval when they respond to aggression with efforts to repair the situation. Feelings of shame and guilt complicate their feelings about third-party revenge. The current study documents how adolescents interpret and understand the frequently contradictory feelings and judgements about their behavior.

4.1   Benevolence, Competence, Security and Power

Adolescents explained their self-affirming emotions and judgements following third-party repair as being due to the action's promotion of important goals. Emotions and judgements were often explained by a strong triad linking goal promotion, benevolence, and competence. The following example illustrates this common pairing, showing benevolent goal rationales underlined and competent goal rationales in bold.

> It was an argument. I just told them that you're sorry to the other person for saying that and try to work it out with them…<u>They worked it out and there wasn't going to be no fight or anything. And they became good friends.</u> [Felt good about actions] **because I made those two stop thinking about fighting and it worked out pretty well.**

Benevolent actions (underlined) sometimes promoted a feeling of security (bolded) in young peoples' relationships. Friendships might feel more intimate after adolescents undertook benevolent but risky action on behalf of a peer. These efforts were not always successful, hence the link to goal threat. This young person describes a successful effort:

> My friend confronted this girl about talking bad about her, and then I tried to make a joke out of it… <u>make her laugh and make her forget what she was doing, so nothing happened between them</u>…Because she actually took it as a joke, not as me trying to hype her up or something… **Now we have a good joke to laugh about…Whenever we find out about her [antagonist] talking bad, we just make that joke, and she really wouldn't do anything.**

Benevolence goals were also strong elements of self-evaluative rationales following third-party revenge, but revenge threatened that goal as strongly as promoting it. In this example, the first phrase describes how the actions promoted a benevolent goal (underlined) while the last phrase describes how the actions ultimately threatened the actor's attempt to be helpful and benevolent in the eyes of the victim. "<u>I felt that I really helped him out</u>…It was a mini fight. **Felt like I won that fight**…<u>but I stopped him from dealing with it in his own way.</u>"

This youth also cited competence as a goal promoted by his actions (bolded), but pride in his fighting ability accompanied guilt over his friend's unwilling involvement in a physical fight. The specificity of the described threat to benevolence suggests that the friend may have complained to the actor.

Links to power emerged in the third-party revenge network, but both the overall low frequency of the citation and weak links within the network suggest that it was a secondary goal in third-party efforts, despite its links to goal promotion. Like benevolence and security, power was linked to goal threat as much as to goal promotion. Aggression, especially of the physical sort, is highly risky. As shown in the following example, it does not always go well. "I was angry that he was making jokes at the fact that I tried to fight him later on…" Conversely, successful attempts to scare aggressors were generally considered helpful at reducing a friend's victimization.

This guy was super weird and rude to this girl, my best friend. She's the best. She was like, "Can you like stop," because he had been so like disgusting and rude for like the longest time. So I told his best friend what he was doing and now they're not friends at all. [It helped a lot] Because he realized I have more power than he thought; then he stopped.

4.2   Self-directed Growth and Shame

Self-directed growth differs from the other goals in its future orientation. While helping someone out or dominating them are short-term goals, self-directed growth refers to the person one wants to be now and in the future. So it may seem counterintuitive that it was cited most often following third-party revenge than third-party repair. As ENA reveals, references to self-directed growth occur most commonly in situations that threaten adolescents' desired identities. Those links suggest that the young person is stepping back to evaluate past actions in light of important standards of behavior. Threats to self-directed growth were often given as rationales for feelings of shame. The self-critical emotions speak to the tension between self-focus and other-focus. As the following example illustrates, guilt tends to be an other-focused emotion related to a specific action [34]. People feel badly due to the harm caused to another. Shame is a more generalized reaction to feeling inauthentic or deep-seated personal inadequacies. Adolescents' self-reflections were sometimes painful. This young man shifts repeatedly between shame he felt about being inauthentic, and guilt about threats to a peer.

I felt disappointed because **that isn't really me, really**. I felt ashamed because I was, I am talking all that talk and stuff. And I felt guilty because I was talking about hurting him.

This example reminds us that an intense self-focus is not necessarily a narcissistic indulgence in adolescence. Reflecting on mistakes and envisioning a positive future self is a developmentally appropriate task. This example appears to conflict with research indicating that frequent experiences of shame are negatively associated with constructive intentions [34] and positive psychosocial outcomes [35]. Motivation to change oneself, however, may be a more common response in hindsight, after defensive and avoidant reactions have subsided [36].

4.3   Relationships Between Reparative Efforts and Revenge

We also saw threats to self-directed growth in the third-party repair condition. These are challenging situations for adolescents and adults to manage—they do not always go well. As the observations by Hawkins and colleagues revealed, unsuccessful attempts at reconciliation are often followed by aggression. Here, benevolent attempts to solve the problem are frustrated by the aggressor and replaced by non-benevolent actions that threatened the actor's feelings of self-direction.

I was trying to talk, "Can you please stop?" But the person wouldn't listen the first time and so I started being a little bit mean and when the person rolled her eyes at me and then I kind of got mad and I said 'I hope your eyes stay like that because you didn't listen… [Action didn't help at all] Because **I should've just talked, um and not try to be rude**, but some people just make me mean. Like **I'm in a good mood the first time, trying to talk nicely but then once they get me angry, then they get me angry.**

This example illustrates the complex relationship between revenge and reconciliation. Failed attempts to be accommodating and generous are likely to engender increased anger and frustration. This may heighten the intensity of a vengeful response. Sometimes revenge precedes reconciliation attempts, also a fraught proposition. On the one hand, adolescents suggest that reconciliation may be easier after they've avenged the harm [37]. Cohen and colleagues [38] noted that southern US males who retaliated for the insults they experienced appeared genuinely enthused to shake hands with their offender, while northern males, who refrained from retaliation, held a grudge. Of course, avengers may be willing to make up after "evening the score," but their adversaries may not share the perspective that the score has been evened. Avengers who are angry typically escalate their response slightly [39]. This may require counter-revenge in the mind of their adversary. The difficulty of reconciling the desire for revenge with the desire for peace has led to interesting and adaptive strategies in some cultures. Some highland tribes in Papau New Guinea, for example, used to wear cassowary plumes over their eyes when they went to war. The purpose of masking the warrior's face was so that eventual peace initiatives would not be derailed if negotiators were to identify a representative of the other side as the one who killed their brother, for example. Other cultural practices, such as councils of elders (*shuras, jurgas*) betray a similarly sophisticated understanding of human nature. Third-party intervention, while often inflammatory, can also provide face-saving rationales for de-escalation when adversaries feel that retaliation is required to maintain a respected reputation [38].

There were few ethnic or gender differences in our data. Almost all participants responded with complex moral analyses of their behavior. They celebrated when actions made them feel like good people and mourned when actions did not. Poignantly, many young people offered that they had never told anyone about these feelings before. They sometimes wanted to tell us about other events because, as one boy reported, he had been haunted for years by remorse he had felt for bullying an overweight boy. He said he had thought about it every day. Way's [40] interviews spanning multiple years in the lives of adolescents, indicates that boys become increasingly reluctant to reveal vulnerable emotions to their friends. Many are equally reluctant to speak to parents about misdeeds and moral concerns. These concerns can be heavy burdens, and adolescents need support. Enlisting youth in social norms interventions that provide accurate information and engage discussion about what their peers actually value [41] might reduce some of the barriers youth feel about revealing themselves. Illustrating their peers' support for positive behavior may also enable more constructive actions in the heat of the moment—if not from the aggrieved parties, perhaps from their friends, "cuz we all care about our friends and want to make sure they do something good."

4.4 Limitations

As an initial examination of the goals and self-understandings associated with third-party revenge and third-party reparative efforts, this study makes an important contribution, but limitations deserve notice. This was not an experimental design. Our reliance on actual events in adolescents' lives meant that youths were able to relay personally significant events, but the actions that they chose to execute in each case may have been influenced by the severity or type of the initial aggression. Events that were amenable to repair, for example, may have been less intense than those that elicited revenge, and therefore more likely to resolve successfully. A second limitation is that we examined only a single item for each type of action. Thus, we do not know whether the actions described were representative of typical responses or relatively unique events. Although inferior psychometrically to multi-item measures, work with adolescents has established the validity of single items when participant burden is an issue [42]. A related limitation is the variation in specific actions within each category (revenge and repair). For example, revenge could be physical or relational, and peaceful resolutions could be confrontational or conciliatory. Such variation may elicit different victim emotions judgements, and rationales. While the variation probably elevated error terms in our statistical analyses, our effect sizes were nevertheless robust. Future research might examine victim responses to specific strategies within these types of action to identify the most effective ways that adolescents can support victimized peers.

4.5   Utility of ENA for Mixed Methodologies

In our work, we frequently have data that consists of counts, such as the number of times aggression happens on the playground, and the number of times that interviewees spontaneously cite benevolence as a motivating value. These counts provide important, real world indications of the magnitude of problems and concerns, as well as illuminating the factors that may support resilience and remediation. Coupled with experimental designs, inferential statistics provide useful summaries when evaluating different psychological and educational practices. The resultant models measure how influential increases or decreases in the frequencies of each factor are on outcomes. They are limited, however, in the ability to identify patterns of relationships between multiple factors. In this study, we see adolescents weighing the sociomoral and personal implications of their actions, sometimes rapidly alternating between satisfaction and shame—patterns that provide a unique insight into youths' moral development. Knowing, for example, that a young man's feelings of satisfaction with his fighting competence was tempered by his belief that he had overstepped his rights in the defense of his friend provides a more nuanced understanding of his value-based self-evaluation. Simple analyses of counts would have missed such associations. We might have concluded that *particular actions* led to satisfaction at some times or for some people, and to self-criticism in other circumstances. Instead, we were able to graphically display the sophisticated moral analyses that young people offered of events and their part in them. At the same time, our counts provided easily apprehended summaries of the overall value that youth placed on their actions, as reflected in the rates of goal promotion and goal threat associated with each type of action. Thus, it is clear that each form of analyses, ENA and inferential statistics, enhanced our ability to interpret the results from the other. These data reveal, as Shaffer [31] has argued, that the *connections* between behaviors provide information that is non-redundant and complementary to the analyses of levels or rates of behavior.

Acknowledgements. This research was supported by a grant from the National Institute of Justice (2015-CK-BX-0022). Opinions or points of view expressed are those of the authors and do not necessarily reflect the official position or policies of the U.S. Department of Justice, other funding agencies, cooperating institutions, or other individuals.

References

1. Frey, K.S., Hirschstein, M.K., Snell, J.L., Edstrom, L.V., MacKenzie, E.P., Broderick, C.J.: Reducing playground bullying and supporting beliefs: an experimental trial of the *steps to respect* program. Dev. Psychol. **41**, 479–491 (2005)
2. O'Connell, P., Pepler, D., Craig, W.: Peer involvement in bullying: Insights and challenges for intervention. J. Adolesc. **22**, 437–452 (1999)
3. Nishina, A., Juvonen, J.: Daily reports of witnessing and experiencing peer harassment in middle school. Child Dev. **76**, 435–450 (2005)
4. Salmivalli, C.: Participant role approach to school bullying: implications for intervention. J. Adolesc. **22**, 453–459 (1999)
5. Salmivalli, C., Voeten, M., Poskiparta, E.: Bystanders matter: associations between reinforcing, defending, and the frequency of bullying behavior in classrooms. J. Clin. Child Adolesc. Psychol. **40**, 668–676 (2011)
6. Casper, D.M., Card, N.A., Bauman, S., Toomey, R.: Overt and relational aggression with participant role behavior: measurement and relations with sociometric status and depression. J. Res. Adolesc. **27**, 661–673 (2017)
7. Kärnä, A., Voeten, M., Poskiparta, E., Salmivalli, C.: Vulnerable children in varying classroom contexts: bystanders' behaviors moderate the effects of risk factors on victimizaton. Merrill-Palmer Q. **56**, 261–282 (2010)
8. Gini, G., Albiero, P., Benelli, B., Gianmarco, A.: Determinants of adolescents' active defending and passive bystanding behavior in bullying. J. Adolesc. **31**, 93–105 (2008)
9. Lambe, L.J., Cioppa, V.D., Hong, I.K., Craig, W.M.: Standing up to bullying: a social ecological review of peer defending in offline and online contexts. Aggression Violent Behav. **45**, 51–74 (2019)
10. Reijntjes, A., Vermande, M., Olthof, T., Goossens, F.A., Aleva, L., van der Meulen, M.: Defending victimized peers: opposing the bully, supporting the victim, or both? Aggressive Behav. **42**, 585–597 (2016)
11. Hawkins, D.L., Pepler, D.J., Craig, W.M.: Naturalistic observations of peer interventions in bullying. Soc. Dev. **10**, 512–527 (2001)
12. Frey, K.S., Newman, J.B., Onyewuenyi, A.: Aggressive forms and functions on school playgrounds: profile variations in interactive styles, bystander actions, and victimization. J. Early Adolesc. **34**, 285–310 (2014)
13. Frey, K.S., Higheagle Strong, Z., Onyewuenyi, A.C., Pearson, C.R.: Third-party intervention in peer victimization: Moral emotions and judgements of a diverse adolescent sample (2019). Manuscript submitted for review
14. Jarvinen, D.W., Nicholls, J.G.: Adolescent's social goals, beliefs about the causes of social success, and satisfaction in peer relations. Dev. Psychol. **32**, 435–441 (1996)
15. Crick, N.R., Dodge, K.A.: A review and reformulation of social information-processing mechanisms in children's social adjustment. Psychol. Bull. **115**, 74–101 (1994)
16. Lemerise, E.A., Arsenio, W.F.: An integrated model of emotion processes and cognition in social information processing. Child Dev. **71**, 107–118 (2000)

17. Schwartz, S.H., Bardi, A.: Value hierarchies across cultures: taking a similarities perspective. J. Cross-Cult. Psychol. **32**, 268–290 (2001)
18. Fontaine, R.G., Dodge, K.A.: Real-time decision-making and aggressive behavior in youth: a heuristic model of response evaluation and decision (RED). Aggressive Behav. **32**, 604–624 (2006)
19. Opotow, S.: Conflict and morals. In: Thorkildsen, T.A., Walberg, H.J. (Ed.) Nurturing Morality, vol. 5, pp. 99–115. Springer, Boston (2004). https://doi.org/10.1007/978-1-4757-4163-6_6
20. Leung, A.K.Y., Cohen, D.: Within- and between-culture variation: individual differences and the cultural logics of honor, face, and dignity cultures. Pers. Process Individ. Differ. **100**, 507–526 (2011)
21. Pederson, E.J., McAuliffe, W.H.B., McCullough, M.E.: The unresponsive avenger: more evidence that disinterested third parties do not punish altruistically. J. Exp. Psychol. Gen. **147**, 514–544 (2018)
22. Cappadocia, M.C., Pepler, D., Cummings, J.G., Craig, W.: Individual motivations and characteristics associated with bystander intervention during bullying episodes among children and youth. Can. J. Sch. Psychol. **27**, 201–216 (2012)
23. Henry, D.B., Dymnicki, A.B., Schoeny, M.E., Meyer, A.L., Martin, N.C.: MVPP: middle school students overestimate normative support for aggression and underestimate normative support for nonviolent problem-solving strategies. J. Appl. Sch. Psychol. **43**, 433–445 (2013)
24. De Castro, B.O., Verhulp, E.E., Runions, K.: Rage and revenge: highly aggressive boys' explanations for their responses to ambiguous provocation. Eur. J. Dev. Psychol. **9**, 331–350 (2012)
25. Thomaes, S., Sedikides, C., van den Bos, N., Hutteman, R., Reijntjes, A.: Happy to be "me?" authenticity, psychological need satisfaction, and subjective well-being in adolescence. Child Dev. **88**, 1045–1056 (2017)
26. Schwartz, S.H.: A theory of cultural value orientation: explications and applications. Comp. Sociol. **5**, 137–182 (2006)
27. Schwartz, S.H.: Values and religion in adolescent development: cross-national and comparative evidence. In: Trommsdorff, G., Chen, X. (Eds.) Values, Religion, and Culture in Adolescent Development, pp. 97–122. Cambridge Press, New York (2012)
28. Knafo, A., Daniel, E., Khoury-Kassabri, M.: Values as protective factors against violent behavior in Jewish and Arab high schools in Israel. Child Dev. **79**, 652–667 (2008)
29. McDonald, K.L., Benish-Weisman, M., O'Brien, C.T., Ungvary, S.: The social values of aggressive-prosocial youth. J. Youth Adolesc. **44**, 2245–2256 (2015)
30. Silfver, M., Helkama, K., Lönnqvist, J.E., Verkasalo, M.: The relation between value priorities and proneness to guilt, shame, and empathy. Motiv. Emot. **32**(2), 69–80 (2008)
31. Shaffer, D.W.: Quantitative Ethnography. Cathcart Press, Madison (2017)
32. Marquart, C.L., Hinojosa, C., Swiecki, Z., Eagan, B., Shaffer, D.W.: Epistemic Network Analysis (Version 1.5.2) (2018). http://app.epistemicnetwork.org
33. Shaffer, D.W., Collier, W., Ruis, A.R.: A tutorial on epistemic network analysis: analyzing the structure of connections in cognitive, social, and interaction data. J. Learn. Anal. **3**(3), 9–45 (2016)
34. Tangney, J.P., Stuewig, J., Mashek, D.J.: Moral emotions and moral behavior. Ann. Rev. Psychol. **58**, 345–372 (2007)
35. Irwin, A., Li, J., Craig, W., Hollenstein, T.: The role of shame in the relation between peer victimization and mental health outcomes. J. Interpersonal Violence **34**, 156–181 (2019)
36. Lickel, B., Kushlev, K., Savalei, V., Matta, S., Schmader, T.: Shame and the motivation to change the self. Emotion **14**, 1049–1061 (2014)

37. Recchia, H., Wainryb, C., Inigo, F., Restrepo, A., Gonzalez, O.L., Posada, R.: Coordination of revenge and forgiveness in Columbian adolescents' narrative accounts of peer injury. Paper presented at the Biennial Meeting of the Society for Child Development, Baltimore, Maryland, US, 21–23 March 2019

38. Cohen, D., Vandello, J.A., Puente, S., Rantilla, A.K.: "When you call me that, smile!" how norms for politeness, interaction styles, and aggression work together in southern culture. Soc. Psychol. Q. **62**, 257–275 (1999)

39. Carlsmith, K.M., Wilson, T.D., Gilbert, D.T.: The paradoxical consequences of revenge. J. Pers. Soc. Psychol. **95**, 1316–1324 (2008)

40. Way, N.: Deep secrets: Boys' friendships and the crisis of connection. Harvard University, Cambridge, MA, US (2011)

41. Henry, D.B., Shinn, M., Yoshikawa, H.: Changing classroom social settings through attention to norms. In: Shinn, M., Yoshikawa, H. (Ed.) Toward Positive Youth Development: Transforming Schools and Community Programs, pp. 40–57. Oxförd University Press, New York (2008)

42. Robin, R., Hendin, H., Trzesniewski, K.: Measuring global self-esteem: construct validation of a single-item measure and the rosenberg self-esteem scale. Pers. Soc. Psychol. Bull. **27**, 151–161 (2001)

Using Epistemic Networks with Automated Codes to Understand Why Players Quit Levels in a Learning Game

Shamya Karumbaiah[1]([⊠]), Ryan S. Baker[1], Amanda Barany[2],
and Valerie Shute[3]

[1] University of Pennsylvania, Philadelphia, PA 19104, USA
shamya@upenn.edu
[2] Drexel University, Philadelphia, PA 19104, USA
[3] Florida State University, Tallahassee, FL 32306, USA

Abstract. Understanding why students quit a level in a learning game could inform the design of appropriate and timely interventions to keep students motivated to persevere. In this paper, we study student quitting behavior in Physics Playground (PP) – a Physics game for secondary school students. We focus on student cognition that can be inferred from their interaction with the game. PP logs meaningful and crucial student behaviors relevant to physics learning in real time. The automatically generated events in the interaction log are used as codes for quantitative ethnography analysis. We study epistemic networks from five levels to study how the temporal interconnections between the events are different for students who quit the game and those who did not. Our analysis revealed that students who quit over-rely on nudge actions and tend to settle on a solution more quickly than students who successfully complete a level, often failing to identify the correct agent and supporting objects to solve the level.

Keywords: Learning game · Quitting behavior · Epistemic network analysis · Automated codes · Interaction log

1 Introduction

Digital games are increasingly used as a learning platform and are designed to keep students engaged in a fun experience while learning [1]. Such serious games need to balance the difficulty level to promote both learning and engagement – goals which may be contradictory at times [2]. In increasing the difficulty level to improve learning, there is a risk of frustrating students or even causing them to give up [2]. A student may quit a game level for many reasons. For instance, a student may find themselves unable to make progress due to a lack of conceptual understanding, difficult game mechanics or interface design (see [3] for the impact of conceptual understanding and game mechanics on student frustration). Alternatively, a player may quit a level in order to search for an easier level, a behavior also seen in intelligent tutoring systems that give the learner choice (e.g. [4]). Understanding why students quit a game informs the design of relevant and timely interventions that could keep the student motivated and

© Springer Nature Switzerland AG 2019
B. Eagan et al. (Eds.): ICQE 2019, CCIS 1112, pp. 106–116, 2019.
https://doi.org/10.1007/978-3-030-33232-7_9

prevent frustration from leading the student to give up. Thus, in this paper, we study why students quit a learning game, with a focus on the student cognition that can be inferred from their interaction with the game.

We study this question using the methods of Quantitative Ethnography [5]. Quantitative Ethnography (QE) attempts to better understand learning and learner choice using big data for thick description – a qualitative explanation of the how and why of human experiences. QE relies on creating meaningful codes – elements that are important to understand the phenomena being studied [5]. Finding ways to create good quality automated coding techniques becomes essential when working with huge volumes of learner interaction data. One solution is to explore events that are automatically generated in the interaction log data. Event-driven logging is a popular technique in software systems. In a well-designed logging system, user interaction events are defined based on the user behaviors that are most meaningful and crucial in the context of the system [6]. In this paper, we use the events logged by the learning game as automated codes relevant to student cognition, to study how the temporal interconnections between the events are different among the students who quit the game and those who did not to understand why students quit.

2 Context

Physics Playground (PP) is a learning game designed for secondary school students to understand qualitative physics [1]. In Physics Playground, students are given a sequence of two-dimensional puzzles (levels) where they need to guide a green ball to hit a red balloon (See Fig. 4). Players analyze the objects they see on the screen and draw the solution on the screen in the form of objects - often simple machines or agents (See Fig. 2). Laws of physics relating to gravity and Newton's laws apply to all objects given and drawn on the screen. Successfully completing levels includes understanding concepts such as Newton's law of force and motion, potential and kinetic energy, and conservation of momentum. There are 74 sketching levels in the game (see Fig. 1 for examples) and each level is designed to be optimally solved by particular agents. PP is nonlinear; students have complete choice in selecting levels.

Participants in this study consisted of 137 students (57 male, 80 female) in the 8th and 9th grades enrolled in a public school with a diverse population in a medium-sized city in the southeastern U.S. The game content was aligned with state standards relating to Newtonian Physics. The study was conducted in a computer-enabled classroom with 30 desktop computers over four consecutive days. The software logged all the student interactions in a log file.

A quit prediction model was previously built for this game with the goal to identify potential moments where cognitive support could support a struggling student in developing their emerging understanding of key concepts and principles [7]. While predicting when a student is likely to quit is important, it is also crucial to understand why the student is likely to quit in order to inform the design of supports that address students' individual needs. Studying why students quit could also give insights on how to improve the design of the features used in the model, potentially improving model performance. Since our eventual goal is to use our insights to improve a prediction model (and validate that model on new data), we restrict this paper's analysis to an arbitrarily chosen 20% of our students.

3   Method

The unit of analysis in this study is a level in the learning game. We have selected five levels with a high percentage of quitting. These five levels vary across Physics concept involved in the level, the agent expected to be used to solve the level (See Fig. 2), and difficulty (see Fig. 1, Tables 1 and 2).

Table 1. Description of the five levels chosen for analysis. The Physics concepts involved are N1stL (Newton's First Law), EcT (Energy can Transfer) and PoT (Properties of Torque).

Level	Playground (PG)	Agent	Primary concept	Secondary concept	Solution video
Flower Power	PG #4	Ramp	N1stL	EcT	tiny.cc/flwr
Big Watermill	PG #4	Ramp	N1stL	EcT	–
Caterpillar	PG #4	Springboard	EcT	–	–
Need Fulcrum	PG #3	Lever	PoT	EcT	tiny.cc/flcrm
Shark	PG #3	Lever	PoT	EcT	tiny.cc/shrk

Fig. 1. The five levels of Physics Playground analyzed in this paper. Top row (left to right) – Flower Power, Big Watermill, Caterpillar. Bottom row (left to right) - Shark, Need Fulcrum

Table 2. Quit levels and difficulty levels of the five levels chosen for analysis.

Level	Game mechanics	Physics understanding	#Students	%Quit
Flower Power	4	2	20	35.45
Big Watermill	3	2	23	43.70
Caterpillar	4	2	22	40.24
Need Fulcrum	1	4	25	42.41
Shark	4	4	22	44.14

3.1 Epistemic Network Analysis Using Automated Codes of Student Gameplay

We study learning within this serious game by interpreting students' activity as expressed in the record of their interaction with the game. Event-based logging automatically captures how and when students use the human-created codes within the tool. For instance, in PP, students draw symbolic representations of various objects like agents (e.g. lever, ramp), fulcrums, mass, and pins. An agent identification system was developed in the game to infer the type of agent drawn based on features like the presence of an arm, number of pins attached, and direction of its movement (see [1] for more details). This system has a 95% accuracy when compared with human ratings. The logging mechanism uses this system to automatically detect the creation and modification of agents in real-time from the student's sketch and logs these as events with timestamps. Studying the temporal interconnection between events using Epistemic Network Analysis (ENA; [8]) gives us a way to better understand students' cognition. We examine how the events are related differently to one another in the group of students who quit a game level as compared to the ones who did not. In this analysis, we are focusing on the following five automatically coded events, detected as indicated below:

1. *Agent Creation* (see Fig. 2)
   a. *Ramp* – detection of an object that a ball rolls along, across the screen
   b. *Springboard* – detection of an object that is attached to two or more pins and rotates to propel the ball upward
   c. *Pendulum* – detection of an object that rotates on a single pin
   d. *Lever* – detection of a secondary object that falls on a primary object, which in turn rotates on a fulcrum (a support on which a lever pivots) to launch the ball.
2. *Draw.Freeform* – the creation of a freeform object (including the agents)
3. *Draw.Pin* – the creation of a pin object
4. *Erase* – the erasure of a freeform or pin object
5. *Nudge* – the player clicked on the ball to move it to the left or right

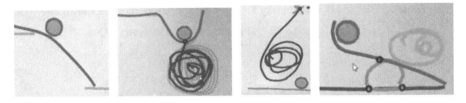

Fig. 2. The four agents used to solve PP levels. *Left to right* – ramp, springboard, pendulum, lever.

An epistemic network for a game level is created from the temporally sequenced one-hot encodings of the events that occurred during the students' gameplay. We segment the data by the student since a single students' attempt in a level is an appropriate unit of interconnected behaviors. On an average, PP logs an event once every 2 s. We chose a moving window size of 10 which on average corresponds to 20 s of gameplay. Due to the fine-grained nature of the event logs, we chose a relatively high moving window size than in many ENA analyses (e.g. [9, 10]) to get an appropriate temporal context to identify relevant co-occurrences of events (e.g. mass drawn and erased to find the needed weight, pins drawn and erased to place a springboard at a precise position, mass dropped on a lever to launch the ball).

4   Results

Figure 3 presents the difference networks for the five levels examined. These networks highlight the most salient difference between the epistemic networks of students who quit the level and those who did not. Along the X-axis (dimension 1 after means rotation), a two-sample t-test assuming unequal variance showed the group who quit was statistically significantly different from the group who did not quit at alpha = 0.05 with effect size of $d = 1.38$ for *Flower Power*, $d = 1.31$ for *Big Watermill*, $d = 0.84$ for *Caterpillar*, $d = 1.29$ for *Shark* and $d = 0.84$ for *Need Fulcrum*. The difference is not statistically significant along the Y-axis ($p = 1$, $d = 0$). Next, we present four key themes that provide us insights on why students quit a level in PP.

4.1   Missing Agent Identification

One of the key steps in solving a level in PP is to identify the agent needed. Across all the five networks in Fig. 3, the students who quit the level without solving it did not use the agent involved - a ramp for *Flower Power* and *Big Watermill*, a springboard for *Caterpillar*, and a lever for *Shark* and *Need Fulcrum* – and other events. By contrast, students who solved the level successfully used the agent involved - except in *Need Fulcrum* (this is explained further in the next theme). Since the event *Draw.Freeform* encompasses the drawing of any freeform object including all agents, it is expected to be the event with the most connections. Apart from *Need Fulcrum*, the strongest connection out of *Draw.Freeform* for the students who did not quit a level is to the agent expected to be needed in that level. Quitting in this case can be attributed to the lack of conceptual understanding of physics.

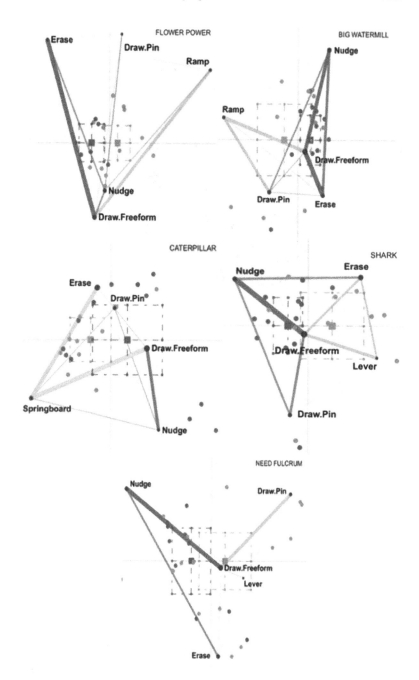

Fig. 3. The difference networks corresponding to the five levels examined. From top to bottom and left to right – *Flower Power, Big Watermill, Caterpillar, Shark, Need Fulcrum.* In these networks, red connections are made more frequently by students who eventually quit the level without solving it, while green connections are made more frequently by those who complete the level. (Color figure online)

4.2 Missing Identification of Supporting Objects

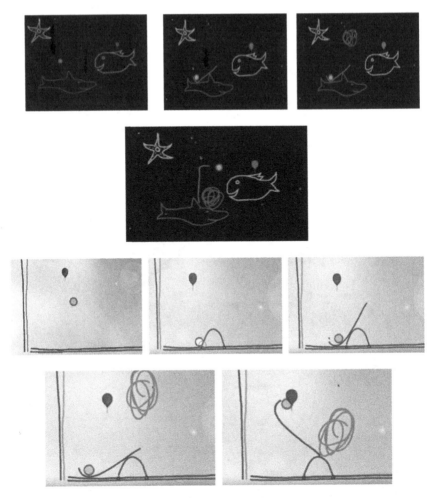

Fig. 4. Solution for *Shark* (top) as compared to *Need Fulcrum* (bottom). Note the extra action (second step) of adding a fulcrum in *Need Fulcrum*. (Color figure online)

Comparing connections in the pairs of networks involving the same agent helps us highlight another missing cognitive connection among the students who quit. The first pair is *Flower Power* and *Big Watermill* – both of which use a ramp in their solutions. However, whereas the ramp can rest on one of the flowers in *Flower Power* (See Fig. 1), *Big Watermill* needs pins on the watermill to hold the ramp from falling off the screen. This cognitive connection can be observed in the students who did not quit *Big Watermill*. Along with a stronger connection between *Draw.Freeform* and *Ramp*, these students also have a strong connection between *Ramp* and *Draw.Pin*. Failure to hold the ramp on the screen using pins results in the ramp being undetected even for the students who may have identified the right agent but eventually quit the level due to missing pins. Quitting in such cases can be attributed to the difficult game mechanics instead of the student lacking conceptual understanding in Physics.

The second pair is *Shark* and *Need Fulcrum* – both of which needs a lever to solve. In *Shark* (See Fig. 4), a lever resting on the blue Shark's fin can catch the falling ball and launch it to the balloon when a mass is dropped. By contrast, *Need Fulcrum* (as the name suggests) needs an additional object – a fulcrum (the red arch in Fig. 4 bottom row). While the identification of the agent needed may not be difficult, the idea of using a fulcrum by fixing it on the plane using pins is the missing cognitive connection among the students who quit the level without solving it. This can be seen in the difference network as the lack of a strong connection between *Draw.Freeform* and *Draw.Pin* among the students who quit the level without solving it as compared to the students who successfully complete the level.

4.3 Over-Reliance on Nudge

Four out of five networks (See Fig. 3) show a stronger association between *Draw. Freeform* and *Nudge* among the students who did not solve the game level. Nudging the ball with little connection to agent or pin creation events indicates repeated attempts at controlling the ball without creating meaningful objects that correspond to the physics concepts involved. In some cases, this could indicate wheel spinning [11] where students play for substantial amounts of time without making progress and eventually quit the level unsolved. The only other event that is closely associated with *Nudge* is *Erase*. This includes cases where the students are trying solutions that are completely incorrect or where they may have identified the agent conceptually but are unable to implement the solution in the game due to the missing identification of supporting objects or complex game mechanics. It is also interesting to observe the little to no connection to *Nudge* from any of the events in the students who solve the five levels successfully.

4.4 Limited Early Action Expansion and Later Action Convergence

There are also differences in the early and late behaviors between students who quit a level and those who didn't. The average and standard deviations (in parenthesis) of time spent per level (in minutes) for the group that quit and that which didn't are comparable - 3.73 (0.59) vs 4.08 (0.65) for *Flower Power*, 3.97 (0.75) vs 4.07 (0.76) for *Big Watermill*, 3.95(0.66) vs 3.90 (0.81) for *Caterpillar*, 4.25 (1.07) vs 4.65 (0.97) for *Shark*, 4.20 (1.15) vs 4.21 (1.14) *Need Fulcrum*. However, when students' attempts were divided into different time quartiles, we see that students who eventually quit a level do limited action exploration in the beginning (often not involving agents) and continue to produce the same limited event set for the rest of the attempt. By contrast, the students who eventually solve the level start with an expanded search for possible actions and continue to converge on a smaller subset of actions as their gameplay goes on. Figure 5 illustrates this by comparing two students – one who solved *Need Fulcrum* successfully and one who did not. As we see in the three red networks, the student who quit started by frequently connecting three events – *Draw.Freefrom, Erase* and *Nudge*

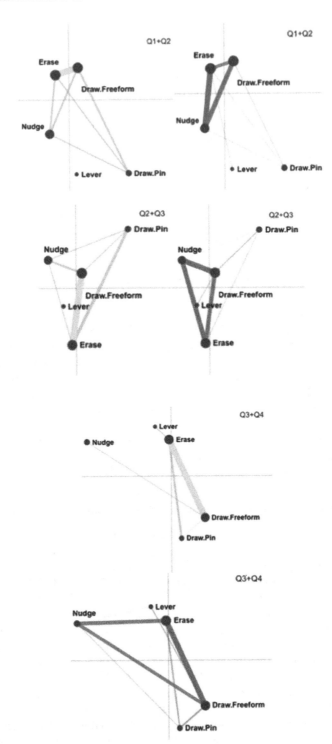

Fig. 5. Comparing early (left) to late (right) epistemic networks of a student who successfully solved *Need Fulcrum* (in green) to a student who quit (in red) (Color figure online)

and continued to follow the same approach until they quit. In contrary, the student who solved the level began with more distributed exploration and converged to just *Draw*. *Freeform* and *Erase* – in this case indicating the final attempts to find the correct weight to drop on the lever to launch the ball.

5   Discussion

In summary, we used Epistemic Network Analysis (ENA) to generate thick description of students' quitting behavior in a Physics learning game called Physics Playground (PP). We do so using the automatically generated events from the interaction log as codes for the quantitative ethnography analysis. Across the five levels investigated, our analysis revealed a set of themes which point at some potential root causes for why students quit levels unsolved in a learning game. In some cases, students may not identify the agent needed to correctly solve the level, indicating their lack of conceptual understanding of Physics. In other cases where students appear to have identified an appropriate agent, they may struggle with the difficult game mechanics around placing supporting objects or timing and precision in the placement of the objects. Other students may display wheel spinning behavior where they just nudging the ball repeatedly without making any progress. These insights are valuable to design appropriate interventions for individual students' needs. For instance, cognitive supports could be provided for the needed conceptual understanding or supports could be incorporated into the play interface for difficult game mechanics.

One limitation in this analysis is that the students in this dataset are of similar ages and live in the same region. Hence, it will be important to test the generalizability of our findings on data from a broader and more diverse range of students. In our future work, we also plan to use ENA to improve the quit prediction model which was built to deliver the interventions in a timely manner. This analysis can inform the engineering of new features that capture the behavior differences like use of a relevant agent, use of nudge and action convergence over time. We also may be able to use the epistemic networks directly for quit prediction. When a new student works on a level, we could look for whether their gameplay converges to the previously observed networks of students who solved the level successfully or those who did not.

In this paper, we have demonstrated an approach of using ENA with automated codes that has the potential to be applied in other intelligent tutoring systems with well-designed event-based logging mechanism to study constructs related to student learning, engagement, and experience in the system. It is worth noting that while this paper has primarily focused on student cognition inferred through their interaction, this does not eliminate the possibility that there could be other broader social and cultural reasons influencing factors such as students' interest, motivation, and their perceptions or beliefs about competence, that might in turn lead to students' quitting behavior. No model is perfect, but what we have learned from ENA on interaction data has the potential to focus our efforts to enhance how we support students within the scope of the game.

References

1. Shute, V.J., Ventura, M., Kim, Y.J.: Assessment and learning of qualitative physics in newton's playground. J. Educ. Res. **106**(6), 423–430 (2013)
2. Lomas, D., Patel, K., Forlizzi, J.L., Koedinger, K.R.: Optimizing challenge in an educational game using large-scale design experiments. In: Proceedings of the SIGCHI Conference on Human Factors in Computing Systems, pp. 89–98. ACM, New York (2013)
3. Karumbaiah, S., Rahimi, S., Baker, R.S, Shute, V.J., D'Mello, S.: Is student frustration in learning games more associated with game mechanics or conceptual understanding? In: Kay, J., Luckin, R. (eds.) 13th International Conference of Learning Sciences, London, UK, vol. 3, pp. 1385–1386 (2018)
4. Baker, R.S.J., Mitrović, A., Mathews, M.: Detecting gaming the system in constraint-based tutors. In: De Bra, P., Kobsa, A., Chin, D. (eds.) UMAP 2010. LNCS, vol. 6075, pp. 267–278. Springer, Heidelberg (2010). https://doi.org/10.1007/978-3-642-13470-8_25
5. Shaffer, D.W.: Quantitative Ethnography. Cathcart Press (2017)
6. Owen, V.E.: Capturing in-game learner trajectories with ADAGE (assessment data aggregator for game environments): a cross-method analysis. Doctoral dissertation, University of Wisconsin-Madison, Madison, WI (2014)
7. Karumbaiah, S., Baker, R.S., Shute, V.: Predicting quitting in students playing a learning game. In: Boyer, K.E., Yudelson, M. (eds.) Proceedings of the 11th International Conference on Educational Data Mining, pp. 167–176 (2018)
8. Shaffer, D.W., Collier, W., Ruis, A.R.: A tutorial on epistemic network analysis: Analyzing the structure of connections in cognitive, social, and interaction data. J. Learn. Anal. **3**(3), 9–45 (2016)
9. Arastoopour, G., Shaffer, D.W., Swiecki, Z., Ruis, A.R., Chesler, N.C.: Teaching and assessing engineering design thinking with virtual internships and epistemic network analysis. Int. J. Eng. Educ. **32**(2) (2016)
10. Knight, S., Arastoopour, G., Williamson Shaffer, D., Buckingham Shum, S., Littleton, K.: Epistemic networks for epistemic commitments. In: Polman, J.L., et al. (eds.) Learning and Becoming in Practice: The International Conference of the Learning Sciences, Boulder, CO, vol. 1, pp. 150–157 (2014)
11. Beck, J.E., Gong, Y.: Wheel-spinning: students who fail to master a skill. In: Lane, H.C., Yacef, K., Mostow, J., Pavlik, P. (eds.) AIED 2013. LNCS (LNAI), vol. 7926, pp. 431–440. Springer, Heidelberg (2013). https://doi.org/10.1007/978-3-642-39112-5_44

Use of Training, Validation, and Test Sets for Developing Automated Classifiers in Quantitative Ethnography

Seung B. Lee[(✉)], Xiaofan Gui, Megan Manquen,
and Eric R. Hamilton

Pepperdine University, Malibu, CA 90263, USA
seung.lee@pepperdine.edu

Abstract. Using automated classifiers to code discourse data enables researchers to carry out analyses on large datasets. This paper presents a detailed example of applying training, validation and test sets frequently utilized in machine learning to develop automated classifiers for use in quantitative ethnography research. The method was applied to two dispositional constructs. Within one cycle of the process, reliable and valid automated classifiers were developed for Social Disposition. However, the automated coding scheme for Inclusive Disposition was rejected during the validation stage due to issues of overfitting. Nonetheless, the results demonstrate the beneficial potential of using preclassified datasets in enhancing the efficiency and effectiveness of the automation process.

Keywords: Qualitative coding · Automated classifiers · Quantitative ethnography

1 Introduction

1.1 Background

Quantitative ethnography (QE) is a research methodology designed to integrate "the thick, in-depth analysis of ethnographic approaches with statistical tools to handle large amounts of data" [16, p. 382]. It is an approach that is particularly well-suited for finding meaning in today's world of big data [17, 18, 20]. There has been significant progress in the development of tools and techniques within QE in recent years, including updates to the epistemic network analysis (ENA) web-tool to enhance its capacity to handle large datasets [14]. The coding of qualitative data, however, has mostly remained a time-consuming process [2], limiting the broader application of QE in research involving substantial amounts of data.

The use of automated classifiers offers one solution to this challenge. Shaffer [16] provides extensive discussion on the topic, including a description of nCoder, a method involving the formulation of regular expressions (regex) for automatic classification of qualitative data. nCoder is based on an iterative process of manually coding small sets of data. In addition to the Cohen's kappa statistic to calculate the agreement between two raters, the method also utilizes the Shaffer's rho statistic, which estimates the

© Springer Nature Switzerland AG 2019
B. Eagan et al. (Eds.): ICQE 2019, CCIS 1112, pp. 117–127, 2019.
https://doi.org/10.1007/978-3-030-33232-7_10

probability of Type I error for a given interrater reliability (IRR) measure and thereby provides a means to statistically test for its generalizability [5].

The nCoder tool utilizes an interactive process that allows the researcher to review specific pieces of data and refine the regex list as needed. The nCoder process generally begins with the development of an initial set of regular expressions to determine the presence of a chosen construct within the data. Based on this regex list, nCoder automatically codes a new dataset for the construct, then creates a test set of a specified length that contains a minimum number of lines in which the construct is present. This is referred to as a baserate-inflated test set, where the baserate is the frequency of occurrence for a given Code [16]. For example, a test set of 40 lines with an inflated baserate of 0.2 would include at least 8 lines that have been positively coded. The remaining 32 lines are randomly selected from the dataset. After the test set has been manually coded by a human rater, kappa and rho values are calculated to determine the level of reliability and generalizability. While possible, it is highly unlikely that the kappa and rho values will have reached acceptable levels in the first iteration. Therefore, this process is typically followed by iterations of the following procedures until the thresholds for the two measures have been met: (1) review of the discrepancies in the coding of the test set between the human rater and the automated classifier; (2) adjustment of the regex list; (3) manual coding of a new baserate-inflated test set; and (4) calculation of kappa and rho statistics.

However, several practical challenges exist in the implementation of the current nCoder methodology. First is that the researcher will need to undertake multiple rounds of manual coding during the process, as a new test set will be generated for each iteration of the regex list. This can become repetitive and time-consuming particularly for complex constructs, which may require numerous cycles to identify and fine-tune the regular expressions that can consistently and accurately distinguish its presence in the data. Furthermore, developing automated classifiers for additional Codes will require that the researcher to go through the entire process for each Code, which will result in even more manual coding.

Another challenge arises from the fact that the nCoder tool utilizes baserates derived from the computer-coded dataset to create the baserate-inflated test sets. This poses a problem especially during the early stages of the iterative process, when the data is automatically classified using an incomplete or under-developed set of regular expressions. As a consequence, the preliminary regex list may result in a computer-coded dataset with a considerable number of false negatives or Type II errors (i.e. lines in which the Code is present, yet not recognized by the regex list used). Although the human rater needs to be able to identify such false negatives and revise the regex list to correctly classify them in later iterations, it is left to chance that this will happen. This is because the baserate-inflated test set, by design, is over-represented by lines that have been positively coded by the regular expressions. The false negatives will only be included if they are randomly selected as part of the remaining lines in the test set. While one of the main goals of the automation process is to minimize the occurrence of false negatives in the computer-coded dataset, the application of preliminary regex lists to create baserate-inflated test sets may actually be hindering this objective. As such, it may be argued that only regex lists that have been sufficiently developed and validated should be considered for generating baserate-inflated test sets.

In an effort to improve the efficiency and validity of the automation process, an alternate method of using preclassified data was adopted for this paper by drawing on approaches frequently utilized in machine learning. This paper presents a detailed description of using training, validation and test sets for developing automated classifiers in quantitative ethnography. In specific, it focuses on the effort to automate the coding of two dispositional constructs, Inclusive Disposition and Social Disposition. The analysis is based on conversational discourse data from video conference sessions, referred to as meet-ups, involving participants from fifteen digital makerspace clubs located in Africa, Europe, and North America [3, 11]. During the sessions, participants ages 10–19 work collaboratively to create media artifacts on STEM-related topics. Finally, the analysis presented in this paper utilized the following R packages: ncodeR (ver. 0.1.2) [13] and rhoR (ver. 1.2.1.0) [4].

1.2   Codes for Automation

Two dispositional constructs were selected for the initial process of developing automated classifiers: Inclusive Disposition and Social Disposition. These two Codes were selected mainly for their relatively high baserates in the meet-up data. The high baserates for these Codes can be attributed to the nature of the online synchronous meet-ups, during which participants share their projects and provide each other comments and feedback [6]. Inclusive Disposition is reflected in utterances that encourage the participation of specific individuals in the discussion. Social Disposition is present when an utterance demonstrates pro-social tendencies of the speaker, especially in expressing appreciation, acknowledgement or validation [10]. The codebook for the two constructs is presented in Table 1.

Table 1. Codebook for Inclusive Disposition and Social Disposition.

Code	Description	Examples
Inclusive Disposition	Encouraging participation of specific individuals in the discussion	*"I would also like to invite Student A to say something about Student B's presentation on food and health"* *"Student C, can you tell us more about a project you've done?"*
Social Disposition	Demonstrating pro-social tendencies, especially in expressing appreciation, acknowledgement or validation	*"I just really look forward to working with you and everybody else and to meet all of you too"* *"That was a wonderful video. I really like the way you have done it"*

2   Methodology

Training, validation and test sets are used prevalently in machine learning contexts. The process involves the division of preclassified data into the respective sets to serve different objectives at each stage [1, 8]. Using three independent sets allows the researcher to assess whether a particular model, or network, is generalizable beyond the examples contained in the training set [15]. Bishop [1] elaborates that:

Various networks are trained by minimization of an appropriate error function defined with respect to a *training* data set. The performance of the networks is then compared by evaluating the error function using an independent *validation* set...Since this procedure can itself lead to some over-fitting to the validation set, the performance of the selected network should be confirmed by measuring its performance on a third independent set of data called a *test* set. [1, p. 372]

Figure 1 provides an overview of the methodology used in this paper to utilize training, validation and test sets in the development of automated classifiers for Inclusive Disposition and Social Disposition. In this analysis, the different iterations of the regex lists created during the training stage will constitute the "models" that are validated and tested at later stages. As such, the automated classifiers developed during this process are rule-based models, distinct from predictive models that are often utilized in machine learning contexts [15]. Kappa and rho statistics are used in all three stages to assess the effectiveness and generalizability of each regex list in classifying the data.

2.1   Dataset Construction

The first consideration in the construction of the datasets was to determine the sizes for each set. In machine learning, an 80/20 split is commonly used to divide labeled data into subsets for training and evaluation [12]. This ratio was applied for the training and validation sets used in this analysis. As the data was collected from online meet-ups, lines of data from the same session were kept together. Therefore, rather than applying the ratio to individual lines of data, it was applied to whole meet-up sessions. Four meet-ups were included in the training set (776 lines) and one meet-up was reserved for the validation set (168 lines). The resulting ratio of the number of lines in the training set to the validation set was 82% to 18%.

To create the preclassified datasets, a total of 944 lines from the five meet-ups used for the training and validation sets were coded separately by two human raters for Inclusive Disposition and Social Disposition as well as several other salient constructs identified through a grounded analysis of the data. This was followed by a process of social moderation undertaken for each line, which allowed the raters to come to an agreement on the coding of the data [7, 9]. The baserates for Inclusive Disposition were 0.13 and 0.15 for the training and validation sets, respectively. For Social Disposition, they were 0.13 and 0.24.

The test set utilized data from a new meet-up of 167 lines that had not been previously coded. Applying the regex list selected in the validation stage, the "autocode" function in ncodeR was used to classify all lines in the new dataset. Then, a baserate-inflated test set of 40 lines was constructed for each Code to be assessed, consisting of at least 8 lines (20%) in which the Code was identified to be present by the regex list.

2.2   Training Stage

The objective of the training stage was to develop regex lists that can automatically identify the presence of the two Codes in the data. To begin, a preliminary list of regular expressions was produced by searching for patterns in words, phrases and sentence structures in all utterances in the training set that had been coded for the

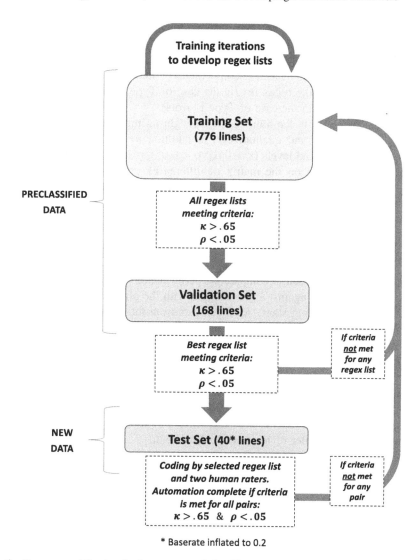

Fig. 1. Process used for developing automated classifiers using training, validation and test sets.

construct. Once the preliminary list was constructed, it was used to automatically code the entire data in the training set, resulting in the first training iteration. Kappa and rho values were calculated to assess the interrater reliability between the computer-coded dataset and the training set.

This was followed by a comparison of the automated and preclassified codes, which allowed the researchers to review and refine the regex list. To facilitate the review process, several new functions were written in R to supplement those included in the ncodeR package. The functions were used to create a single matrix containing the utterance along with the preclassified and automated codes for the entire training set.

For utterances with positive codes that were automatically identified, the matrix also provides the specific regular expression that was matched in the text. Having the full matrix of utterances and codes provided the researchers the ability to sort and search through the coded training data in a comprehensive manner—to identify additional patterns for inclusion in the regex list (in the case of Type II errors) and to adjust or exclude certain phrases (in the case of Type I errors).

Subsequent iterations in the training stage sought to improve upon previous regex lists, both quantitatively and qualitatively. In addition to achieving kappa and rho values beyond the threshold levels ($\kappa > 0.65$, $\rho < 0.05$), researchers continued to fine-tune the regex list based on the matrix of differences produced for each iteration. Training iterations were carried out until the researchers determined qualitatively that no further adjustments could be made to the regex list without undermining the generalizability of the automated classifier.

2.3   Validation Stage

The validation stage was used to compare the performance of the regex lists developed during the training phase against a new dataset, with the objective to select the single best regex list for the testing stage. All regex lists from the training stage meeting the minimum criteria for kappa and rho values ($\kappa > 0.65$, $\rho < 0.05$) were included in the analysis. This allowed the researchers to account for any overfitting that may have occurred during the later parts of the iterative training process. The selection of the best classifier was determined based on the kappa and rho statistics—the regex list that produced the highest kappa and lowest rho measures against the validation set was chosen for the next stage. In the case that none of the regex lists met the thresholds for both kappa and rho, the process of automation for the particular Code was returned to the training stage for further revisions and adjustment.

2.4   Testing Stage

The testing stage applied nCoder's method of using baserate-inflated test sets [8]. The regex list selected in the validation stage was used to code utterances from a new dataset consisting of 167 lines. Several functions were created in R based on the ncodeR package to produce a test set of 40 lines, of which at least 20% had been identified by the automated classifier as having the particular Code present in the utterance. After the test set was individually coded by two human raters, the kappa and rho statistics were computed for each pair of raters, i.e. Computer vs. Human 1, Computer vs. Human 2, Human 1 vs. Human 2. The automated classifier was considered to be valid and reliable for the Code if the kappa and rho thresholds were met for all three pairs of raters [16, 19]. Otherwise, the automation process was returned to the training stage.

3   Results

3.1   Training Stage

The kappa and rho values resulting from each iteration in the training stage are displayed in Table 2 for Inclusive Disposition and Social Disposition. Four of the regex lists for Inclusive Disposition (Iterations #3, #4, #5, and #7) achieved sufficient levels of kappa and rho, while two regex lists satisfied the criteria for Social Disposition (Iterations #3 and #4).

A total of seven iterations were carried out for Inclusive Disposition, the first Code to be applied to the automation process. The relatively high number of iterations was mainly due to the adjustments made to the regex list to account for some irregularities encountered in how quantifiers and punctuations were handled in the nCodeR package. First, it was observed that the quantifiers "*" and "?" used in the regular expressions were not being properly recognized. A work-around was to rewrite the regular expression containing one of these quantifiers with two regular expressions removing the quantifier. For example, the regular expression "(could|can) (you) (just)* (share| say|explain)" was replaced with "(could|can) (you) (share|say|explain)" and "(could|-can) (you) (just) (share|say|explain)." This adjustment, which was made between the third and fourth iterations, may have resulted in the slight loss in the kappa and rho values. During the fifth iteration, it was also noticed that punctuations were not being consistently processed. Modifications were made to the R commands to remove all punctuations from utterances in the matching. The regex list was also revised accordingly. The reduction in kappa and increase in rho values seen in the sixth iteration may be due to this correction. The highest kappa-lowest rho combination for Inclusive Disposition was achieved in the seventh and final iteration, resulting in a kappa of 0.7162 and a rho of 0.0038.

The insights gained during the training iterations for Inclusive Disposition were applied to the process for Social Disposition. As a result, only four iterations were required to obtain kappa and rho values of 0.8245 and 0.0000, respectively.

Table 2. Kappa and rho values for each iteration in the training stage for the two codes.

Training Iteration No.	Inclusive Disposition		Social Disposition	
	Kappa	Rho	Kappa	Rho
# 1	0.5233	1.0000	0.5482	0.4525
# 2	0.6465	0.0725	0.6132	0.1538
# 3	0.7138	0.0088	0.7959	0.0000
# 4	0.6957	0.0125	0.8245	0.0000
# 5	0.6930	0.0150		
# 6	0.6116	0.1513		
# 7	0.7162	0.0038		

3.2   Validation Stage

Four regex lists for Inclusive Disposition and two for Social Disposition were utilized to automatically classify the utterances in the validation set. The results are displayed in Table 3. For Inclusive Disposition, none of the regex lists met the criteria for the kappa and rho values. Kappa remained within a low range of 0.34 and 0.41 and the rho statistic for all four regex lists was 1. However, both regex lists for Social Disposition performed well against the validation set. Iteration #4 was ultimately selected for the testing stage based on its relatively high kappa (0.7284) and low rho (0.0188) levels.

Table 3. Kappa and rho values for select regex lists included in the validation stage.

Training Iteration No.	Inclusive Disposition		Social Disposition	
	Kappa	Rho	Kappa	Rho
# 3	0.3493	1.0000	0.7235	0.0275
# 4	0.3615	1.0000	0.7284	0.0188
# 5	0.4099	1.0000		
# 7	0.4099	1.0000		

3.3   Testing Stage

The only regex list to reach the testing stage was the one developed during Iteration #4 for Social Disposition. The baserate-inflated 40-line test set was coded individually by the automated classifier and two human raters. The kappa and rho statistics between each pair of raters are presented in Table 4. The kappa and rho values between the computer and the first human rater were 0.8065 and 0.0400, respectively. Coding by the computer and the second human rater resulted in a kappa of 0.8750 and a rho of 0.0138. The kappa value was the highest between the two human raters at 0.9390, while the rho value was the lowest at 0.0013.

Based on the high levels of agreement between all three pairs of raters, it was concluded that the automated classifier was valid and reliable in coding the data for Social Disposition. The final regex list used to obtain these results is presented in Table 5.

Table 4. Kappa and rho values for each pair of raters in the testing stage for Social Disposition.

Raters	Kappa	Rho
Computer & Human 1	0.8065	0.0400
Computer & Human 2	0.8750	0.0138
Human 1 & Human 2	0.9390	0.0013

Table 5. Final list of regular expressions for Social Disposition.

Regex list for Social Disposition (training iteration #4)	
"(look|looking) forward"	"(enjoy|love|like|enjoyed|loved|liked) (your|how)"
"(like|liked) the presentation"	
"(I|(I really)) liked"	"I (wanna|want) say (thanks|(thank you))"
"very exiciting"	"I (wanna|want) to say (thanks|(thank you))"
"that (s|is) awesome"	"(enjoy|love|like|enjoyed|loved|liked) (that|it) because"
"really well done"	
"well done"	"(nice|happy|good|love|(((my)|(our)) pleasure)|excited) to (meet|see)"
"I just loved"	
"I (agree|agreed) completely"	"((thank you)|(thanks)|(thanks so much)| (thank you so much)|(thank you all)|(thank you everyone)) for"
"(appreciate|appreciated|appreciates)"	
"(creative|creativeness|creativity)"	
"(video|presentation|idea) because"	"you made a (good|cool|nice| interesting|great) (presentation|video)"
"(I think it was)|(I feel like this was)"	
"really great topic"	"((That is) |(that s)|(that was)) (a|an) (good|cool|nice|interesting|great)"
"(video|presentation) (was|is)"	
"(good|cool|nice|interesting|great)"	"((That is)|(that s)|(that was)) (a|an) (really| very|so) (good|cool|nice|interesting|great)"
"(that|video|presentation) (s|was|is)"	
"((I am)|(I m)) (glad|appreciative|excited)"	"(really|very|so) (good|cool|nice| interesting|great)"
"thank you for (participating|sharing) (enjoy|love|like|enjoyed|loved|liked) (collaborating|working) (with|for) (you| everyone|this)"	"(inspirational|wonderful|amazing|(well put together))"
"(good|great|nice) (job work) ((it s)|(it is)|you| on|because)"	"((I am)|(I m)) really (glad|appreciative| excited)"

4   Discussion

Using automated classifiers to code discourse data enables researchers to carry out analyses on large datasets. This paper presented a detailed example of applying training, validation and test sets frequently utilized in machine learning to the develop automated classifiers for use in quantitative ethnography research. The development and refinement of regex lists for identifying the presence of particular constructs in the data was facilitated by the use of preclassified data in the training and validation sets. In addition to enhancing the efficiency of the automation process, the preclassification of the data also allowed the researchers to become more familiar with the qualitative data and to gain a better understanding of the meanings contained within them.

This paper focused on one cycle of the training-validation-testing process for two dispositional constructs. While the automated coding scheme for Social Disposition was able to reach acceptable levels of reliability and validity, the automated classifiers for Inclusive Disposition did not produce sufficient IRR levels when tested against the validation set. This may be due to a number of factors, including the learning curve the researchers experienced during the first implementation of the automation tools and techniques as discussed above. At the same time, it may be indicative of the more difficult nature of the automating the coding for this particular construct. Coding for

Inclusive Disposition within an utterance is largely dependent on the "invitation for participation" that is extended by the speaker to specific individuals or groups. As such, it is often associated with the names of individuals or groups, which is difficult to generalize in any regular expression. For this reason, overfitting may have resulted in the development of regex lists for Inclusive Disposition in the training stage, thereby leading to their rejection in the validation stage. While beyond the scope of this paper, this raises the question of whether a larger training set will be required for Inclusive Disposition in the second cycle of training. Additional research will be needed to look into the issue of overfitting and how it might be resolved in subsequent cycles of training and validation.

Nonetheless, the relatively efficient and effective automation of the coding for Social Disposition points to the beneficial potential of using the training-validation-testing process. Analysis of discourse data often involves coding for several salient constructs. While the creation of preclassified training and validation sets may require substantial investment of resources in the preparation phase, a key advantage is that the same training and validation sets may be used for the automation of all of the Codes which were included in the preclassification process. A follow-up study will examine in detail the automation of several other Codes included in the preclassified data used for this paper. It is expected that the findings of this subsequent analysis will provide important insights on the broader applicability of the approaches outlined in this paper. In addition, further research will be required to determine the most effective lengths of training, validation and test sets for use in the development of automatic classifiers for discourse data.

Acknowledgements. The authors gratefully acknowledge funding support from the US National Science Foundation for the work this paper reports. Views appearing in this paper do not reflect those of the funding agency.

References

1. Bishop, C.M.: Neural Networks for Pattern Recognition. Oxford University Press, Oxford (1995)
2. Dönmez, P., Rosé, C., Stegmann, K., Weinberger, A., Fischer, F.: Supporting CSCL with automatic corpus analysis technology. In: Proceedings of the 2005 Conference on Computer Support for Collaborative Learning (CSCL), pp. 125–134. International Society of the Learning Sciences (2005)
3. Eagan, B.R., Hamilton, E.: Epistemic network analysis of an international digital makerspace in Africa, Europe, and the US. Paper presented at the annual meeting of the American Education Research Association (AERA), New York (2018)
4. Eagan, B.R., Rogers, B., Pozen, R., Marquart, C., Shaffer, D.W.: rhoR: rho for inter rater reliability (version 1.2.1.0) (2019)
5. Eagan, B.R., Rogers, B., Serlin, R., Ruis, A.R., Arastoopour Irgens, G., Shaffer, D.W.: Can we rely on IRR? Testing the assumptions of inter-rater reliability. In: Proceedings of the 12th International Conference on Computer Supported Collaborative Learning, Philadelphia (2017)

6. Espino, D.P., Lee, S.B., Eagan, B.R., Hamilton, E.R.: An initial look at the developing culture of online global meet-ups in establishing a collaborative, STEM media-making community. In: Proceedings of the 13th International Conference on Computer-Supported Collaborative Learning (CSCL), pp. 608–611. International Society of the Learning Sciences (2019)

7. Frederiksen, J.R., Sipusic, M., Sherin, M., Wolfe, E.W.: Video portfolio assessment: creating a framework for viewing the functions of teaching. Educ. Assess. 5(4), 225–297 (1998)

8. Haykin, S.S.: Neural Networks and Learning Machines, 3rd edn. Prentice Hall, New York (2009)

9. Herrenkohl, L.R., Cornelius, L.: Investigating elementary students' scientific and historical argumentation. J. Learn. Sci. 22(3), 413–461 (2013)

10. Katz, L.G., McClellan, D.E.: Research into practice series, vol. 8. Fostering children's social competence: the teacher's role. National Association for the Education of Young Children, Washington, D.C. (1997)

11. Lee, S.B., Espino, D.P., Hamilton, E.R.: Exploratory research application of epistemic network analysis for examining international virtual collaborative STEM learning. Paper presented at the annual meeting of the American Educational Research Association (AERA), Toronto (2019)

12. Lever, J., Krzywinski, M., Altman, N.: Points of significance: model selection and overfitting. Nat. Methods 13(9), 703–704 (2016)

13. Marquart, C., Swiecki, Z., Eagan, B.R., Shaffer, D.W.: ncodeR: techniques for automated classifiers (version 0.1.2) (2018)

14. Marquart, C., Hinojosa, C., Swiecki, Z., Eagan, B., Shaffer, D.W.: Epistemic network analysis (version 1.5.2) (2018)

15. Sebastiani, F.: Machine learning in automated text categorization. ACM Comput. Surv. 34(1), 1–47 (2002)

16. Shaffer, D.W.: Quantitative Ethnography. Cathcart Press, Madison (2017)

17. Shaffer, D.W.: Big data for thick description of deep learning. In: Millis, K., Long, D., Magliano, J., Wiemer, K. (eds.) Deep Comprehension, pp. 265–277. Routledge, New York (2018)

18. Shaffer, D.W., Ruis, A.R.: Epistemic network analysis: a worked example of theory-based learning analytics. In: Lang, C., Siemens, G., Wise, A.F., Gasevic, D. (eds.) Handbook of Learning Analytics, pp. 175–187. Society for Learning Analytics Research (2017)

19. Swiecki, Z., Ruis, A.R., Farrell, C., Shaffer, D.W.: Assessing individual contributions to collaborative problem solving: a network analysis approach. Comput. Hum. Behav. (2019, in press)

20. Wise, A.F., Shaffer, D.W.: Why theory matters more than ever in the age of big data. J. Learn. Anal. 2(2), 5–13 (2016)

Theme Analyses for Open-Ended Survey Responses in Education Research on Summer Melt Phenomenon

Haiying Li[(✉)], Joyce Zhou-Yile Schnieders, and Becky L. Bobek

ACT, Inc., Iowa City, IA 52243, USA
{haiying.li,joyce.schnieders,becky.bobek}@act.org

Abstract. Summer melt is a phenomenon when college-intending students fail to enroll in the fall after high school graduation. Previous research on summer melt utilized surveys, typically consisting of Likert scale questions and open-ended response questions. Open-ended responses can elicit more information from students, but they have not been fully analyzed due to the cost, time, and complexity of theme extraction with manual coding. In the present study, we applied the topic modeling approach to extract topics and relevant themes, and evaluated model performance by comparing model-generated topics and categories with the human-identified topics and themes. Results showed that the topic model allows for extracting similar topics as the survey questions that were investigated, but only extracted part of the themes classified by the human. Discussion and implications focus on potential improvements in automated topic and theme classification from open-ended survey responses.

Keywords: Topic modeling · Summer melt · Automated theme analyses

1 Introduction

Quantitative ethnography bridges the gap between quantitative and qualitative research to facilitate better understanding data and the people who provide the data [14]. The present study introduces "topic modeling", a statistical model using an unsupervised machine learning method and one quantitative ethnographic approach.

Topic modeling is a popular method that has been applied to automatically extract themes from open-ended survey responses in a few studies, especially in the fields of political science [12] and social science [13], to reveal the opinions and interests of respondents who participate in surveys. Automated theme extraction enhances studies with open-ended survey responses allowing for investigation of respondents' thoughts, opinions, and interests, which may not be ascertained from forced-choice surveys, and further provides researchers more nuanced information to test and advance more complex theoretical frameworks. We tested this method in the field of education, automatically extracting the topics and themes from open-ended responses in surveys about the summer melt phenomenon, which is described in the next section.

© Springer Nature Switzerland AG 2019
B. Eagan et al. (Eds.): ICQE 2019, CCIS 1112, pp. 128–140, 2019.
https://doi.org/10.1007/978-3-030-33232-7_11

1.1 Summer Melt

The term "summer melt" is defined as a phenomenon when "college-intending students fail to enroll at all in the fall after high school graduation" [5]. Previous studies have revealed that in the United States, the percentage of high school graduates who intend to go to college but do not enroll in the following fall semester after graduation ranges from 10%–40% [4]. The enrollment rate is likely to be even lower for students who are first-generation, socio-economically disadvantaged, or underrepresented minority compared to their peers [4]. The lack of higher education may lead to the loss of employment opportunities, as well as financial and societal benefits [16, 17].

Diverse intervention programs have been developed for students at risk of summer melt as a way to improve college enrollment rates among these students, some of which include highlighting the benefits of postsecondary education [10], and providing information about college costs and how to complete financial aid applications [1]. However, approximately one-quarter of high school graduates who received these types of personalized intervention supports in one school system still failed to enroll in college on the first day [11].

Potential factors that contribute to or protect students from summer melt have been identified through surveys [11]. Surveys in such studies typically consist of two types of question formats: Likert scale questions and open-ended response questions. Open-ended responses were used to elicit more personalized information relative to Likert scale questions, further unpacking the possible contributors to the summer melt phenomenon and providing insights to potential interventions that reduce the summer melt phenomenon.

The traditional approach to analyzing open-ended survey responses is through manual coding. Due to the cost, time, and complexity of theme extraction with manual coding, open-ended survey responses in many cases are not fully analyzed. The purpose of this study is to explore the feasibility of using the topic modeling approach to analyze open-ended survey responses as an accurate and more efficient method to extract relevant themes on this education topic.

1.2 Topic Modeling in Theme Extraction

Recent research has shown that topic modeling is a promising approach to automated extraction of hidden themes from open-ended survey responses [6, 7, 12, 13]. Latent Dirichlet allocation (LDA), a generative probability model, is a popular algorithm for topic modeling that involves three levels of words, documents, and corpora [2]. Documents are represented as a series of latent topics, where each topic is represented by a distribution over words [2]. Specifically, if a document contains N words, $\{w_1, w_2, \cdots, w_N\}$, the LDA model probabilities $pk = (pk(w_1), pk(w_2), \cdots, pk(w_N))$ form a representation of the k_{th} topic ($k = 1, 2, \cdots, K$). The words with the highest probability in each topic inform the potential topics.

Recently, researchers have explored the application of the automated approach to extract themes from open-ended survey responses. For example, a few studies have adopted a semi-automated approach – the structural topic model (STM) – to extract themes from open-ended responses in the fields of political science [12], social science [13], and even from multilingual textual datasets [7]. The STM extracted themes utilizing probabilistic topic modeling algorithms – LDA [2]. Recently, Li, Zhou, and Bobek [6] utilized LDA to extract the same number of broad themes as human coders from open-ended responses in nine surveys related to summer melt. Results showed that the LDA model extracted four themes that were similar to human-identified themes (e.g., college talks, college stressors, future career, supports), even though two were different (e.g., help, uncertainty) [6]. This study suggests that topic modeling might be a potential approach to automatically extract themes from open-ended survey responses. This initial study, however, extracted 6 themes from all the responses in the 9 surveys as a whole rather than extracted themes from responses to survey questions in each survey.

The present study expanded on previous studies to investigate the potential for extracting survey question topics and underlying themes from open-ended responses to each survey. This study aims to seek a method that allows for automatically retrieving important hidden topics and themes from open-ended responses to survey questions. The contribution of this study is to provide a potential approach to automated topic and theme extraction from open-ended survey responses so that we could more fully use students' responses to identify the reasons for summer melt. Additionally, this approach has the potential to be applied to open-ended survey responses in other fields, saving the time and labor required for manual coding.

1.3   Quantitative Ethnography Theories and Topic Modeling

Quantitative ethnography is a new approach that combines quantitative and qualitative research methods to generate meaningful insights using big data [14]. In the information age, quantitative data mining techniques have been widely applied to spot patterns in large human behavioral datasets. However, it is risky to simply identify patterns from big data, without going further to make sense of the patterns [15]. Ethnography, on the other hand, aims to create thick descriptions of the structures of societies and cultures through in-depth observations of small groups. Thus, with the merging of ethnography and quantitative approaches, the mechanisms in ethnography can inform how to make sense of patterns found in big data. At the same time, "statistical approaches can expand the tools of ethnography" as well [15].

An important part of quantitative ethnography is taking the basic process of constructing quantitative models and applying these models to organized qualitative data [14]. Before modeling the way people express their understanding about the world, the first step is to identify what people mean as they talk and act, in other words, transforming qualitative data into forms amenable to statistical analysis [15]. And when the dataset is too big to be accurately and efficiently coded by humans, automated coding is necessary. Work on natural language processing provides insights to how to translate human languages into terms that a computer can understand [14]. Shaffer [14] pointed out that topic modeling is a potentially useful tool for automatically identifying the codes in the data. However, he also highlights the necessity to check whether the topics are actually grounded in the original meaning of the data. To accomplish this, the topics need to be validated by getting good interrater reliability between human raters and the machine [14]. The current study tested out this process in the context of "summer melt".

2 Method

2.1 Participants and Surveys

Participants were graduating seniors from five tuition-free public charter high schools. Based on students' responses in a high school exit survey, a pre-interview, teacher feedback, and demographic data, 54 students (52% male, 48% female) identified as at risk for summer melt were chosen to participate in the study. The racial composition of participants included Hispanic (47%), Asian (26%), multi-racial (13%), White (11%), and Native Hawaiian (2%).

Participants completed nine bi-weekly online surveys during summer 2017. Table 1 displays the response rate for each survey, ranging from 72% to 89%. Multiple areas that might interfere with students' college attendance in Fall 2017 were investigated, including college readiness perceptions, belonging, self-efficacy, self-regulated learning, academic needs, goal setting, intention, persistence, core self-evaluation, barriers, supports, career goals, and educational attitude. Each survey consisted of two types of questions, Likert scale questions and open-ended response questions. The topics of each survey were designed to match students' experiences relevant to college preparation in the corresponding time period.

Table 1. Response rate of each online survey.

	1	2	3	4	5	6	7	8	9
Total	48	44	44	43	43	43	43	40	39
Percentage	89%	81%	81%	80%	80%	80%	80%	74%	72%

2.2 Human Coding

Qualitative analysis was conducted for the majority of open-ended survey response questions. Using the "summer melt" literature [4, 5] and the Education & Career Navigation domain of the ACT Holistic Framework [3] as the theoretical framework, a coder began data analysis by reading all the responses for each survey question and highlighting segments of data that were relevant to that question. The segments were then grouped and compared together. Tentative categories were constructed for each question through this analytical coding process. To increase the trustworthy and consistency of the analysis, peer examination was used [9]. After the tentative categories were constructed, a second coder scanned all the raw data, assessed the occurrences of each category across all the participants, and verified the plausibility of each category. The categories for each question were finalized afterwards. Table 2 shows the topical focus for each survey question, and the themes that human coders categorized from each open-ended response question across nine surveys.

Table 2. Topics and themes from responses to questions (Q) in each survey (S).

S	Q	Type	Topic	Theme
1	7 (N = 46)	Human	**Support Types**	People who are close, advice, decisions/choice made, academic, financial, family support, emotional support, career literacy
		Model	**Support**	Support, life, family, people, friend, move, academic, decision, live, career
	9 (N = 47)	Human	**Anticipated College Stressors**	Pressure, expectation uncertainty, fear of failure, finance, difficult adjustment, moving away from home, making friends/meeting new people
		Model	**College Stress**	College, stress, year, finance, graduation, success, future, father, community, meeting
	11 (N = 47)	Human	**Financing College Stressors**	Parent financing, self-financing, finding ways to pay for college, pay for loans after college, lot of money, not qualify for certain types of aid
		Model	**Financial Stress**	Pay, parent, money, job, loan, learn, worry, summer, scholarship, debt
	12 (N = 47)	Human	Summer (pre-college) Thoughts	Work, future academic or career (e.g., class), friends (time spent), college experience, college preparation, negative emotion
		Model	Work, Time, and Class	School, work, time, hard, class, student, home, good, preparation, experience
2	3 (N = 44)	Human	**Summer Plan**	Working, relaxing, spending time with family/friends, vacation, summer classes, preparing for college
		Model	**Summer**	Summer, job, college, family, work, friend, school, week, relax, class
	6 (N = 44)	Human	**Career Plan Fit**	Undecided/indecisive, identified general direction, specified goal, driven by interest/passion/skill, committed to a goal, information/experience, potential barrier to career path /goal achievement, identified what will motivate goal attainment
		Model	**Career**	Career, choice, hard, major, path, life, goal, people, hope, change
	11 (N = 4)	Human	**Plan Change**	Family, college logistics, distance from job, future goal direction
		Model	**Plan**	Work, time, plan, money, year, day, college, home, read, fall
3	6 (N = 44)	Human	**College Success Confidence**	Support self in coursework/belief in capability, finances might interfere, open to opportunities/ challenges, focused, other obligations/ responsibilities might interfere, resources /supports, new environment concerns, academic demands, self-doubt

S	Q	Type	Topic	Theme
		Model	**Confidence**	Support, work, great, people, complete, skill, confidence, nervous, education, campus
	8 (N = 44)	Human	**College Talk**	Rigor expected of coursework, manage time, hopes, cost, if right college, living on campus differs from living at home, responsible for self, how classes will be, college experience
		Model	**College Talk**	College, talk, major, family, friend, attend, prepare, pay, money, plan
	10 (N=44)	Human	**Anticipated College Challenges**	Rigorous course content, lots of studying, fast paced quarter system, higher level of education
		Model	**College challenging**	College, school, challenging, year, student, choice, big, community, mind, level
	11 (N = 43)	Human	**Personal Strengths**	Time management, plan, goal driven, determination, resourceful, ask for help, build relationships, way of thinking
		Model	**Strength**	Success, class, learn, habit, academic, strength, experience, living, easy, people
	12 (N = 43)	Human	**Personal Limitations**	Money, coursework too much, procrastination, lack of perseverance, time management, distractions
		Model	**Limit**	Time, hard, limit, life, ability, effort, procrastination, order, management, coursework
4	7 (N = 40)	Human	**My College Knowledge**	Diversity, programs, majors, clubs, organizations, sports, overview of the college (history, background, location)
		Model	**College**	College, attend, community, orientation, transfer, plan, support, time, education, decision
	8 (N = 40)	Human	College Choice	Location, rating, size, financial reasons, program offering
		Model	Decision	School, friend, year, great, decision, change, community, transfer, finance, complete
	14 (N = 39)	Human	**Student Similarities to You**	Same goal (e.g., transfer from community college), similar interest, background, area, religion, nervous/stressed
		Model	**Student Similar**	Student, similar, class, interest, work, diversity, hard, sport, challenge, share

S	Q	Type	Topic	Theme
	15 (N = 39)	Human	Student Differences from You	Different background, academic/career goal, values/opinions/views, interests, living habits
		Model	People	School, people, great, goal, life, meet, experience, friend, background, view
5	7 (N = 43)	Human	**Family Supports**	Financially help, encouragement, pride and positivity, involved in process, guidance and asks others who have children in college
		Model	**Supportive**	Supportive, choice, excited, decision, education, application, question, major, encouragement, class
	8 (N = 43)	Human	**Family Lack of Supports**	Finance, college/major/career choice, lack of involvement
		Model	**Family Push**	Family, job, pay, hard, care, money, long, push, child, young
	10 (N = 43)	Human	**Friend Supports**	Discuss goals and how to achieve, decision support for a degree, proud of me, encouraging, advice and help
		Model	**Supportive People**	Supportive, people, time, life, happy, living, offer, close, success, moving
	11 (N=)43	Human	**Friend Lack of Supports**	College choice, discouraging
		Model	**Unsupportive Friends**	Friend, school, year, change, bad, unsupportive, proud, transfer, future, complete
	13 (N = 9)	Human	College Plan Shift	Finance, location, environment, offer rescinded, changed to community college
		Model	College Attendance	College, great, attend, community, process, stay, understand, waste, remind, environment
6	3 (N = 43)	Human	**Future Career Options**	More than 1 option, general or vague direction, undecided
		Model	**Future Career**	Career, future, decision, time, path, people, enter, success, plan, opportunity
	5 (N = 43)	Human	College supports Future Career	Expand network, find work opportunities, having the degree, develop skills and knowledge for career, find future
		Model	College	College, learn, engineer, knowledge, interest, science, prepare, computer, school, skill
	7 (N = 43)	Human	**Career Choice Confusion**	Difficult to find what to enjoy, too many options to navigate, not sure what will fit, lack of understanding about choices and paths, not sure about doing something for a long time, experience helped

S	Q	Type	Topic	Theme
		Model	**Confusion**	Major, passion, life, business, class, great, rating, confusing, idea, explore
	10 (N =43)	Human	**Career Choice Difficulties**	So many choices, need to consider interest/pay, big decision
		Model	**Difficulties**	Option, job, difficult, field, enjoy, hope, question, figure, interest, school
7	8 (N = 41)	Human	**College Beliefs /Feelings**	Commute difficult, school priority, confident about passing classes, independent, stressed, excited for what is new, pressure to do well, getting along with people, accountable
		Model	**College Feelings**	College, excited, enjoy, homework, job, nervous, life, learn, home, stress
	10 (N = 3)	Human	College Going Shift	Finance, transfer, major
		Model	Class and School	Class, school, great, friend, time, difficult, experience, people, week, day
8	9 (N = 2)	Human	**College Preparedness Certainty**	Providing sufficient preparation and challenging courses
		Model	**Learning Challenge**	Class, time, learning, professor, rating, lecture, challenging, test, coursework, math
	17 (N = 37)	Human	**College Preparation**	Self-directed/motivated learning, independent, time management, skills
		Model	**School Preparation**	School, job, prepare, student, teach, skill, homework, deadline, university, content
	19 (N = 22)	Human	**College Surprises**	Diversity, small class size, heavy workload, professor with freedom to cancel classes, more responsibility, hard to make friends, people are nice
		Model	**Surprises (in College)**	College, surprise, school, people, complete, great, teacher, friend, assignment, essay
9	7 (N = 38)	Human	**Advice to Students**	Time management important, set priorities, make friends, have fun, have new experience
		Model	**Advice (to) Students**	College, time, school, job, student, class, advice, friend, learning, experience

2.3   Procedure of LDA Theme Extraction

The present study used Mallet topic modeling [8] inbuilt within Genism in Python 3.6. To extract themes more accurately, we removed the stop words after slight modification to the stop words in the NLP package, including removing the high frequency words that do not deliver rich meanings (e.g., thing, know). We also added a bi-gram and tri-gram model in order to extract semantic related keywords. Moreover, we considered the synonyms that highly occurred in the data set to avoid two words with similar meaning being included in the same topic which leads to two repeated keywords or keywords in two topics which yield an "overlapped topic" [6]. Take survey question 7 in Survey 1 (see Table 2) as an example. This question asked about support types, so "help" and "support" occurred frequently in students' responses. To avoid the occurrence of the overlapped topic "help" and "support", we replaced the term "help" with "support".

The number of responses to each survey question ranged from 2 to 47 (see Table 2), so it is impossible to achieve better performance of LDA model for survey questions with few responses, such as 2, 3, or 4 responses. Fewer survey responses, however, is typical for qualitative studies and must be considered. One possible method for researchers is to remove the smaller number of responses. This method, however, may eliminate important information that should be included in the analysis. We proposed an alternative approach: running nine topic models, each model for each survey. Specifically, we mixed the responses to all the survey questions within one survey and thus generated nine text data sets. The number of topics in each data set was determined based on the number of survey questions in the survey. For example, if Survey 1 contained four questions, four topics were selected in the model. In this way, we were able to examine whether LDA model could retrieve themes similar to the investigated survey questions. This allowed for investigating whether key words in the topics demonstrated relevant themes compared to the categories found through manual coding. Another advantage with this approach is that it is possible to merge similar themes across different survey questions that consist of similar topics.

3   Results and Discussion

Table 2 displays the topics that the human predetermined based on the question itself as well as the topics that the LDA model generated based on students' responses. Table 2 also displays themes for each survey question that human coders categorized through manual coding and themes from all the open-ended responses in a particular survey (nine in total).

3.1   Extracted Survey Topics

Results showed that the topics generated from the LDA model from students' responses were similar to those predetermined by the human based on the content of survey questions. Due to the paper limit, we provide Survey 1 as an example to demonstrate the way that we interpreted the results. Survey 1 included four open-ended survey questions: Question 7 (What more support would you want in your life?); Question 9 (Explain why you rated yourself at this stress level, based on the previous question "How stressed out are you about going to college?"); Question 11 (Explain why you rated yourself at this stress level, based on the previous question "How stressed out are you about paying for college?"); and Question 12 (What is on your mind now that you've graduated? What kinds of things are you thinking about?). Therefore, four topics were extracted from responses to these four questions in Survey 1. The humans predetermined short labels (hereafter called topics) for each question: Question 7 (Support types), Question 9 (Anticipated college stressors), Question 11 (Financing college stressors), and Question 12 (Summer (precollege) thoughts). LDA analyses showed that the topic model generated similar topics as the predetermined topics: supports, college stress, financial stress, and work-time-class. We labeled the model-generated topics based on the high-weighted keywords and the corresponding semantic context.

Results indicated that 25 out of 31 topics that the model identified were matched with the predetermined topics with the accuracy reaching Kappa = 0.87. In some surveys, all the model-coded topics were matched with the predetermined topics, such as Survey 2 (summer plan, career plan fit, and plan change), Survey 3 (college success confidence, college talk, anticipated college challenge, personal strengths, and personal limitations), Survey 8 (college preparedness certainty, how school prepared you, and surprises in college), and Survey 9 (advice to students) (see Table 2). In other surveys, model-coded topics were partially matched with the predetermined topics. Some surveys have more topics matched, such as Surveys 1 & 6 (3 out of 4) and Survey 5 (4 out of 5). Others had half of the topics matched, such as Survey 4 (2 out of 4) and Survey 7 (1 out of 2) (see Table 2).

These findings implied that the LDA model allowed for extracting similar topics in each survey when the same number of topics was selected in the model as the number of open-ended survey questions. This step is critical because its accuracy will determine whether the corresponding themes related to each survey question could be extracted. As we mentioned in the Method section, the recommended way is to extract topics from responses to each survey question. The prerequisite for this method, however, is a large number of responses to each survey question. In our survey data, four questions had less than 10 responses ($N = 2, 3, 4$, and 9 respectively). The present study provides a potential approach for a limited number of responses to each survey question: instead of removing the smaller sized data that could not be included in the analyses, we could combine all the responses in each survey to enlarge the sample size and then extract the same number of topics based on the number of survey questions. This approach is likely to address the issue of smaller sample sizes.

3.2   Extracted Themes

The themes that human coders identified from the responses to each survey question are listed in Table 2. This table also displays 10 high-weighted keywords (hereafter called themes) for each survey question generated from the LDA model. The themes consistent with human-coded categories were underlined with different types of lines. The model-generated themes matched some categories that human coders identified, but the frequencies of similarity varied across survey items. The model-extracted themes in the majority of surveys were well matched with human-identified themes, a match rate above 50% (see Table 2). Sometimes, the model extracted the exact words as humans identified, such as "finance" in Question 9 in Survey 1. Sometimes, the model extracted words with close semantic meaning as humans identified, such as synonyms like "unsupportive" and "discouraging" in Question 11, Survey 5. For some survey questions, the model did not identify as many themes that matched with humans, such as Question 10 in Survey 7.

One possibility is that the survey question had different numbers of responses. The fewer responses (e.g., N equals to 2 to 9) could not display varied topics, which led to low model accuracy. On the other hand, some themes generated by the model were categorized into another survey question. For example, in Survey 5, the human coder identified the category "proud of me" from responses to the survey question "Friend Supports", whereas the model generated the theme "proud" related to unsupportive. This mis-classified phenomenon is likely due to the usage of negation rather than antonyms to express the opposite meaning. In the future, not only synonyms but also antonyms and negation need to be considered before model generation. Another interesting finding is that the more narrow the scope of the topic, the more overlap between the human-coded themes and the model-extracted themes, allowing for specific types of responses to be more identifiable. For example, Question 10 in Survey 7 was relatively broad, "explain how your plan changed and why", which might elicit more diverse responses due to the different struggles that individual seniors confronted. Moreover, this question only received 3 responses. Fewer responses plus the broadness of the question might lead to lower accuracy in the model. Alternatively, more responses to the broader question might lead to higher accuracy in the model, such as Question 12 in Survey 1: "What is on your mind now that you've graduated? What kinds of things are you thinking about? Provide as much detail as you can". Or fewer response plus the more narrow question might also yield higher accuracy in the model, such as Question 9 in Survey 8: "Explain how well Summit has prepared you for college coursework?". This may suggest ways to represent open-ended survey questions to potentially increase the likelihood that model results would be more comparable to human coders.

4   General Discussion, Conclusions and Implications

This paper employed topic modeling to automatically extract topics and themes from open-ended survey responses related to the summer melt phenomenon. This study considered the close semantic meanings, such as synonyms, hypernyms, and antonyms, which were not included in a previous study [6]. Findings indicated that this modification yielded higher agreement between human-coded themes and model-generated themes. Moreover, the method that extracted topics/themes from the responses to the questions within one survey enabled the extraction of similar topics as predetermined human-coded topics, and the ability to generate themes similar to the themes humans categorized even though the agreement varies across questions. To sum up, this paper provides a potential approach to automated topic and theme generation for open-ended response surveys with a smaller sample size. The model allows for greater accuracy of topic generation based on the content of the survey questions and partial themes related to the corresponding survey questions with limited sample sizes.

This study contributed to research on quantitative ethnography in two ways. First, it made the connection between the quantitative model, topic modeling and qualitative data, open-ended survey responses [14]. This method allows for large-scale analyses with open-ended survey responses so as to maximally retrieve and utilize information from these open-ended responses. Second, this study included the validation of topic modeling performance through human raters' coding [14], which demonstrates that automated coding using topic modeling can reflect humans' understanding and interpretation of open-ended survey responses. This validation is a critical step in quantitative ethnography, which requires transforming qualitative data for use in quantitative analyses [15].

Future studies need to focus on improving the model by considering the use of negation or other semantic-related conditions in the corpus. For example, using n-gram demonstrates the nouns or noun phrases and their modifiers that provide more concrete information than that with adjectives as keywords. In the future, it will also be important to compare the topic modeling method with other automated topic/theme extraction models to further demonstrate which model is a better fit for theme extraction with open-ended survey responses.

References

1. Bettinger, E.P., Long, B.T., Oreopoulos, P., Sanbonmatsu, L.: The role of application assistance and information in college decisions: results from the H&R Block FAFSA experiment. Q. J. Econ. **127**, 1205–1242 (2012)
2. Blei, D.M., Ng, A.Y., Jordan, M.I.: Latent dirichlet allocation. J. Mach. Learn. Res. **3**(1), 993–1022 (2003)
3. Bobek, B.L., Zhao, R.: Education and career navigation. In: Camara, W., O'Connor, R., Mattern, K., Hanson, M.A. (eds.) Beyond Academics: A Holistic Framework for Enhancing Education and Workplace Success, pp. 39–51. ACT Inc, Iowa City (2015)
4. Castleman, B.L., Page, L.C.: A trickle or a torrent? Understanding the extent of summer melt among college-intending high school graduates. Soc. Sci. Q. **95**(1), 202–220 (2014)
5. Castleman, B.L., Page, L.C., Snowdon, A.L.: Strategic data project summer melt handbook: a guide to investigating and responding to Summer Melt. Center for Education Policy Research, Harvard University (2013)
6. Li, H., Zhou, Y., Bobek, B.L.: Automated identification of open-ended survey response themes in education research: a summer melt study. Poster presented at Twenty-Ninth Annual Meeting of the Society for Text & Discourse, New York, NY (2019)
7. Lucas, C., Nielsen, R.A., Roberts, M.E., Stewart, B.M., Storer, A., Tingley, D.: Computer-assisted text analysis for comparative politics. Polit. Anal. **23**(2), 254–277 (2015)
8. McCallum, A.K.: MALLET: a machine learning for language toolkit (2002). http://mallet.cs.umass.edu
9. Merriam, S.B., Tisdell, E.J.: Qualitative Research: A Guide to Design and Implementation. Wiley, Hoboken (2015)

10. Oreopoulos, P., Dunn, R.: Information and college access: evidence from a randomized field experiment. Scand. J. Econ. **115**, 3–26 (2013)
11. Paek, P.L., Bobek, B.L.: Unpacking the factors contributing to summer melt and impacting college readiness. Paper presented at the Annual Conference for the American Research Education Association (AERA), New York, NY (2018)
12. Roberts, M.E., et al.: Structural topic models for open-ended survey responses. Am. J. Polit. Sci. **58**(4), 1064–1082 (2014)
13. Roberts, M.E., Stewart, B.M., Airoldi, E.M.: A model of text for experimentation in the social sciences. J. Am. Stat. Assoc. **111**(515), 988–1003 (2016)
14. Shaffer, D.W.: Quantitative Ethnography. Cathcart Press, Madison (2017)
15. Shum, S.B.: Book review: quantitative ethnography by David Williamson Shaffer. J. Learn. Anal. **6**(1), 99–101 (2019)
16. Trostel, P., Smith, M.C.: It's Not Just the Money: The Benefits of College Education to Individuals and to Society. Lumina Foundation, Indianapolis (2015)
17. Vandenbroucke, G.: Lifetime benefits of an education have never been so high. Reg. Econ. **25**(1), 10–11 (2015)

Computationally Augmented Ethnography: Emotion Tracking and Learning in Museum Games

Kit Martin$^{(\boxtimes)}$, Emily Q. Wang, Connor Bain ,
and Marcelo Worsley

Northwestern University, Evanston, IL 60208, USA
{kitmartin, eqwang, connorbain}@u.northwestern.edu,
m.worsley@northwestern.edu

Abstract. In this paper, we describe a way of using multi-modal learning analytics to augment qualitative data. We extract facial expressions that may indicate particular emotions from videos of dyads playing an interactive table-top game built for a museum. From this data, we explore the correlation between students' understanding of the biological and complex systems concepts showcased in the learning environment and their facial expressions. First, we show how information retrieval techniques can be used on facial expression features to investigate emotional variation during key moments of the interaction. Second, we connect these features to moments of learning identified by traditional qualitative methods. Finally, we present an initial pilot using these methods in concert to identify key moments in multiple modalities. We end with a discussion of our preliminary findings on interweaving machine and human analytical approaches.

Keywords: Multimodal learning analytics · Affect tracking · Game-based learning · Physical traces

1 Introduction

Informal learning environments can provide rich educational experiences without the need for an instructor or classroom [3]. While these sorts of learning environments have become more common, museum exhibits have long utilized these environments as the main way of engaging their visitors [6]. In order to assess the effectiveness of these exhibits, designers need to understand how participants' interactions with the environment inform their learning experience. However, it is often difficult to identify and track participants' learning through these open-ended interactions [16]. Traditionally, such work has been done through ethnographic practice, requiring researchers to be either *in situ* or sort through enormous amounts of video data manually. With the advent of multimodal sensing, many researchers have moved to a model of observation that focuses on instrumenting participants and the environment itself in order to capture many types of interaction streams. However, these automated multi-modal protocols lose the depth and detail that an experienced ethnographer can capture [15].

© Springer Nature Switzerland AG 2019
B. Eagan et al. (Eds.): ICQE 2019, CCIS 1112, pp. 141–153, 2019.
https://doi.org/10.1007/978-3-030-33232-7_12

In this paper, we are working to augment ethnographic practice (e.g., observations and semi-structured interviews) with multimodal sensing. Instead of viewing the two skill sets as opposing models for data collection and analysis, we envision them working in concert. Multi-modal sensing could provide a lens through which to identify specific learning moments for more in-depth qualitative analysis. At the same time, qualitative analysis could reveal moments where social phenomena, such as side-by-side collaboration, might be better understood through a multi-modal lens. In the following sections, we show one possible use of computationally augmented ethnographic practice by analyzing a single dyad's learning trajectory as they interact with a table-top museum exhibit focused around the biology of ants. Specifically, we focus on how the relationship between affect and cognition can be utilized to better analyze learning in this environment.

2 Prior Work

This work is informed by prior research at the intersection of neurobiology, ethnography, and artificial intelligence. More broadly, the work builds on a growing body of research in multimodal learning analytics [16–18]. Research on neurobiological parts of memory shows that emotionally arousing stimuli consolidate and preserve more often over the long term [13]. Both positive and negative emotions are associated with memorable moments controlled selectively within the basolateral amygdala. The brain region regulates the consolidation of memory for various experiences through projections from the amygdala to many other regions involved in storing newly acquired information [14].

More recent work has more explicitly explored the relationship between cognition and emotions. For example, D'Mello and Graesser [5] advance a model of affect dynamics that describes the complex interactions that exist among different emotional states, tracked through facial expressions, and how those afford learning. They are particularly concerned with how students transition in and out of moments of confusion. Central to their model are the roles of surprise and joy as indicators of a student transitioning into or out of a moment of confusion. While still discussed in terms of short term cognitive gains, the model implies that examining affective states can be instrumental for better understanding learning.

Worsley, Scherer, Morency, and Blikstein [26] similarly leverage the information embedded within affective states, or facial expressions, to segment multi-modal data streams. Specifically, their paper uses changes in facial expression as a proxy for delineating meaningful changes in the learners cognitive or behavioral state. Put differently, they segment their data stream whenever the user has a change in their most probable facial expression. Underlying this proxy, is the assumption that affective information can be informative for examining complex learning experiences.

Different from prior work, however, we adopt strategies from data mining and qualitative ethnography in order to avoid analyses that are purely computational or purely human-annotated. Our motivation to do to this work is to provide an example of iteration between ethnographic practice and computational techniques to analyze a previously developed learning environment [11]. To this end, we focus on a single dyad interacting with an informal learning environment in order to demonstrate our methodology in a particular informal learning activity. We also articulate future work to continue using these techniques with more participants and other modalities.

3   Data Collection

We collected data from 114 participants in 38 groups as they participated in a museum exhibit called *Ant Adaptation* [8]. Ant Adaptation is a game built from an agent-based model (ABM) implemented in NetLogo [23]. Two players interact with the game by controlling one of two ant colonies, using sliders to adjust ant behaviors and touch-screen gestures to set pheromone trails towards flowers for food. Ants can also "fight" over resources, which sets up a feedback loop that drives the action of the complex system. Through gameplay, players learn about (a) ant colony behavior and (b) entities' interactions in the complex system. More details of the game will be discussed in the following section.

To explore computationally augmented ethnography with Ant Adaptation, we collected two different types of data: (a) synchronous video, audio, and Kinect data streams using the Social Signal Interpretation framework [21] and (b) ethnographic data in the form of written field notes and a pre/post-gameplay semi-structured interview. This paper does not include analysis of the Kinect data.

In this paper, we present data collected on one dyad of the 38 groups. This pilot serves as a study on the potential of this approach before continuing to work with more participants and modalities. The dyad played the game side-by-side on a 52 horizontal touch screen in a research lab. A researcher interviewed the dyad before and after gameplay in order to probe their evolving under-standing of the ant life cycle. We placed the camera and Kinect approximately five feet away from the dyad in order to capture facial expressions and body motions for further analysis. Audio of interviews and gameplay conversation were transcribed verbatim.

4   Ant Adaptation

Agent-based modeling (ABM), and NetLogo ABM in particular, has been shown to be an effective and empowering platform for classroom learning with successful curricula spanning many disciplines, from chemistry [22], to physics [1], to biology [24]. Ant Adaptation builds off this rich history of ABM for learning, but positions itself as a rich tangible interaction form factor for walk-up-and-play use in an informal learning space. As shown in Fig. 1, Ant Adaptation simulates two ant colonies side by side. Leveraging a large multi-touch display mounted horizontally as a table-top exhibit, the model tracks players' touches within the model. Through this interface, players can both interact directly with the gameplay area and with sets of widgets that directly control parameters of the core simulation.

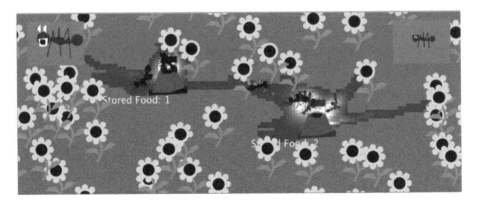

Fig. 1. A screenshot of the Ant Adaptation table-top game.

4.1   Ant Adaptation as an ABM

It is important to note that Ant Adaptation is a fully functioning ABM. Even without user interaction, ant agents go out to collect food and return to the nest. As they return to the nest, ants lay down a pink pheromone that attracts others nearby. When ants find a flower, their food source, they return, lay down more pheromones, and thus reinforce the pink trail. This creates an emergent feedback loop that routes more and more ants to successful sites of forage. As the ants exhaust a food source, they must find new locations and thus repeat a cycle. When two or more ants of opposing colonies encounter each other, they fight or scare each other away also leaving chemicals that attract more ants. For the winner, this works to protect the food source from competing colonies. The ant queen reproduces when the ants in her colony collect enough food, in other words when collection surpasses the current Create Cost. Flowers periodically grow up around the map, adding food to the game.

The player interacts with this complex system by tweaking parameters and effecting the environment. Through interacting with the system, users form a functional understanding of the ants and their mechanisms of action (i.e. agents and their rules) in the model.

4.2   Ant Adaptation as a Digital Table-Top Game

To supplement the original NetLogo software [23], we have developed software designed for touch interaction with the model, NetLogo Touch [10]. The original NetLogo software only allows for a single interaction by a single user at any given time. That model is simply not conducive to cooperative or simultaneous gameplay as in Ant Adaptation. As such, NetLogo Touch allows users to interact with the model simultaneously and via touch instead of via the more traditional mouse-based input.

Each team is given control of three simulation parameters via widgets in addition to two *monitors* that give summary information about a colony. From Fig. 2, element (1) is a monitor of the ant populations labelled Black Ants on the left and Red Ants on the right which simply keeps track of population sizes. Element (2) shows the three

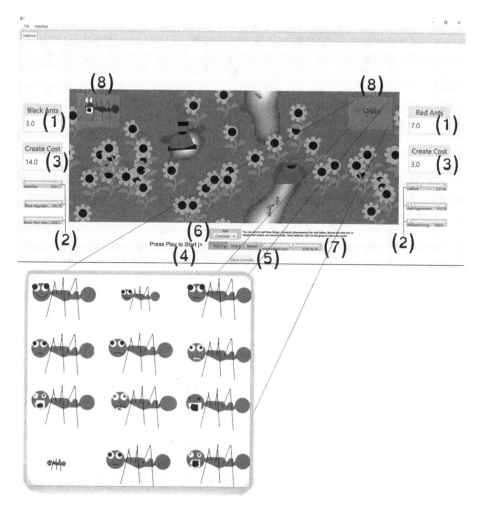

Fig. 2. A view of the Ant Adaptation table-top game interface. (Color figure online)

widgets which are sliders players can use to adjust their ants' size, aggressiveness, and the maximum amount of energy. In this particular model, energy is essentially equivalent to how long ants can walk without eating. These sliders can be adjusted at any time during the game to experiment with different settings for each colony depending on the players' wishes. Element (3) is another monitor that shows the *Create Cost* of an ant or the amount of energy required from the colony to spawn a new ant.

Players decide how big and aggressive their ants are. As the size of ants increases, they become slightly stronger in a fight against other ants. At the highest levels they are 13 times stronger than ants of the smallest size. When players make their ants more aggressive, it increases the radius in which ants detect opposing ants and thus the probability that they will attack. Increasing either the size or the aggressiveness also increases how much energy is required to raise a new ant, so the largest ant requires 13 times as much food to feed to adulthood. Increases in either of these parameters reduces the expected population of the colony by increasing the Create Cost, though it increases their likelihood of fighting and winning through the emergent interactions of parameters (size and aggressiveness) and agent actions (collecting food, leaving trails, and fighting).

The colony produces a new ant when the stored food is greater than the Create Cost. This is an example of what Chi et al. [2] call an opaque summing mechanism that NetLogo designers employ, where the collective mechanism is computed by the NetLogo system itself, thus [left] opaque to the students (p. 21). The cost is calculated by the current value of three sliders, meaning players can change the create cost of their ant using these three dimensions. Because the outcome of the calculation is the current cost for the colony to birth one more ant, this summing mechanism becomes a key element of the gameplay for players to understand in order to strategize.

Lastly, there are two representations of the players' ants in the top right and left of the play space. These show the user how large and aggressive their ants are when born. As shown in element (8) of Fig. 2 the display changes according to the mixture of aggressiveness and size the player chooses for their team. This provides the player immediate feedback for changing slider parameters, giving them a better sense of cause and effect in the model. This is important because adjusting the sliders only affects new ants born, instead of extant ants. So the effect of the interaction is longer than the 30 to 60 Hz, 16.6 to 33.3 ms periods, people associate with cause and effect within games [7].

In addition to the colony controls, the players share five widgets in the bottom center of the screen. As shown in Fig. 2, element (4) Play and Stop buttons, which control the model's time; element (5) a Restart button, which sets the model back to initial conditions; element (6) a drop-down chooser which allows players to select the main action of the game as a series of strategic choices (which will be discussed in more detail in a moment); and element (7) a slider to control the evaporation rate of pheromonesa chemical trail that ants leave behind them as they travel in the world.

Element (6) of Fig. 2 is a main mechanism of the gameplay, allowing users to select the action that takes place when they touch within the game world seen in Fig. 2. Players can choose to add *chemical, flowers,* or *vinegar*. Chemical is a pink pheromone that usually is only laid down by ants. Ants are attracted toward the highest concentration pheromone near their location, which is displayed by whitish-pink shades. Flowers are these ants' main food source. Ants collect and eat the flowers to feed themselves and bring food back to the colony for collective rearing of young. Vinegar erases ants' trails allowing the player to mask pheromones, disrupt communications, and clear the ground by applying vinegar to the chemical trails. In essence, this chooser allows players to decide whether to have their colony focus on collecting food, thereby increasing the population, go on the warpath, forcing colonies to fight for resources, or perhaps purposefully prevent their colony from finding the opposing colony. However, it is important to remember that in order to use these elements effectively, a player must understand how the ants react to each of these different action items. These three options are not directly affecting the ants, but rather the world that the ants live in. As such, players are forced to try to understand how, in this particular complex system, the environment and agents interact.

Each method of play could lead to high populations or the elimination of the opponent through better-controlled food resources. For example, after learning about the consequences of strategic choices through gameplay, players could strategize by increasing ants' size, aggressiveness, or both. This might lead them to win the game by annihilating the other group's ant colony. However, bigger and/or more aggressive ants consume more food to reproduce and potentially reduce the colony's population size. Thus, a player could strategize by adding more flowers and pheromone tracks around the colony to help the larger ants survive. This learning and strategy cycle interweaves the learning into the gameplay. The design scaffolds experimentation and encourages players to interact with emergent phenomenon like feedback loops, local optima, and more.

Ant Adaptation has four main affordances that support two central learning objectives. In Ant Adaptation, playing with parameters allows players to: (1) construct their colony in competition with an opponent; (2) share strategies through comparison; (3) discuss what is happening through observer scaffolding such as parents' intervention or interaction between players, including slapping hands; and (4) learn about the emergent impacts of colony behavior arising from individual ant behavior in a complex system game. This approach allows visitors to learn (1) the impacts of adaptation on ant colony life and (2) how attractants such as pheromones work in ants' organization to increase the population. As such, the game also offers the potential to scaffold learners in switching from a direct to an emergent schema of how the phenomenon we see in everyday life might occur [12]. Next, we analyze and discuss these learning opportunities by looking at participant data via a qualitative and computational lens.

5   Data Analysis

We began our analysis with traditional qualitative methods involving multiple coders following a constructivist dialogue mapping approach [9] a type of concept mapping [15] on transcripts of interviews and gameplay conversation. We extracted emotional evidence from gameplay video to identify segments when learning occurred. Finally, we describe how iterative explorations of emotion informed and augmented our analysis of dialogue, and vice versa.

5.1   Using Joy Values Based on Facial Recognition to Identify Potential Learning Moments

We used FACET [20] to extract the strength of detected emotions through the proxy of facial expressions. The basolateral amygdala is an essential component of the neuro-modulatory system regulating behavioral states and is thought to consolidate experience-dependent plasticity [14]. As a result, both positive and negative emotions are behavioral states that may be involved in vital brain function modulating the neuronal representation associated with memorable moments. To identify reasonable emotional moments that might be associated with moments of learning [13, 14], we used a 2.5th and 97.5th percentile threshold to reveal peaks and valleys of joy values

Participant Joy During Tabletop Play Over Time

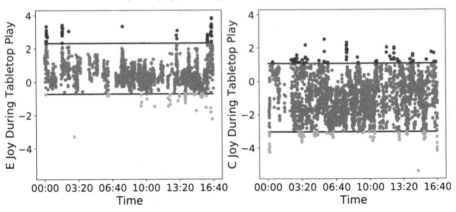

Fig. 3. A visualization of participant joy over time, as detected by FACET analyzing facial expressions extracted from videos. Emotional peaks and valleys outside of the middle 95% of the data are colored. (Color figure online)

computed via FACET. The percentile thresholds are visualized as horizontal line boundaries in Fig. 3. This search for peaks and valleys is informed by McGaugh's previous work, which argued high stimulus periods lead to higher memory encoding. We then extracted the dialogue from the transcript that occurred approximately ten seconds before each peak or valley to compare and further analyze whether learning occurred at these time segments. We call these segments *learning windows*. Note we do

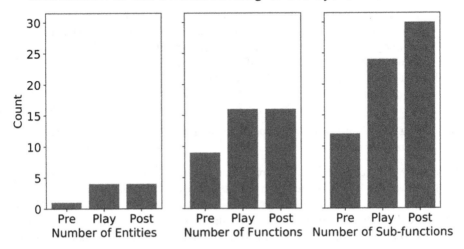

Fig. 4. A plot of how many entities, functions, and sub-functions the dyad verbalized during the pre-gameplay interview, gameplay, and post-gameplay interview.

not directly compare FACET values between participants, but instead focus on FACET values for each participant individually.

Cognitive Mapping for Learning Concepts. We define learning in this dataset as how the participants elaborate their understanding of the ant life cycle through informal play. For example, if initially they say "ants walk", but after play "ants walk to follow paths to reach food to feed the colony and themselves", we would interpret this as an elaboration of their understanding. This is based on how Ant Adaptation is designed to embed learning into gameplay as participants manipulate ant parameters (e.g., aggression, size, distance to food) [11].

With this in mind, each coder created a concept map for transcripts of the interview before gameplay, during gameplay conversation, and of the interview after gameplay.

Each map consisted of entities, functions, and sub-functions discussed by the dyad. Entities included different agents and resources in the game, such as ants and flowers. Functions are the processes that engage entities, such as leaving a trail. Sub-functions are the motivations or results for functions, such as collecting food or directing the paths of other ants. Two coders created cognitive maps, and had 96% inter-rater reliability on the functions and sub-functions they identified.

6 Results

In this section, we present results from our cognitive mapping analysis and three examples of learning windows identified at the timestamps of peaks and valleys in Fig. 3. In addition to providing more insights on learning processes that we already manually coded with cognitive mapping analysis, emotion tracking unveils other learning processes that we did not notice previously. We see this interlacing of human coders and computational techniques having potential in iterative qualitative research processes. When we used facial expressions to locate moments we find three main benefits: (1) Through indexing moments of high joy, we find moments we would not otherwise notice; (2) Tracking facial expressions can provide addition insight into learning moments; (3) Using affective tracking we can see past epistemological bias to see moments of learning in a new way. This third benefit focuses on the main purpose of multimodal learning analytics in computational ethnography: to see the interactions with additional subjective lenses to gain a fuller understanding of learning in the space.

6.1 Cognitive Mapping

As shown in Fig. 4, we found that during the pre-gameplay interview which focused on questions of ants, players described one entity (ants), with nine functions (e.g., placing paths to signal food) and twelve subfunctions. Participants elaborated both the entities and functions of those entities through playing the game. During gameplay, players mentioned three additional entities (flowers, queen ants, and a GUI element in the interface) as well as sixteen functions, such as "ants hiding in their colony" and "organizing society". They also discuss twenty-four sub functions (e.g., food collection through leaving attractant pheromones). By the end of the intervention, the participants expanded to thirty sub-functions. In other words, most of the "learning" (75% of the entities and 40% of the subfunctions) were elaborated on during play.

While cognitive mapping provides a structured set of codes about participant sense making, one key limitation is that it extracts insight solely from transcribed audio and not necessarily capture evidence of learning in other modalities. Next, we share three examples of how emotion tracking extended our analyses of participant knowledge evolving throughout the session.

6.2 Seeing Past Epistemological Bias: Learning Occurs During Instruction Before Gameplay

This example shows how emotion tracking revealed potential moments of learning that we missed in our qualitative coding. In Fig. 3, we noticed a peak joy value with Emma (pseudonym) and a valley joy value with Chris (pseudonym) at timestamp 00:04, showing that this is a high-stimulus moment for both participants. Revisiting the transcript, this moment corresponded with when the facilitator described how to play the game, "On your side of the board, you have three sliders. One of them changes the size of new ants formed, one of them changes the aggressiveness...".

This evidence of potential learning during instruction was overlooked because when coding interview transcripts, the analyst cannot confirm whether participants are learning when they are not verbalizing. Additionally, given our interest in informal learning that occurs in museums through games, we primarily focus on the sponta- neous, emergent learning that occurs rather than learning that happens during instruction. As such, during the cognitive mapping analysis, we did not identify learning during the moments when the facilitator instructed participants about the game.

Ultimately, emotion tracking enabled us to detect a potential learning moment when participants were silently listening to instructions on how to play the game. We see the potential to augment qualitative coding beyond what can be detected with a single analytic frame and transcription techniques.

6.3 Learning While Experimenting with Gameplay

Both Chris and Emma had a peak joy value (3.5 and 2.1, respectively) near timestamp 13:30 on Fig. 3. We did not initially code this as a learning moment because functions and sub-functions were not clearly articulated in the transcripts. Revisiting the video near this peak's timestamp, we noticed participants were comparing their different slider conditions to draw conclusions about ant behavior as Chris says "Wait, how aggressive are you?... Wait, are you are a lot...you're slightly bigger than I am." while noticing that Emma's population count had risen to 77. As they watch the action unfold on screen, they noticed how different settings led to the victory or loss of their ant colonies.

While the first example revealed learning while silently listening to the facilitator's instructions, this example reveals how participants elaborated their understanding as they experimented with the game interface's sliders but did not verbalize an updated understanding of ant behavior. Using Constructionist Dialogue mapping during these moments lets us more carefully chronicle learning during moments of higher memory encoding. This example is another instance of how emotion tracking revealed moments of learning that we glossed over in our initial cognitive mapping analysis. By triangulation we came to better understand our data.

6.4 Joy During Previously Identified Learning Moments

Looking at Fig. 3 at timestamp 7:00, Emma and Chris have peak joy values of 3.8 and 1.8–2.2, respectively, and we revisited the video at this segment for further analysis. Chris and Emma verbalized that the flowers at close proximity to the ant hill increases their ants' population. Chris says, "Can I have more flowers?" Emma responds, "Yes. Ring of flowers." They place a ring of flowers around their ant hills and notice ants picking up the food. Both watch the table-top intently to observe the resulting ant behavior and Chris says, "Ooh, now I've got lots of ants." The dyad discovered a powerful relationship they can use to manipulate the environment. In contrast to previous examples, this example shows how emotion tracking provided additional insight on a learning moment identified in the cognitive mapping analysis.

As the dyad continues tinkering with parameters, they laugh through their trial and error attempts, and cemented the concept that food close to the nest increased the population of ants. This moment was coded in the cognitive mappings as one of the flowers' primary functions and our emotion analysis adds an additional understanding to this moment. That is, this discovery led to a sense of joy or what might be interpreted as "satisfaction" which is important to cultivate in informal learning environments and gameplay. Though we selected many of the same moments using emotion logging as we did through manual cognitive mapping, emotion logging also drew our attention to moments we would not have analyzed otherwise. In other words, the approach both reinforced our prior units of analysis and added to our approach to analyzing the interaction. We aim to continue this back-and-forth between qualitative methods and computational techniques as we collect data on more dyads and extract insights from other modalities.

7 Future Work

Applying computational techniques to understand ethnographic datasets has a lot of potential [4, 19, 25]. Researchers are examining means for human and artificial intelligence to interact by both bootstrapping human analysis with artificial intelligence and using human inference in the computational data analysis pipeline. As we continue collecting and analyzing data from different dyads, we will renew our pipeline and iteratively update our findings. For example, we want to correlate emotions with body positions. If we could correlate emotions to taxonomies of gestures identified in prior work [18], then body position could also be a proxy for emotion and reveal learning windows. We aim to develop tools to integrate exploratory analyses of data captured in different modalities to better understand learning and interaction.

8 Conclusion

This paper presented a preliminary approach to augment qualitative analysis of an informal learning environment. Using techniques from multimodal learning analytics, we were able to expand our analysis of learning while participants interacted with a multitouch environment. Our methodological approach required us to extract emotions from the low-level logs of facial action units using FACET and then revisit video corresponding to particular FACET values to identify moments of high emotional stimulation theoretically implicated in learning. Working towards a whole-body analysis, we are continuing to refine our pipeline to identify proxies of learning that are useful in museums and other informal learning environments. While our use of constructivist dialogue mapping showed that the users learned during their interaction with Ant Adaption, emotional logging identified alternative moments of learning outside of our analytic framework. These approaches augment each other by creating an iterative analysis approach where computational and traditional coding feed into each other, so the machine can learn from us, and we can learn from the machine. This work has the potential to allow ethnographers in informal learning environments to leverage computational techniques without losing the depth of information that ethnographic field work captures.

References

1. Blikstein, P.: An atom is known by the company it keeps: a constructionist learning environment for materials science using multi-agent simulation (2006)
2. Chi, M.T.H., Roscoe, R.D., Slotta, J.D., Roy, M., Chase, C.C.: Misconceived causal explanations for emergent processes. Cogn. Sci. **36**(1), 1–61 (2012). https://doi.org/10.1111/j.1551-6709.2011.01207.x. https://onlinelibrary.wiley.com/doi/abs/10.1111/j.1551-6709.2011.01207.x
3. Council, N.R.: Learning Science in Informal Environments: People, Places, and Pursuits. National Academies Press (2009)
4. D'Mello, S., Dieterle, E., Duckworth, A.: Advanced, analytic, automated (AAA) measurement of engagement during learning. Educ. Psychol. **52**(2), 104–123 (2017). https://doi.org/10.1080/00461520.2017.1281747
5. D'Mello, S., Graesser, A.: Dynamics of affective states during complex learning. Learn. Instr. **22**(2), 145–157 (2012). https://doi.org/10.1016/j.learninstruc.2011.10.001. http://www.sciencedirect.com/science/article/pii/S0959475211000806
6. Falk, J.H., Dierking, L.D.: The Museum Experience Revisited. Routledge (2016)
7. Gregory, J.: Game Engine Architecture. AK Peters/CRC Press (2017)
8. Martin, K., Wilensky, U.: Netlogo ant adapatation model (2019). http://ccl.northwestern.edu/netlogo/models/AntAdaptation
9. Martin, K.: Constructivist dialogue mapping: evaluating learning during play of ant adaptation, a complex interactive tabletop museum game. In: 31st Annual Visitor Studies Association, p. 20 (2018)
10. Martin, K.: Multitouch NetLogo for museum interactive game. In: Companion of the 2018 ACM Conference on Computer Supported Cooperative Work and Social Computing, CSCW 2018, Jersey City, NJ, USA, pp. 5–8. ACM, New York (2018). https://doi.org/10.1145/3272973.3272989. http://doi.acm.org/10.1145/3272973.3272989

11. Martin, K., Horn, M., Wilensky, U.: Ant Adaptation: A Complex Interactive Multitouch Game about Ants Designed for Museums, August 2018

12. Martin, K., Horn, M., Wilensky, U.: Prevalence of direct and emergent schema and change after play. Inf. Educ. **18**(1), 183 (2019)

13. McGaugh, J.L.: Memory and Emotion: The Making of Lasting Memories. Columbia University Press (2003)

14. McGaugh, J.L.: Make mild moments memorable: add a little arousal. Trends Cogn. Sci. **10** (8), 345–347 (2006). https://doi.org/10.1016/j.tics.2006.06.001. http://www.sciencedirect.com/science/article/pii/S1364661306001355

15. Miles, M.B., Huberman, A.M., Saldaa, J.: Qualitative Data Analysis: A Methods Sourcebook, 3rd edn. Sage, Thousand Oaks (2014)

16. Ochoa, X., Worsley, M.: Editorial: augmenting learning analytics with multimodal sensory data. J. Learn. Anal. **3**(2), 213–219 (2016). https://doi.org/10.18608/jla.2016.32.10. https://learning-analytics.info/journals/index.php/JLA/article/view/5081

17. Oviatt, S., Cohen, A., Weibel, N.: Multimodal learning analytics: description of math data corpus for ICMI grand challenge workshop. In: Proceedings of the 15th ACM on International Conference on Multimodal Interaction, ICMI 2013, Sydney, Australia, pp. 563–568. ACM, New York (2013). https://doi.org/10.1145/2522848.2533790. http://doi.acm.org/10.1145/2522848.2533790

18. Schneider, B., Blikstein, P.: Unraveling students' interaction around a tangible interface using multimodal learning analytics. J. Educ. Data Mining **7**, 89–116 (2015)

19. Spikol, D., et al.: Exploring the interplay between human and machine annotated multimodal learning analytics in hands-on stem activities. In: Proceedings of the 6th International Learning Analytics & Knowledge Conference, vol. 6, pp. 522–523. Association for Computing Machinery (2016)

20. Taggart, R.W., Dressler, M., Kumar, P., Khan, S., Coppola, J.F.: Determining emotions via facial expression analysis software. In: Proceedings of Student-Faculty Research Day, CSIS, Pace University, May 2016

21. Wagner, J., Lingenfelser, F., Baur, T., Damian, I., Kistler, F., Andr, E.: The social signal interpretation (SSI) framework: multimodal signal processing and recognition in real-time. In: Proceedings of the 21st ACM International Conference on Multimedia, MM 2013, Barcelona, Spain, pp. 831–834. ACM, New York (2013). https://doi.org/10.1145/2502081.2502223. http://doi.acm.org/10.1145/2502081.2502223

22. Wilensky, U.: Netlogo gaslab gas in a box model (1997). http://ccl.northwestern.edu/netlogo/models/GasLabGasinaBox

23. Wilensky, U.: NetLogo (and NetLogo user manual) (1999). https://ccl.northwestern.edu/netlogo/

24. Wilensky, U., Reisman, K.: Thinking like a wolf, a sheep, or a firefly: learning biology through constructing and testing computational theories–an embodied modeling approach. Cogn. Instr. **24**(2), 171–209 (2006)

25. Worsley, M., Blikstein, P.: A multimodal analysis of making. Int. J. Artif. Intell. Educ. **28**(3), 385–419 (2018). https://doi.org/10.1007/s40593-017-0160-1

26. Worsley, M., Scherer, S., Morency, L.P., Blikstein, P.: Exploring behavior representation for learning analytics. In: Proceedings of the 2015 ACM on International Conference on Multimodal Interaction, ICMI 2015, Seattle, Washington, USA, pp. 251–258. ACM, New York (2015). https://doi.org/10.1145/2818346.2820737. http://doi.acm.org/10.1145/2818346.2820737

Using Process Mining (PM) and Epistemic Network Analysis (ENA) for Comparing Processes of Collaborative Problem Regulation

Nadine Melzner$^{(\boxtimes)}$ ⓘ, Martin Greisel ⓘ, Markus Dresel ⓘ, and Ingo Kollar ⓘ

Universität Augsburg, Universitätsstraße 10, 86159 Augsburg, Germany
nadine.melzner@phil.uni-augsburg.de

Abstract. Learning Sciences research often concerns the analysis of data from individual or collaborative learning processes. For the analysis of such data, various methods have been proposed, including Process Mining (PM) and Epistemic Network Analysis (ENA). Both methods have advantages and disadvantages when analyzing learning processes. We argue that a concerted use of both techniques may provide valuable information that would be obscured when using only one of these methods. We demonstrate this by applying PM and ENA on data from a study that investigated how students regulate collaborative learning when faced with either motivational or comprehension-related problems. While PM showed that collaborative learners are more incoherent (i.e. more heterogeneous in their chosen activities) when regulating motivational problems than comprehension-related problems at the beginning, ENA revealed that in later stages of their learning process, they focus on fewer activities when being confronted with motivational than with comprehension-related problems. Thus, a combination of the two approaches seems to be warranted.

Keywords: Epistemic network analysis · Process mining · Self-regulation · Collaborative learning · Co-regulation · Shared regulation

1 Problem Statement

Learning Sciences research is often concerned with the analysis of how learning processes emerge over time [1]. Traditionally, research typically used a coding-and-counting approach to analyze such processes (e.g., summing up frequencies by which learners employ certain strategies).

However, the problem with this routine is that it does not account for the dynamics of the learning process, i.e. for the fact that learners' engagement in different learning processes may change over time. Researchers have thus called for methods that consider learning processes in their temporal sequence [2]. One approach to do this is to use process mining (PM). PM uses mathematical algorithms to inductively discover sequences of processes in event traces by visualizing them in process models. Based on Petri nets, process models are illustrations of systematically connected codes and

B. Eagan et al. (Eds.): ICQE 2019, CCIS 1112, pp. 154–164, 2019.
https://doi.org/10.1007/978-3-030-33232-7_13

transitions between codes and serve to uncover hidden information on the processes of interest.

For instance, using PM, [3] found that successful self-regulators initially prepare their learning before deeply processing information, whereas less successful learners did not show the described shift towards in-depth information processing. Yet, PM also has limitations such as partially producing "spaghetti-like" models that run the danger of becoming too complex for visual comparison. Additionally, PM does not provide statistical tests that check for differences between processes of different groups on a global level [4]. Furthermore, PM includes the individual activities of all subjects as an influence on the same process model with equal weighting. For example, if we want to investigate how groups regulate motivational as opposed to comprehension-related problems, it might be that one single person may be accountable for most loops on a single code in one situation (e.g., the person repeatedly applies an elaboration strategy), whereas in the other situation, such loops might be more evenly distributed across persons. If these loops are not weighted, this may lead to a distorted picture of the regulatory differences between situations resp. between different groups.

An approach that may help overcome these challenges is Epistemic Network Analysis (ENA) – a network analysis method based on a dimensional reduction procedure for tagging, extracting, and plotting meaningful compounds of activities by considering regulation processes as a network of coherent activities [5]. Some of its advantages are that ENA provides global statistical tests to compare models from different conditions (e.g., regulation processes in groups that experience motivational vs. regulation processes in groups that experience comprehension-related problems), that it provides information on the relatedness of codes within a specific window size that can consider more than just two successive codes (as it is the case in PM), and routines such as rotating networks in space for visually highlighting group differences, or normalizing vectors to check whether differences between two models are caused by single individuals within a group, but rather by a concerted (i.e., more or less evenly distributed) effort of the group.

ENA has lately been used to analyze data from a wide variety of different contexts [6, 7]. Despite its advantages, though, ENA still faces challenges: Since the networks drawn by ENA are based on so called adjacency matrices that include sums of counted code-code connections, it ignores start and end points and self-loop information. Additionally, it simply highlights connections between certain activities, rather than the direction of transitions. Thus, when visually and statistically comparing two models with ENA, these characteristics are not considered. Given the mutual strengths and weaknesses of the two approaches, we argue that a concerted use of PM and ENA might help to better understand the temporal structure of learning processes. We test this assumption by applying both methods to the analysis of data from a study on how learners cope with different kinds of collaborative regulation problems.

2 Method

2.1 Participants and Design

$N = 82$ students (61 female, $M_{Age} = 21.79$, $SD_{Age} = 4.86$) who were on average in their 2nd semester ($M_{Stud} = 2.12$, $SD_{Stud} = 0.57$) of studies participated in this study. They received a booklet with four vignettes (in randomized order) that described a self-organized study group preparing for an exam that faced different kinds of regulation problems. One of the four vignettes described the group as experiencing no regulation problems at all, another one described the group as experiencing solely motivational problems, a third one said the group would experience solely comprehension-related problems, and a fourth one describing the group as experiencing both motivational and comprehension-related problems. That way, we established a 2×2-factorial within-subjects design with the independent factors "motivational problems" (with vs. without) and "comprehension-related problems" (with vs. without).

For example, in the condition "motivational problems", the vignette read: "*Imagine you are part of a study group with three fellow students. You meet regularly and are a well-rehearsed team. Currently, you prepare with your group for an exam that is in three weeks. Concerning the content to be learnt for the exam, all group members have high knowledge and low learning motivation*". In the vignette "comprehension-related problems", for example, "high knowledge" and "low learning motivation" were turned into "*low knowledge*" and "*high learning motivation*".

Due to lack of space, in this paper we focus our analysis on the conditions "with motivational problems/without comprehension-related problems" and "with comprehension-related problems/without motivational problems".

2.2 Variables

After each vignette, students received open-ended questions that asked them to indicate (a) what *types of strategies* they would apply if they were a member of the group, and (b) at what *social level* they would apply each of those strategies.

To measure the *types of strategies*, after each vignette, participants had to write down the exact sequence of actions they would perform to ensure high quality of learning in each situation (1. At first …, 2. After that …, 3. After that …, After that …, and so on). Open answers were coded by means of a coding scheme based on strategy classification schemes of [8, 9]. This coding scheme differentiated between 1. elaboration strategies, 2. surface-oriented strategies, 3. metacognitive strategies, 4. resource-oriented motivational strategies, 5. resource oriented-non motivational strategies, 6. other strategies, and 7. no strategies (see Table 1). Two independent coders rated ten percent of the data and reached a sufficient level of interrater reliability (Cohen's Kappa = 0.73).

To measure the *social level* at which participants would apply those strategies, we provided three tick boxes after each strategy that asked them to indicate whether they would apply the respective strategy to (a) regulate their own learning ("self-level"), (b) to regulate some other group member's learning ("co-level"), or whether the person would negotiate about that strategy with all group member ("shared level"; [10]).

Table 1. Coding scheme for regulation activities along with examples.

Strategy type code	Example (in brackets the social level at which the answer was mentioned)
Elaborative	"[After that] I try to understand my part" (Self), "[After that], other members ask their questions" (Co), "[After that] joint elaboration of a summary" (Shared)
Surface oriented	"[After that] I skim through the material independently" (Self), "[After that], everyone learns their notes by heart" (Co), "[After that] everyone repeats the content independently" (Shared)
Metacognitive	"[After that] I also check if I am more motivated" (Self), "[After that] I ask who needs help with topics which the others perceived to be difficult" (Co), "[After that], we'll see if we've completed all that we had planned to learn." (Shared)
Motivational	"[First], I formulate a bond between knowledge and my life" (Self), "[After that], I try to bring humor into the learning situation" (Co), "[After that], the contents are asked together in plenary and made playful" (Shared)
Non motivational	"[After that] I start to prepare independently: I structure my learning materials" (Self), "[After that] I ask the group what thoughts about it they had" (Co), "[After that] we make fixed dates so that we are "forced" to come" (Shared)
Other	"[First] I make an appointment" (Self), (no example provided for Co), "[After that], everyone goes home" (Shared)
No	"[After that] I write the exam with my already collected knowledge" (Self), (no examples provided for Co and Shared)

2.3   Data Preparation

Strategy type codes were paired with social level codes to generate meaningful codes (e.g., "Motivational Shared" indicates a motivational strategy that a participant reported to apply at the shared level) for each condition. Thus, from each of the seven strategy codes mentioned above, 3 "strategy type"—"social level" pair codes (= 18 codes in total) were generated.

Since sample size was insufficient to perform dimensionality reduction through ENA with all 18 codes, the aforementioned code pairs with their absolute and relative frequency were listed in descending order so that 7 codes, each accounting for at least five per cent of all pairs, could be selected for data analysis (see below). By choosing this threshold (= selection criterion), we arrived at almost complete models.

For example, of the codes that met this condition in the "motivational problems" condition, the *Elaboration Shared* code had the highest relative frequency of 0.21, while the *Elaboration Self* code reached the lowest relative frequency of 0.07. In the "comprehension-related problems" condition, also the *Elaboration Shared* code was most frequent (with a relative frequency of 0.25), but different to the other problem condition, also the *Metacognitive Self* code met the selection criterion with a relative frequency of exactly 0.05. It is noteworthy that in both conditions, five times the same of these seven codes fulfilled the inclusion criterion (the two exceptions: *Motivational*

Shared in the motivational problem condition (relative frequency = 0.16 in this condition) and *Metacognitive Self* in the comprehension-related problem condition) (since PM is based on event logs, artificial timestamps with identical time intervals between all consecutive codes were added to two event log files we created before conducting PM).

Process Mining. For plotting regulation sequences with PM, we used the R package "bupaR" (version 0.4.2; [11]). The PM algorithm generated one precedence matrix per condition by using the absolute frequencies of antecedent and consequent codes (activities) and flow of each person within a condition (= "absolute_case"). As the data was stored in the data.frame format it had to be transformed into an eventlog object before the process map could be computed based on this object.

Epistemic Network Analysis. For plotting the regulation sequences with ENA, we used the ENA 0.1.0 online tool [12] and included the following codes: *Elaboration Self, Elaboration Co, Elaboration Shared, Motivational Shared, Non motivational Shared, Metacognitive Self*, and *Metacognitive Shared*. We defined the units of analysis as the lines associated with a single value per condition (i.e., motivational problems, comprehension-related problems) associated with each participant's case ID (= subset). Resultantly, one unit consisted, as an example, of the lines associated with the "motivational problems" condition and the participant with Case ID 42. The ENA algorithm counted the frequencies of each of two "strategy type"—"social level" pairs (= binary summation) based on a moving stanza window size of three (each of three lines plus the two previous ones) within a given conversation [5]. That way, each person received one value within the 28 dimensional vector space (for seven codes the space is calculated $7 + 6 + 5 + 4 + 3 + 2 + 1$) represented by the matrices per condition.

To represent these values in the lower-dimensional vector space (= dimensional reduction), only the first seven dimensions were used as descriptors $svd_i = 1.7$. Equally, the respective node positions were calculated based on the summed adjacency matrices within each condition, $N_j = 1$–7, while the centroid values of the network graphs were calculated based on the weighted connections of the nodes. A final optimization routine served to minimize the difference between the plotted points and the corresponding network centroids (Σ_i (p_i–c_i)), while an additional means rotation minimized the network's distance towards the x-axis in order to make possible group differences visible.

The projection of all subsequent dimensions, on the other hand, was done using a singular value decomposition, which produces orthogonal dimensions that maximize the variance explained by each of these dimensions.

3   Results

Process Mining. Process models (see Figs. 1 and 2) show that students with motivational problems tend to start off with one of two kinds of strategies, both at the shared level: *Motivational* and *Metacognitive Shared*. In the comprehension-related problems condition, students clearly prefer starting off with *Metacognitive Shared* regulation, while Motivational Shared regulation does not play a large role in that condition at all. In both conditions, *elaborative shared regulation* seems to particularly be chosen later in the process.

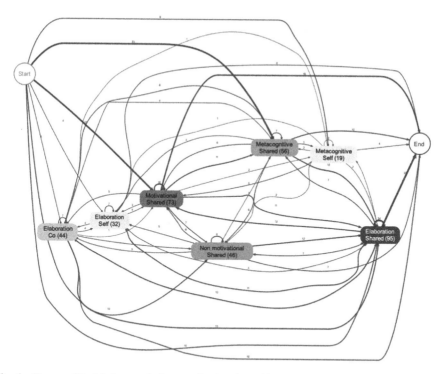

Fig. 1. Process Model for regulating motivational problems with absolute frequencies of all codes (boxes), as well as all observed directional code-code connections (arrows). Darker box colors indicate higher absolute code frequencies which means that the corresponding activities were observed more frequently, indicating that several persons have progressed from the corresponding first to the corresponding second activity (or, in the case of self-loops, that one person has performed the same activity several times in succession). (Color figure online)

When experiencing motivational problems, students appear to switch more often between *Motivational* and *Metacognitive Shared* regulation, between *Non Motivational* and *Elaborative Shared* regulation, and between *Elaborative* and *Motivational Shared* regulation, as compared to situations with comprehension-related problems.

When experiencing comprehension-related problems, in turn, students seem to switch more often between *Metacognitive Shared* and *Elaborative Self*-regulation, between *Metacognitive* and *Elaborative Shared* regulation, and between *Elaborative Co-* and *Shared* regulation.

The fact that weaker and stronger connections are more distinct in this process model might give rise to the interpretation that students tend to regulate comprehension-related problems in a more coherent way than they tend to regulate motivational problems. Further, it is noticeable that the temporal arrangement of codes in both conditions appears to be the same, and that only the "Motivational Shared" code comes in earlier in case of motivational problems.

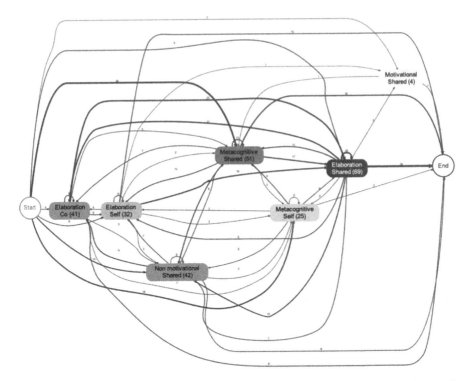

Fig. 2. Process Model for regulating comprehension-related problems with absolute frequencies of all codes (boxes), as well as all observed directional code-code connections (arrows). Again, darker box colors indicate higher absolute code frequencies which means that the corresponding activities were observed more frequently, indicating that several persons have progressed from the corresponding first to the corresponding second activity (or, in the case of self-loops, that one person has performed the same activity several times in succession). (Color figure online)

Epistemic Network Analysis. The first ENA model (see Fig. 3a) shows that the first component mr1 represented by the x-axis explained 10.20% of the variance in the ENA parameter space, while the second component svd2 represented by the y-axis accounted for 15.10% of the variance.

Visualization of subtracted networks (see Fig. 3b) shows that with the relatively stronger connections between metacognitive, elaboration, and non-motivational shared strategies with motivational shared strategies, the center of mass of the motivational problems condition network shifts to the right, while the relatively stronger links between elaboration and metacognitive strategies at all levels place the center of mass of the comprehension-related problems condition network to the left quadrants.

In addition, the subtracted network retains the differences found by PM: When encountering motivational problems, students show higher scores along the x-axis (mr1 can be seen as representative for motivational shared regulation) than when encountering comprehension-related problems. It also reveals higher relative co-occurrences between *Metacognitive Self*-regulation and *Elaborative Shared* regulation in the

a)   b)

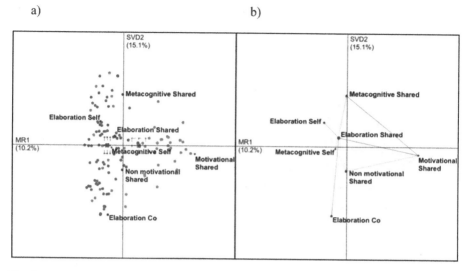

Fig. 3. (a) Networks of students in the conditions "motivational problems" (red) and "comprehension-related problems" (blue) with mean values (squares) and confidence intervals (boxes around squares). The x-axis is based on the descriptor mr1: values on this axis increase as participants demonstrate a higher emphasis on motivational regulation. The y-axis is based on svd2 and primarily focuses on (meta-)cognitive regulation. (b) Subtracted (= contrasted) networks for the conditions „motivational problems" (red) and „comprehension-related problems" (blue) which were generated by subtracting both networks' nodes and connection weights from each other. They serve to represent the differences between the two network graphs and illustrate what makes regulation of motivational problems in collaborative learning different to the regulation of comprehension-related problems. (Color figure online)

condition with comprehension-related problems, but does not retain the higher frequencies between *Non-Motivational* and *Elaboration Shared* in the motivational problem condition any more that was shown by PM.

Moreover, students in both conditions scored similarly on the y-axis (svd2 is representative for meta-cognitive activities). This was statistically confirmed by a paired t-test along the y-axis that failed to reject the null hypothesis as no statistical differences were found between the centroids in the condition with motivational ($M = 0.00$, $SD = 0.60$) and with comprehension-related problems ($M = 0.00$, $SD = 0.66$, $t(81) = 0.00$, $p = 1.00$). Nonetheless, a t-test for paired samples along the x-axis revealed that the centroid of the motivational problem condition ($M = 0.29$, $SD = 0.56$) was significantly different from the centroid of the comprehension-related problems condition ($M = -0.29$, $SD = 0.24$, $t(81) = 8.76$, $p = .00$). On a more general level, these results illustrate that there are differences in how students in groups regulate motivational problems and in how they regulate comprehension-related problems. On a more specific level, they illustrate the shift of the regulation focus to motivational group activities in situations with motivational problems and to (meta-)cognitive activities at different social levels in situations with comprehension-related problems.

Additional Analyses to Converge Findings from PM and ENA. Apart from the fact that ENA, unlike PM, cannot consider start and end points as codes, ENA also lacks to consider self-loop frequencies that may differ between conditions. To make sure taking the loops into account would not have resulted in completely different results of the t-tests, we proved that at least the self-loops of codes that were not plotted close to the x-axis by ENA did not significantly differ between conditions. Thus, we used exact Fisher's tests for count data to compare the cell frequencies of self-loops of all codes between both conditions that were already revealed by PM for significant differences (as the cell frequencies for the Motivational Shared code in the comprehension-related problems condition had zero counts, we have corrected for all cell frequencies based on a proposed procedure by Dureh, Choonpradub, and Tongkumchum [13]).

Since results showed no significant differences of self-loop frequencies between conditions except for the Motivational Shared code which was higher in the motivational than in the comprehension-related problem condition (this code was already plotted close to mr1 in the ENA), $M_{Mot} = 32$, $M_{Comp} = 1$, $p = .00$, $OR = 36.33$ (95% CI: 6.02, 1472.99), we take this as an indication that the group differences we found regarding mr1 would have maximally been even larger if the global test had also taken into account the self-loops on the motivational shared code beside the higher frequencies of this code in the motivational compared to the comprehension-related problems condition.

4   Discussion

This paper intended to demonstrate a procedure for comprehensively testing differences between regulation processes by aid of PM and ENA. At the same time, it intended to depict ways to bypass the drawbacks of each technique.

When performing PM and ENA individually, we encountered some of the problems of the two methods that are already discussed in literature. For example, when preparing the data for analysis, it turned out that our sample size was appropriate for PM, but too small for ENA. Therefore, less frequent codes had to be excluded from ENA (we also excluded these codes from PM as to better demonstrate the extent results of both methods converge). Also, PM created rather confusing models which were barely visually comparable due to the representation of all observed paths. The visual comparison in PM is also generally impeded by the fact that process models include all person-specific regulation paths with same weight irrespective to the person specific activity rate (for what ENA offers a solution). In addition, with PM, global differences between the models could not be verified by a statistical test. Interestingly enough, the arrangement of the codes in both models showed differences only in terms of the "Motivational Shared" code, which was positioned earlier in the process when motivational problems were present. Thus, PM showed that motivational problems are primarily regulated motivationally and metacognitively in the beginning, whereas for comprehension-related problems, the initial focus is on metacognitive regulation.

These findings – which ENA failed to reveal – might indicate that students more or less automatically activate different motivational strategies to solve motivational problems, whereas they seem to be more analytical (and coherent) when faced with comprehension-related problems (see [14]). However, that elaborative shared regulation was rather chosen at the end of the process (in both process models), which seems to be in line with Boekaerts' [8] three-layered model, claiming that goals and resources need to be regulated before learning and which is adheres to the findings of [3] which are described above.

However, ENA allowed for a global statistical verification of the differences between the compared processes that could not be gained by PM. Additionally, while PM would have required further reductions of codes or code-connections to clearer visualize regulation processes, ENA was able to clearly visualize group differences through differently weighted code-code-connections. The clear visualization of group differences was further due to the subtraction of networks and to the provided rotation of networks, which cannot be done in PM. Even though the shift towards joint motivational group efforts for motivational and towards activities closer to the learning process for comprehension-related problems were already observed in PM, ENA revealed that students with comprehension-related problems frequently control their own regulation when acquiring knowledge: because students constantly switched to motivational shared regulation in the regulation of motivational problems, but did not that much switch to one specific cognitive activity in the comprehension-related problems condition, ENA's visual comparison of conditions revealed that students regulate comprehension-related problems comparatively incoherently.

This latter observation would not have been apparent from PM. We specifically revealed this through the ENA's normalization of adjacency vectors. This routine, which is not implemented for PM, ensured that the number of each participant's activities did not affect the structure and thickness of networks. In return, however, it removed the effect of the person-specific regulatory length, which PM considers, but does not clearly visualize. Furthermore, as ENA disregards start and end points of processes, as well as the order of selected paths and self-loop information (all information provided by PM), exact Fisher's tests for count data based on the self-loop information provided by PM complied with the findings of the global comparison.

Overall, when critically appraising the use of each of the two methods and what they add to our understanding of the differences between motivational vs. comprehension-related problem regulation in groups, we argue that the concerted use of PM and ENA has high potential in comparing regulatory processes and is superior to using only either technique. As an outlook for further research with process data remains to say that researchers are already working on the development and refinement of so-called directed ENA [15].

References

1. Csanadi, A., Eagan, B., Kollar, I., Shaffer, D.W., Fischer, F.: When coding-and-counting is not enough: using epistemic network analysis (ENA) to analyze verbal data in CSCL research. Int. J. Comput.-Support. Collaborative Learn. 13(4), 419–438 (2018). https://doi.org/10.1007/s11412-018-9292-z

2. Hadwin, A.F., Järvelä, S., Miller, M.: Self-regulated, co-regulated, and socially shared regulation of learning. In: Zimmerman, B., Schunk, D. (eds.) Handbook of Self-regulation of Learning and Performance, pp. 65–84. Routledge, New York (2011)

3. Bannert, M., Reimann, P., Sonnenberg, C.: Process mining techniques for analysing patterns and strategies in students' self-regulated learning. Metacognition Learn. 9(2), 161–185 (2014). https://doi.org/10.1007/s11409-013-9107-6

4. Bolt, A., van der Aalst, W.M.P., de Leoni, M.: Finding process variants in event logs. In: Panetto, H., et al. (eds.) On the Move to Meaningful Internet Systems. Lecture Notes in Computer Science, vol. 10573, pp. 45–52. Springer, Cham (2017). https://doi.org/10.1007/978-3-319-69462-7_4

5. Shaffer, D.W.: Quantitative Ethnography. Cathcart Press, Madison (2017)

6. Ruis, A.R., Rosser, A.A., Quandt-Walle, C., Nathwani, J.N., Shaffer, D.W., Pugh, C.M.: The hands and head of a surgeon: Modeling operative competency with multimodal epistemic network analysis. Am. J. Surg. 216(5), 835–840 (2018). https://doi.org/10.1016/j.amjsurg.2017.11.027

7. Zhang, S., Liu, Q., Cai, Z.: Exploring primary school teachers' technological pedagogical content knowledge (TPACK) in online collaborative discourse: an epistemic network analysis. Br. J. Edu. Technol. (2019). https://doi.org/10.1111/bjet.12751

8. Boekaerts, M.: Self-regulated learning: where we are today. Int. J. Educ. Res. 31(6), 445–457 (1999). https://doi.org/10.1016/S0883-0355(99)00014-2

9. Friedrich, H.F., Mandl, H.: Lernstrategien: Zur Strukturierung des Forschungsfeldes. In: Mandl, H., Friedrich, H.F. (eds.) Handbuch Lernstrategien, pp. 1–23. Hogrefe, Göttingen (2006)

10. Hadwin, A., Oshige, M.: Self-regulation, coregulation, and socially shared regulation: exploring perspectives of social in self-regulated learning theory. Teachers Coll. Rec. 113(2), 240–264 (2011)

11. Janssenswillen, G.: bupaR: Business Process Analysis in R. R package version 0.4.2 (2019)

12. Marquart, C.L., Hinojosa, C., Swiecki, Z., Shaffer, D.W.: Epistemic network analysis version 0.1.0 (2018)

13. Dureh, N., Choonpradub, C., Tongkumchum, P.: An alternative method for logistics regression on contingency tables with zero cell counts. Songklanakarin J. Sci. Technol. 38(2), 171–176 (2016). https://doi.org/10.14456/sjst-psu.2016.23

14. Melzner, N., Greisel, M., Dresel, M., Kollar, I.: Effective regulation in collaborative learning: an attempt to determine the fit of regulation challenges and strategies (long paper). In: Lund, K., Niccolai, G., Lavoué, E., Hmelo-Silver, C., Gweon, G., Baker, M. (eds.) A Wide Lens: Combining Embodied, Enactive, Extended, and Embedded Learning in Collaborative Settings: Proceedings of the 13th International Conference on Computer Supported Collaborative Learning, CSCL, vol. 1, pp. 312–319. International Society of the Learning Sciences, Lyon (2019)

15. Marquart, C.L., Swiecki, Z., Collier, W., Eagan, B., Woodward, R., Shaffer, D.W.: rENA: epistemic network analysis. R package version 0.1.6.1 (2019)

Students' Collaboration Patterns in a Productive Failure Setting: An Epistemic Network Analysis of Contrasting Cases

Valentina Nachtigall[1](✉) and Hanall Sung[2]

[1] Ruhr-University Bochum, Universitätsstr. 150, 44780 Bochum, Germany
valentina.nachtigall@rub.de
[2] University of Wisconsin–Madison, Madison, WI 53706, USA

Abstract. In this paper, we aim at uncovering collaborative problem-solving patterns associated with students' successful learning of social sciences research methods in a Productive Failure (PF) setting. We report an epistemic network analysis (ENA) of PF students' conversations. Conversations are compared between PF groups that generated high quality solution ideas (HQ groups) and groups that developed low quality solution ideas (LQ groups). The ENA results demonstrate significantly different patterns. The collaborative problem solving of four HQ triads in a PF setting is characterized by debates and elaborations related to canonical contents of the targeted learning concept. The collaborative problem solving of four LQ triads is featured by task-pursuance actions and elaborations related to the instructions and contents stated in the worksheet. We also compared the eight groups based on their learning outcome (i.e., performance on a knowledge test). The comparison of four groups with a high learning outcome and of four groups with a low learning outcome revealed similar ENA results as the comparison of the HQ and LQ groups. These findings offer empirical evidence for the often hypothesized but rarely supported notion of certain collaborative problem-solving processes being important for the effectiveness of PF. The potential relevance of the collaborative problem-solving patterns of HQ groups for learning in a PF setting is discussed in light of mechanisms hypothesized to underlie the PF effect.

Keywords: Productive failure · Collaborative learning · Problem solving

1 Theoretical Background

1.1 Productive Failure

In a Productive Failure (PF) setting, students explore the solution of a complex problem by the generation of intuitive (often erroneous) solution ideas [1]. Afterwards, the features of typical erroneous solution ideas are compared and contrasted to the components of the canonical solution by the instructor [2]. Thus, in a PF setting, students receive delayed instruction after they have attempted to solve a problem on their own. These initial problem-solving attempts usually result in students' failure to solve the problem canonically during the initial problem-solving phase. However, this failure is

© Springer Nature Switzerland AG 2019
B. Eagan et al. (Eds.): ICQE 2019, CCIS 1112, pp. 165–176, 2019.
https://doi.org/10.1007/978-3-030-33232-7_14

hypothesized to be productive for students' development of a deep understanding of the targeted learning concept during the subsequent instruction phase [3].

Previous research indeed has demonstrated that PF is more effective for students' development of a deep understanding of a targeted learning concept than approaches that engage students in instruction prior to problem solving [2]. This effectiveness of PF settings is hypothesized to be enhanced by asking students to collaboratively invent their solution ideas. By collaboratively inventing solution ideas, PF students are provided with opportunities to "share, elaborate, critique, explain, and evaluate" their proposed solution ideas ([4], p. 51). By collaboratively discussing and elaborating their solution ideas, students may be enabled to attend to the features of the canonical solution which should prepare them for developing a deep understanding of the components of the canonical solutions that are explained and discussed during subsequent instruction [4]. In addition, these collaborative processes, such as sharing and discussing each other's solution ideas, could lead to group solutions with a higher quality [5].

But how can we examine what role collaboration plays to get deep understanding of the learning contents in PF settings? The answer to that question clearly depends on collaboration processes: how students take part in collaboration, how they generate solutions, and how they learn from this experience. In this paper, we investigate how the process of students' collaboration in a PF setting can be related to their learning achievement. To accomplish this, we used epistemic network analysis (ENA; [6]) to analyze students' collaborative problem solving in a PF setting. Specifically, we identified (a) high quality (HQ) solution groups and (b) low quality (LQ) solution groups by the quality of solutions students generated through collaborative discussion. We then coded discussion data and used the resulting ENA model to examine whether and to what extent HQ and LQ groups showed different collaboration patterns.

1.2 Collaboration in a PF Setting

Empirical evidence regarding the relationship between collaborative problem solving and learning achievement in a PF setting is still inconclusive. So far, few studies (e.g., by Weaver, Chastain, DeCaro, and DeCaro [5] or by Mazziotti, Rummel, Deiglmayr, and Loibl [7]) have systematically examined the effect of collaborative versus individual problem solving prior to instruction on learning. These studies revealed no beneficial effect of collaborative problem solving on students' acquisition of knowledge in a PF setting. However, given the findings from research on collaborative learning more generally, the findings from Weaver et al. [5] and Mazziotti et al. [7] are not surprising. That is, Dillenbourg, Baker, Blaye, and O'Malley claim that "collaboration is in itself neither efficient nor inefficient" ([8], p. 196), which could be a reason why research that has compared the effects of collaborative versus individual learning has revealed inconsistent findings. Instead, according to the "interactions paradigm", research should focus on analyzing learners' collaboration processes and interaction patterns and in turn the role of certain collaboration processes for students' learning outcomes [8]. In this perspective, Kapur and Bielaczyc conducted a qualitative contrasting-case analysis and compared the collaborative problem-solving processes of two PF groups: a group that had invented solution ideas with highly different quality levels and had reached a high learning outcome and a group that had generated solution ideas with less different quality levels and had reached a low learning outcome [4]. Their results revealed that the collaboration processes differed between both groups, as students' problem solving in the highly different quality group contained more evaluation, elaboration, and explanation activities than in the less different quality group [4].

Against this background, it is likely that certain collaboration processes (e.g., discussion and elaboration) during problem solving are associated with the quality of PF students' generated solution ideas and in turn with PF students' learning outcome. However, empirical evidence that supports this argument is rare. So far, few investigations (i.e., [9, 4]) indicate that collaboration processes during problem solving may – under certain circumstances – be beneficial for the quality of students' generated solution ideas during problem solving and their knowledge acquisition in a PF setting. To extend these findings of previous PF research, we aim at uncovering collaborative problem-solving patterns linked to students' successful learning in a PF setting. These relationships can be modeled with ENA, a technique for identifying and quantifying connections among epistemic frame elements and representing them in dynamic network models [6]. In this study, we compare students' collaborative problem-solving processes between PF groups that had generated a solution with a high quality (HQ groups) and PF groups that had generated a solution with a low quality (LQ groups). We then compare the relationship between PF groups' solution quality and their learning outcome.

2 Method

2.1 Research Context

The data analyzed here are part of a study conducted by Nachtigall and colleagues [10] with 10th graders from social sciences classes of secondary schools in Germany. 121 students from six classes participated in the implemented PF setting.

During the initial problem-solving phase (45 min), PF students worked in small groups. Most of the groups were triads (38 groups out of 44). The small groups were formed by students, often with seat neighbors. PF students were asked to solve a problem that required them to design experiments related to educational sciences. Students were provided with an illustration depicting the suggestions of three fictitious math teachers on how to improve math learning in 10th grade: One teacher suggested to implement group work in order to improve math learning, a second teacher proposed to use a computer-supported learning program, and a third teacher claimed that students with language problems who just arrived from other countries would be disadvantaged by group work as well as by working on the computer. PF students were asked to collaboratively generate as many study designs as possible to investigate all suggestions mentioned by the teachers. While PF students generated their solution ideas, they did not receive any instructional support on the content or on problem-solving steps. They were only provided with motivational prompts that aimed at encouraging them to persist in solving the task (e.g. "the task is difficult, but together you will definitely come up with some great ideas" or "you are doing a good job together, keep going").

After the problem-solving phase, PF students experienced an instruction phase (45 min). During instruction, the instructor compared and contrasted the features of typical erroneous student solutions to the components of the canonical solution. To keep the instructional material constant between the different classes that participated in the study, the instruction used typical erroneous student solutions that were based on pilot tests of the learning materials. Thus, the contents of instruction were the same in all classes. Moreover, the same experimenter led the instruction in all six classes. PF students' solution ideas typically included at least one of the following four errors: no systematic variation of variables, no inclusion of a control condition, no administration of a pretest next to a posttest, and no measurement of a certain control variable by using a questionnaire. More specifically, students often designed an experiment with two to three conditions (e.g., working in small groups vs. working on computers) and planned to use a posttest in order to investigate the effectiveness of the different learning methods for students' learning in mathematics. As this example demonstrates, students did not systematically vary the independent variables (i.e., group work: yes/no and work with computers; yes/no), they did not include the usual learning method (i.e., working individually) as control condition, they did not plan to use a pretest next to a posttest, and they did not assess students' language skills (which was a concern of one of the three fictitious math teachers in the task) as control variable. These four typical errors were discussed during instruction and compared and contrasted to the components of the canonical solution. The canonical solution, which was presented and explained in a step-by-step procedure, was a 2×2-design with pre- and posttest and a questionnaire to measure a certain control variable. See Fig. 1 for an illustration of the canonical solution presented during instruction.

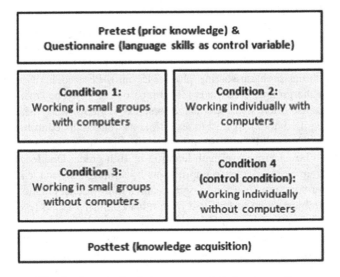

Fig. 1. Canonical solution presented during instruction.

Group conversations during the problem-solving phase were audio recorded and the quality of generated group solutions was coded. To measure the quality of the solution ideas that students had generated in small groups, we assessed the number of canonical components within each solution idea (total score: 0–8). The following eight canonical components were coded: (1) Experimental study design, (2) between-subject-design, (3) correct amount and naming of conditions, (4) inclusion of a posttest, (5) inclusion of a pretest, (6) inclusion of a control group, (7) systematic variation of variables, (8) consideration of a particular control variable. The score of the best idea (highest number of canonical components) that each group had generated was used as measure for solution quality (cf., [11]). Students' individual learning outcome was assessed after the instructional phase by using a knowledge test with eight items (total score: 0–25). Three items asked students to design experimental studies, two items required students to evaluate study designs and to identify difficulties within these designs, and three items asked students to complete a sentence in their own words by recalling the canonical components of the experimental design presented during instruction. The posttest, the coding of the solution ideas as well as the study design are described in more detail in Nachtigall, Serova, and Rummel [12].

2.2   Data Analysis

Selection of Contrasting Cases

We selected eight PF groups and transcribed the respective audio recordings of their initial collaborative problem-solving process. The selection of the eight groups was based on the quality of solutions they generated through collaborative discussion. On average (N = 44), the best solution idea generated by PF groups included 5.07 canonical components (SD = 1.29). We selected four HQ groups that had developed a solution with a rather high quality (i.e., solution quality >5.07) and four LQ groups that had invented a solution with a rather low quality (i.e., solution quality <5.07).

All eight groups consisted of three members. As the small groups were formed by students, we asked the teachers to assess whether the groups were homogenous or not in terms of their prior achievement in class. We, moreover, asked the teachers to evaluate the general achievement level of the groups based on the students' prior performance in class using a five-point scale with "1" indicating low achievement and "5" reflecting high achievement of the group. A teacher rating is missing for one of four LQ groups. The teachers characterized two HQ groups and one LQ group as homogenous and two HQ and two LQ groups as heterogeneous. Hence, there were heterogeneously composed HQ groups as well as homogenously composed HQ groups and the same is true for the selected LQ groups. With respect to the general achievement level, the teachers characterized all groups as rather high achievement groups. More specifically, teachers rated the achievement level of three HQ groups with 4, of one HQ group with 5, and of the three LQ groups with 3, 4, and 5 points, respectively.

Discourse Coding

Based on grounded analysis, we developed a set of codes to represent the key elements of students' collaboration and their discussion contents (see Table 1). To code the discussion data (1457 lines of talk), we used an automated coding process (ncodeR; [13]) based on regular expression matching. We validated all five codes using a series of comparisons between two human raters and the computer with resulting Cohen's kappa scores between 0.82 and 1.00 (see Table 1). The interrater reliability analysis shows that all pairwise agreements among rater 1, rater 2, and the computer meet standards for kappa [14]. We used Shaffer's rho to determine, for each kappa value, the likelihood that it would be found by two coders if their true rate of agreement was kappa < 0.65 [15]. As shown in Table 1, all of the kappa values achieved have Shaffer's rho values less than 0.05, meaning that the Type I error rate for assuming that if the coders were to code the whole data set, they would have a level of agreement over kappa of 0.65.

Table 1. Coding scheme and inter-rater reliability statistics.

Code name	Description	Examples (translated from German to English)	Kappa		
			R1 v. R2	R1 v. ncodeR	R2 v. ncodeR
Canonical content	Students describe features of an experimental study design. These descriptions may include specific details (e.g., sample size)	"One uses the entire semester to test the different methods and after each method one uses a test. Then, the test results could be compared"	1.00**	0.93**	1.00**
Worksheet content	Discussions are oriented on the instructions and information given in the task worksheet	"Here is written: students with a migration background"	0.85**	1.00**	0.94**
Agreement and elaboration	Students agree with proposals or opinions of their group members and/or elaborate on these	"Exactly and then we could as a second idea [...]"	0.90**	0.95**	0.90**
Debate	Students discuss and/or evaluate each other's proposals	"But, when working individually, this assessment does not make sense"	0.82*	0.97**	0.84*
Task-pursuance actions	Students conclude steps for their work progress, summarize ideas and draft a plan on next actions	"Okay, lets write it down: Working in groups is not a good idea, as students may show a lot of off-task behavior"	0.90**	0.93**	0.93**

*indicates $\rho(0.65) < 0.05$; **indicates $\rho(0.65) < 0.01$.

3   Results

3.1   Quantitative Results

Figure 2 shows each student's network location, along with the means and 95% confidence intervals of students in HQ (red) and LQ groups (blue). The unit of analysis is individual, and each point represents the center of mass of the individual networks. There is a statistically significant difference between HQ and LQ students on the first dimension with a moderate effect ($mean$HQ = −0.83, $mean$LQ = 0.83; $t = −7.60$, $p < 0.01$, Cohen's $d = 3.10$).

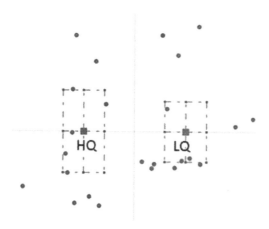

Fig. 2. ENA scatter plot showing HQ (red) and LQ (blue) groups. Each point is a single student; the squares are group means; the dashed boxes are 95% confidence intervals (t-distribution). (Color figure online)

To examine which connections accounted for the differences between HQ and LQ groups, we constructed mean epistemic networks for each group. The mean networks represent the average connections among the epistemic frame elements which students make during the discussion within HQ or LQ groups. As Fig. 3 shows, both HQ (red network, left) and LQ (blue network, right) groups made dense networks of connections but HQ groups (red network, left) made more links from AGREEMENT AND ELABORATION to DEBATE and CANONICAL CONTENT, while LQ groups (blue network, right) made more links to WORKSHEET CONTENT and TASK-PURSUANCE ACTIONS.

To be specific, when we subtract one network from the other (Fig. 4) to identify why there is a significant difference between the two groups, the HQ groups (red) made more links between DEBATE and CANONICAL CONTENT, while LQ groups (blue) made more links between AGREEMENT AND ELABORATION and TASK-PURSUANCE ACTIONS.

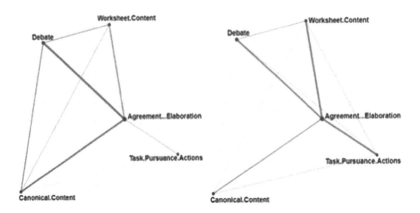

Fig. 3. Mean ENA network diagrams showing the connections made by HQ (red network, left) and LQ (blue network, right) groups. (Color figure online)

Fig. 4. ENA difference graph for conversations in the HQ (red) and LQ (blue) groups. (Color figure online)

3.2   Qualitative Results

To understand the importance of these differences between HQ and LQ groups, we analyzed students' discussions qualitatively. Here, we present a short excerpt of a student discussion in HQ groups (Table 2) and in LQ groups (Table 3), respectively.

The discussion of the HQ group (see Table 2) shows that students evaluated and elaborated on each other's proposals. These discussions related to important canonical content (i.e., crucial criteria of an experimental study). For instance, Julia is concerned about comparing different classes due to different prior-knowledge levels (see line 2).

The discussion of the LQ group (see Table 3) demonstrates that these students are concerned with taking school students' opinions about certain learning methods into

Table 2. Excerpt of high solution quality team's discussion.

Line	Student	Discussion utterance	Codes
1	Lucy	Using different classes would be faster, right?	CANONICAL CONTENT
2	Julia	Using different classes might be faster, but different classes may be not on an equal level of prior knowledge	CANONICAL CONTENT DEBATE
3	Lucy	Yes, yes, right!	AGREEMENT AND/OR ELABORATION
4	Anita	I would do it with one class	CANONICAL CONTENT
5	Julia	Yes, yes. That one class simply [stops talking]	AGREEMENT AND/OR ELABORATION CANONICAL CONTENT
6	Anita	But that its mixed. That there are also students with language issues and so on	DEBATE

Table 3. Excerpt of low solution quality team's discussion.

Line	Student	Discussion Utterance	Codes
1	Lynn	Yes, write down: 1. Step	AGREEMENT AND/OR ELABORATION TASK PURSUANCE ACTIONS
2	Sandra	We could use a questionnaire again for assessing the opinions and preferences of the children	CANONICAL CONTENT
3	Lynn	Maybe we could make it specific and ask who can work best in groups and how they evaluate their own abilities, and how they might contribute to the group work	WORKSHEET CONTENT
4	Agnes	Yes	AGREEMENT AND/OR ELABORATION
5	Lynn	Do you know what I mean?	/
6	Agnes	Yes, but look, we just had the case in which we had an inefficient group. It would make sense, if we were able to split them up	AGREEMENT AND/OR ELABORATION DEBATE

account (see Sandra's proposal in line 2) and with developing strategies for increasing the effectiveness of the learning methods that are stated in the worksheet (see Agnes' proposal in line 6). In addition, their conversation is featured by summarizing or dictating the things that should be written down (see Lynn's statement in line 1).

3.3   Learning Outcome

We also compared the eight groups based on their learning outcome (i.e., performance on the knowledge test). On a group level (i.e., mean posttest scores of each group), PF groups ($N = 44$) reached on average the following learning outcome: $M = 10.95$, $SD = 3.51$. The comparison of four groups with a rather high learning outcome (i.e., learning outcome >10.95) and of four groups with a rather low learning outcome (i.e., posttest score <10.95) revealed similar ENA results (see Figs. 5 and 6) as the comparison of the HQ and LQ groups. These results are not surprising as three out of the four HQ groups also reached a high learning outcome and three out of the four LQ groups reached a low learning outcome. Moreover, as reported in Nachtigall et al. [12], the solution quality correlated positively and significantly with PF students' learning outcome (medium-sized effect).

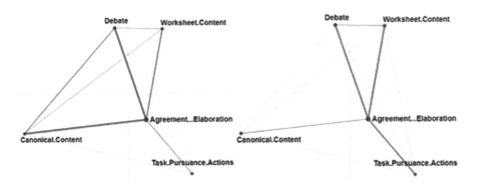

Fig. 5. Mean ENA network diagrams showing the connections made by groups with a rather high learning outcome (red network, left) and with a rather low learning outcome (blue network, right). (Color figure online)

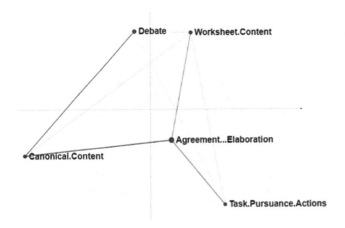

Fig. 6. ENA difference graph for conversations in groups with a rather high learning outcome (red) and a rather low learning outcome (blue). (Color figure online)

4 Discussion

The ENA results presented in this paper are a relevant contribution and extension of previous PF research. More specifically, our findings offer empirical evidence for the often hypothesized but rarely supported notion of collaborative problem solving being important for the effectiveness of PF. Our quantitative and qualitative contrasting-cases analyses demonstrated that PF groups that had generated high-quality solution ideas (HQ groups) evaluated, discussed, and elaborated on each other's proposals and attended to canonical components of the targeted learning concept. PF groups that had generated low-quality solutions (LQ groups) elaborated on task-pursuance actions and on contents stated in the worksheet. In contrast to HQ groups, LQ groups' conversations were less featured by debates (i.e., evaluations of proposals) on canonical content. The collaboration pattern that characterizes the HQ groups (i.e., debating and elaborating on canonical contents) is in line with the following three mechanisms hypothesized to underlie the effectiveness of PF [2]: (1) activation of prior knowledge, (2) awareness of knowledge gaps, and (3) recognition of the deep features of the canonical solution. It is likely that elaboration processes during collaborative problem solving help students in a PF setting to activate more prior knowledge than during individual problem solving [7]. Moreover, debating on each other's ideas may lead to socio-cognitive conflicts, uncover misconceptions, and trigger PF students' awareness of knowledge gaps [9]. Both, elaborating and debating on different solution ideas might support PF students to attend to the deep features of the canonical solution [4]. It is likely that all three mechanisms evolved during the collaborative problem solving of the HQ groups as they (in contrast to LQ groups) elaborated and debated on each other's proposals which related to canonical content. As a consequence, they were seemingly enabled to develop group solutions with a higher quality and, thus, to reach a higher learning outcome on the knowledge test. It would be interesting to investigate whether facilitating certain collaboration processes would promote students' learning in a PF setting. Then, it would be interesting to match the individual's talk during collaboration to their individual learning outcome. This was not possible in the current study, which, nevertheless, revealed interesting insights into the role of collaborative problem solving for the effectiveness of PF.

Acknowledgements. The data analyzed in this paper are part of a project that the first author conducted in cooperation with Prof. Dr. Nikol Rummel and Dr. Katja Serova at the Institute of Educational Research at Ruhr-University Bochum (RUB). We want to acknowledge their input and support with respect to, for instance, the study design. We also want to thank the Research School at RUB for funding a research stay at the Educational Psychology Department at University Wisconsin-Madison. The research stay allowed the first author to visit the lab of Prof. Dr. David W. Shaffer and made this joint publication possible. This work was funded in part by the National Science Foundation (DRL-1661036, DRL-1713110), the Wisconsin Alumni Research Foundation, and the Office of the Vice Chancellor for Research and Graduate Education at the University of Wisconsin-Madison. The opinions, findings, and conclusions do not reflect the views of the funding agencies, cooperating institutions, or other individuals.

References

1. Kapur, M.: Learning from productive failure. Learn. Res. Pract. **1**(1), 51–65 (2015)
2. Loibl, K., Roll, I., Rummel, N.: Towards a theory of when and how problem solving followed by instruction supports learning. Educ. Psychol. Rev. **29**(4), 693–715 (2017)
3. Kapur, M.: Examining productive failure, productive success, unproductive failure, and unproductive success in learning. Educ. Psychol. **51**(2), 289–299 (2016)
4. Kapur, M., Bielaczyc, K.: Designing for productive failure. J. Learn. Sci. **21**(1), 45–83 (2012)
5. Weaver, J.P., Chastain, R.J., DeCaro, D.A., DeCaro, M.S.: Reverse the routine: problem solving before instruction improves conceptual knowledge in undergraduate physics. Contemp. Educ. Psychol. **52**, 36–47 (2018)
6. Shaffer, D.W., Collier, W., Ruis, A.R.: A tutorial on epistemic network analysis: analyzing the structure of connections in cognitive, social, and interaction data. J. Learn. Anal. **3**(3), 9–45 (2016)
7. Mazziotti, C., Rummel, N., Deiglmayr, A., Loibl, K.: Probing boundary conditions of Productive Failure and analyzing the role of young students' collaboration. npj Sci. Learn. **4**(2), 1–9 (2019)
8. Dillenbourg, P., Baker, M., Blaye, A., O'Malley, C.: The evolution of research on collaborative learning. In: Spada, H., Reiman, P. (eds.) Learning in Humans and Machine: Towards an Interdisciplinary Learning Science, pp. 189–211. Elsevier, Oxford (1996)
9. Hartmann, C., Rummel, N., Loibl, K.: Communication patterns and their role for conceptual knowledge acquisition from productive failure. In: Looi, C.K., Polman, J., Cress, U., Reimann, P. (eds.) Proceedings of the 12th International Conference of the Learning Sciences, vol. I, pp. 530–537. International Society of the Learning Sciences, Singapore (2016)
10. Nachtigall, V., Rummel, N., Serova, K.: Authentisch ist nicht gleich authentisch–Wie Schülerinnen und Schüler die Authentizität von Lernaktivitäten im Schülerlabor einschätzen [Authentic Does not Equal Authentic – How Students Evaluate the Authenticity of Learning Activities in an Out-of-School Lab]. Unterrichtswissenschaft **46**(3), 299–319 (2018)
11. Loibl, K., Rummel, N.: The impact of guidance during problem solving prior to instruction on students' inventions and learning outcomes. Instr. Sci. **42**(3), 305–326 (2014)
12. Nachtigall, V., Serova, K., Rummel, N.: When Failure Fails to be Productive – Probing the Effectiveness of Productive Failure for Learning Beyond STEM Domains (submitted)
13. Marquart, C.L., Swiecki, Z, Eagan, B., Shaffer, D.W.: ncodeR: techniques for automated classifiers. R package version 0.1.2 (2018). https://CRAN.R-project.org/package=ncodeR. Accessed 24 July 2019
14. Landis, J.R., Koch, G.G.: The measurement of observer agreement for categorical data. Biometrics **33**(1), 159–174 (1977)
15. Shaffer, D.W.: Quantitative Ethnography. Cathcart Press, Madison (2017)

The Influence of Discipline on Teachers' Knowledge and Decision Making

Michael Phillips[1]([✉]) [ID], Vitomir Kovanović[2] [ID], Ian Mitchell[1] [ID],
and Dragan Gašević[1] [ID]

[1] Monash University, Melbourne, Australia
michael.phillips@monash.edu
[2] University of South Australia, Adelaide, Australia

Abstract. The knowledge required by teachers has long been a focus of public and academic attention. Following a period of intense research interest in teachers' knowledge in the 1980s and 1990s, many researchers have adopted Shulman's suggestion that expert teaching practice is based on seven forms of knowledge which collectively are referred to as a knowledge base for teaching. Shulman's work also offered a decision-making framework known as pedagogical reasoning and action, which allows teachers to use their seven forms of knowledge to make effective pedagogical decisions. Despite the widespread acceptance of these ideas, no empirical evidence exploring the connections between knowledge and decision-making is evident in the research literature. This paper reports on a pilot study in which the connections between knowledge and decisions in science, mathematics and information technology teachers' lesson plans are quantified and represented using epistemic network analysis. Findings reveal and levels of complexity that have been intimated but, until now, not supported with empirical evidence.

Keywords: Teacher knowledge · Teacher decision making · Epistemic network analysis

1 An Introduction to Teachers' Knowledge and Decision-Making

Public scrutiny and commentary on the work of teachers have long been a feature of public and academic discourse. Questioning the value of teachers and their work is exemplified by a maxim in George Bernard-Shaw's play Man and Superman: "He who can does. He who cannot, teaches". Woody Allen's character in the movie Annie Hall extends this to further denigrate teachers claiming "... and those who can't teach, teach gym".

Partly in response to such jibes, developing a clearer sense of what teachers know and how they use their knowledge to enhance their teaching practices has been an area of focus for education researchers, teacher educators and educational policymakers [2]. An examination of documents from the late 19th century reveals debates about the nature of teacher knowledge have long been part of the educational landscape (for example, see the discussion in [3]). The debate about teacher knowledge arguably

© Springer Nature Switzerland AG 2019
B. Eagan et al. (Eds.): ICQE 2019, CCIS 1112, pp. 177–188, 2019.
https://doi.org/10.1007/978-3-030-33232-7_15

reached a crescendo in the United States in April 1983 with the release of results from a Commission established by then-President Ronald Reagan to investigate declining educational outcomes in the nation's schools. Their ominously titled report A Nation at Risk [4]:

> found that not enough of the academically able students are being attracted to teaching; that teacher preparation programs need substantial improvement ... Too many teachers are being drawn from the bottom quarter of graduating high school and college students. The teacher preparation curriculum is weighted heavily with courses in "educational methods" at the expense of courses in subjects to be taught. (p. 22)

The renewed focus that resulted, in large part, from claims such as this produced a great deal of focused research in the 1980s and 1990s that considered teachers' knowledge from differing epistemological viewpoints. For example, Tom and Valli [5] developed a philosophically grounded review of professional knowledge, Grimmit and MacKinnon [6] analysed craft conceptions of teaching, Clandinin and Connelly's [7] considered teachers' personal, professional knowledge and Shulman's [1, 8–12] program of research sought to "show what forms and types of knowledge are required to teach competently" (Fenstermacher, [18], p. 6). It is this extensive program of research that has led Tom and Valli [5] to suggest that Shulman "probably has gone as far as anyone in his thinking about the forms of teacher knowledge" (p. 6).

Shulman [1] suggested that the categories of the knowledge base for teaching included: content knowledge, general pedagogical knowledge, knowledge of learners, knowledge of educational contexts, knowledge of educational ends and pedagogical content knowledge.[1] While articulating this knowledge base for teaching, Shulman [1] recognised that "a knowledge base for teaching is not fixed and final" (p. 12) and that his "'blueprint' for the knowledge base of teaching has many cells or categories with only the most rudimentary place-holders, much like the chemist's periodic table of a century ago" (p. 12). As such, contemporary examinations of teacher knowledge continue to draw on Shulman's [1] knowledge base for teaching (for example, see [13]). Indeed, Tom and Valli's [5] suggestion of the importance of Shulman's work has been validated in the years following the initial publication of Shulman's knowledge base for teaching with Google Scholar currently indicating that Shulman's [1] work has been cited more than 20,000 times.

Of particular note in the broad range of work using Shulman's knowledge base are the investigations undertaken by a range of researchers around Shulman's notion of pedagogical content knowledge (PCK). Initial theoretical development of PCK began in the early 1980s and, as part of his 1985 American Educational Research Association presidential address, "Lee Shulman tossed off the phrase 'pedagogical content knowledge' and sparked a small cottage industry devoted to the scholarly elaboration of the construct" [7, p. 32].

Fenstermacher [18] suggested that although Nelson's "notion of 'tossed off' seems a bit ungenerous, given the amount of scholarly development that went into the

---

[1] See Shulman (1987a) p. 8 for more detailed descriptions of each of these categories.

concept, there is no doubt that [it] has spawned an extensive set of research studies" (p. 14). The PCK work conducted in the years immediately following the construct's introduction has continued (see the corpus work of Loughran for example).

Fifteen years after Shulman proposed PCK, educational technology researchers began to extend the construct to include teachers' technological knowledge (e.g., [14–16]). While the elements and configurations of these new models and the methods recommended to develop the specialised knowledge that they represented differed, all shared PCK and technological knowledge as their conceptual base.

This new framework has become known as technological pedagogical and content knowledge (TPACK) and has been utilized in more than 1,300 publications that have appeared in less than 15 years (http://activitytypes.wm.edu/TPACKNewsletters/), and has impacted on the practice of teachers, school leaders, and other stakeholders looking to develop meaningful educational uses of technology [17]. Shulman's knowledge base for teaching, therefore, is a valuable starting point for investigations such as this one into the forms of knowledge teachers use as part of their classroom practice.

At the same time as he proposed a knowledge base for teaching, Shulman [1] also articulated a series of stages in a decision-making process that he termed pedagogical reasoning and action (PR&A). Shulman [1] argued that PR&A always starts with comprehension of subject matter and ends with new comprehension but the other four stages (transformation, instruction, evaluation and reflection) "are not meant to represent a set of fixed stages, phases, or steps. Many of the processes can occur in a different order. Some may not occur at all during some acts of teaching. Some may be truncated, others elaborated" (p. 19). In other words, the connections between the stages of pedagogical reasoning, and indeed the connections of these stages of reasoning to the knowledge that teachers use to make pedagogical decisions were not part of Shulman's [1] descriptions.

Outlining a vision for a "mature profession" (p. 422), Sachs [19] argues that contemporary conceptualisations of teacher professionalism revolve around "A better understanding of the form and content of teachers' professional knowledge and how teachers arrive at judgements" (p. 433). While there are many examples of empirical accounts of individual aspects of Shulman's knowledge base – for example PCK and more recently TPACK - empirical evidence that reveals co-occurrences of teachers' knowledge and decision-making, however, is not evident in the corpus of research that has followed Shulman's work in the subsequent three decades from the 1980s. This pilot study, therefore seeks to explore the following question: what are the connections between teachers' knowledge and their decision-making as defined in Shulman's PR&A? In doing so, we hope to begin to much needed empirically-based understandings of the relationships between teachers' knowledge and their decision-making processes.

2 Method

One of the substantial challenges when exploring teacher knowledge is to have teachers articulate what they know and consider when making decisions. As Loughran et al. [20] found more than a decade ago that "attempts to articulate links between practice and knowledge have proved to be exceptionally difficult, because, for many teachers, their practice and the knowledge/ideas/ theories that tend to influence that practice are often tacit (Schön 1983)" (p. 371).

Working with participants in a longitudinal study exploring teachers' knowledge, particularly PCK, Loughran et al. [20] found it valuable to draw on tools that provided representations of teachers' knowledge as did Phillips [21] when investigating teachers' TPACK. This pilot investigation drew on teachers' lesson plans – part of the daily practice of recording what knowledge and pedagogical decisions teachers anticipate using – as the starting point for making the tacit explicit. In an attempt to make this tacit knowledge explicit, this investigation began by identifying participants who would be willing to reveal both tacit and explicit aspects of their planned practice.

3   Participants

The participants for this study were all drawn from one specialist government-run Mathematics, Science and Technology secondary school in Melbourne, Australia. This school has a select entry enrolment process and only enrols students in the final three years of their secondary education program (ages 16–18). In addition to being a specialist, select entry school, an additionally noteworthy contextual factor was that all classes were co-taught by two teachers who worked with 50 students in a class in contrast to the typical practice of one teacher working with 25 students in most other government-run schools.

A pair of teachers from each of the specialist areas within the school (Mathematics, Science and Technology) volunteered to participate in the study, and each of these pairs together taught a co-educational class of approximately 50 Year 10 students (around 16-years old). These six participants provided the initial data for this investigation in the form of their shared lesson plans for the first unit of work that was to be taught in the academic year.

4   Data

Sixteen lesson plans were provided by two Mathematics teachers detailing planned instruction in a unit on Linear Relations and Equations, nine lesson plans were provided by two Science teachers for a unit on DNA and Genetics and 20 lesson plans were provided by two Technology teachers for a unit on Coding. Drawing on Shulman's [1] conceptualisation of a knowledge base for teaching, the 45 lesson plans provided by all teachers were coded for evidence of teacher knowledge using the NVivo12 software program. In total, six codes for teacher knowledge were included in the codebook, namely: content knowledge, general pedagogical knowledge, knowledge of learners, knowledge of educational contexts, knowledge of educational ends and pedagogical content knowledge. Additionally, four codes for teacher knowledge were added by the researchers based on theoretical developments in the three decades following Shulman's [1] publication. The additional codes were big ideas, promoting quality learning and engagement, nature of the domain, tactical and strategic thinking, and TPACK.

The development of codes for PR&A followed the same process, initially drawing on Shulman's [1] categorisations of comprehension, transformation, instruction, evaluation and reflection, new comprehension. Consideration was given to the addition of supplementary PR&A codes; however, it was decided that, based on two factors, supplementary PR&A codes would not be added. First, relatively little empirical research has been conducted on PR&A, particularly for in-service teachers and where there have been empirically grounded studies, little consensus has been reached [22].

In one example, some researchers have considered the impact of technology on teachers' PR&A [23, 24] in a similar way to the way technology has been considered as a factor impacting teachers' knowledge [25]; however, the findings from these investigations have been markedly different in many of these examples, and this has led to questions over the validity of the conceptualization of technological pedagogical reasoning and action [26]. As a result of the lack of broadly accepted additions to Shulman's [1] description of PR&A, the PR&A codes for this pilot study reflected Shulman's categories.

This process follows the recommendation from Shaffer, Collier and Ruis ([27], p. 12) that "the elements of the epistemic frame of a particular community of practice can be identified a priori from a theoretical or empirical analysis". Coding was initially undertaken independently by two authors on a sample of 9 Mathematics lesson plans, 5 Science lesson plans and 5 Technology lesson plans. This process produced inter-rater reliability of Cohen's kappa, $\kappa=0.78$. The remainder of the lesson plans were then coded by one of the authors for teacher knowledge and the by another author for aspects of PR&A.

5   Analysis

Following the coding of lesson plans in NVivo, we examined the relationship between forms of knowledge and processes of PR&A through Epistemic Network Analysis (ENA) [28]. ENA is a graph-based analysis technique used to examine qualitatively coded datasets for the association between different *codes*. For each *unit of analysis*, which are in the present study individual lesson plans, a network of relationships between codes was created, by examining the co-occurrence of different codes within chunks of text called *stanzas*. In the study, each lesson was considered a stanza, and a pair of codes was connected if they both appear within the same lesson. ENA is typically used for analysis of qualitatively coded datasets, such as coded conversation transcripts, and within education has been used to examine students' critical thinking [29], collaborative learning [30], knowledge transferability [31], and mentoring [32]. Unlike other graph-based analysis methods which focus on analysing large networks of relatively simple relationships, a unique characteristic of ENA is that it focuses instead on analysing smaller networks with a rich set of interactions among nodes [31].

In more technical terms, ENA works by constructing a code co-occurrence matrix (N x N triangular matrix, where N is the number of codes) for each unit of analysis, by examining co-occurrences of codes across stanzas within each of the units. The co-occurrence matrices are then converted into high-dimensional vectors, where each element represents the number of co-occurrences for a specific pair of codes.

This high-dimensional space is called *analytic space,* and each unit of analysis is represented as a single point in this space. Since analytic space cannot be visualised directly, a two-dimensional representation of the analytic space is created through dimensionality reduction algorithm called singular value decomposition (SVD), which is then used to create *projection space* and plot individual analysis units. Besides plotting all analysis units together, ENA also produces an ENA graph for each unit of analysis together, and for each unit of analysis individually. On such graphs, the size of codes represents their frequency and the strengths on edge between two codes the strength of their relationship.

As indicated above, in our analysis, we used individual lesson plans as units of analysis and lessons within those plans as individual stanzas. Then, we examined the ENA graphs for each lesson plan separately, so that we could compare the differences among them with regards to code relationships. The implementation of ENA was done using the rENA package for the R programming language [33] and all pre-processing of lesson plan data to a format suitable for ENA was done using the Python programming language.

6   Results

The overall ENA network for co-occurrences between codes from knowledge forms and PR&A categories is shown in Fig. 1. The strongest connections were between pedagogical content knowledge, course content, and instruction codes. However, it was

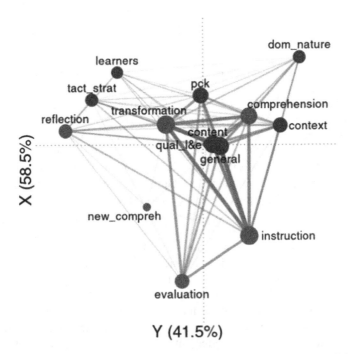

Fig. 1. ENA co-occurrence network between knowledge forms and PR&A for all lesson plans.

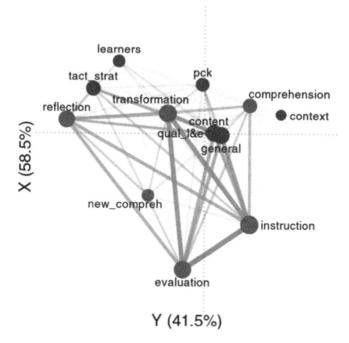

Fig. 2. ENA co-occurrence network between knowledge forms and PR&A in mathematics lesson plans.

even more critical to examine the differences in ENA network graphs for lesson plans of different course types. The analysis revealed substantially different connections between knowledge forms and PR&A stages. In Figs. 2, 3 and 4, nodes representing knowledge forms are coloured purple, nodes representing PR&A stages are coloured green. Figure 2 shows the connections between knowledge forms and PR&A stages for the mathematics lesson plans, while Fig. 3 represents connections evident in the science lesson plans and Fig. 4 represents connections evident in the IT lesson plans.

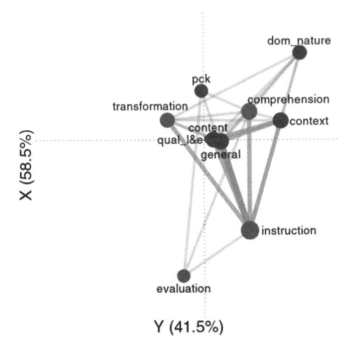

Fig. 3. ENA co-occurrence network between knowledge forms and PR&A in science lesson plans.

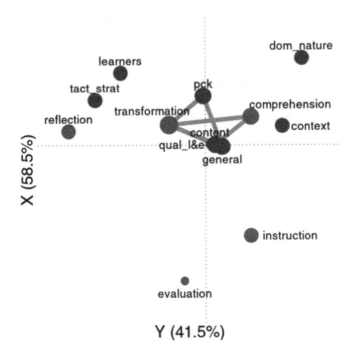

Fig. 4. ENA co-occurrence network between knowledge forms and PR&A in IT lesson plans.

7 Discussion

The ENA representations of co-occurrences of knowledge forms and PR&A stages for Mathematics, Science and IT teachers reveals marked differences in co-occurrences based on subject domain. Figure 1, for example, shows relatively frequent co-occurrences between a number of PR&A stages (particularly transformation, instruction and evaluation) evident in the lesson plans supplied by the Mathematics teachers. Similar co-occurrence frequencies are evident when comparing connections between transformation and instruction with connections between content knowledge and general pedagogical knowledge in the lesson plans supplied by the Science teachers.

However, there are also marked differences in co-occurrences in these two representations; for example, it appears as though the context in which the Science teachers worked was considered more frequently in relation to other forms of knowledge and PR&A stages in comparison to the Mathematics teachers. It is also noteworthy that some forms of knowledge and PR&A stages are absent from one ENA representation while showing co-occurrences in the other. For example, reflection regularly co-occurs with evaluation, transformation and content knowledge in the lesson plans supplied by the Mathematics teachers and yet is not evident in the lesson plans supplied by the Science teachers whereas the nature of the domain regularly co-occurs with four other knowledge forms and stages of PR&A but is absent from the Mathematics teachers' lesson plans. Most strikingly, the lesson plans from the IT teachers showed comparatively few co-occurrences between knowledge forms and stages of PR&A yet had the most codes represented of all three domains.

Despite the small sample size of teachers involved in this project (n = 6), the initial results provide what we believe is the first quantitative examination of the co-occurrence of teachers' knowledge forms and PR&A stages. The initial ENA representations provided in this paper provide new insights into teacher knowledge and decision making that challenge the homogenous nature of these two frameworks presented in Shulman's initial work. The language that Shulman [1] used to describe his knowledge base for teaching was largely singular ("*a* codified or codifiable aggregation of knowledge" (p. 4), "*the* knowledge base" (p. 4), "*an* elaborate knowledge base for teaching" (p. 7) [emphases added]). While Shulman [1] discussed contextual knowledge, the ways in which his knowledge base have been used have, in part, been mostly devoid of contextual considerations suggesting that all effective teachers drew upon all six forms of knowledge irrespective of factors such as discipline taught and age of students.

Shulman's [1] description of stages of PR&A, while helpful in regard to identifying different components of teachers' decision-making processes, did not provide much guidance for researchers or practitioners as these stages "are not meant to represent a set of fixed stages, phases, or steps. Many of the processes can occur in a different order. Some may not occur at all during some acts of teaching. Some may be truncated, others elaborated" (p. 19). The ENA representations presented in this paper provide empirically-based insights into the co-occurrences of the planned PR&A processes of these teachers. While not suggesting that there is a definitive order for these, it is encouraging to see such representations as the process of coding and representing these co-occurrences promises greater insights than have been previously possible.

8   Conclusion

Public scrutiny and debates about the professionalism of teacher knowledge and decision-making have long been a feature of public and academic discourse. Shulman [1] highlighted more than 30 years ago the profession suffered from an "incomplete and trivial definition of teaching … [and] a proper understanding of the knowledge base of teaching, the sources of that knowledge, and the complexities of the pedagogical process" (p. 20) were required to develop expert educators. Despite the work exploring specific aspects of Shulman's knowledge base (in particular, the extensive corpus of work examining PCK and more recently TPACK), developing a more comprehensive picture of the relationships between these knowledge forms and how teachers' use this knowledge to make decisions has remained elusive.

This paper reports on an initial and encouraging attempt to discover the co-occurrences of forms of teacher knowledge and stages of PR&A in STEM teacher lesson plans and reveals levels of complexity that have been intimated but, until now, not supported with empirical evidence. The use of ENA has been particularly beneficial in revealing the differences in co-occurrences of different forms of teacher knowledge and decision making in the lesson plans of teachers from different subject domains. While these initial results provide what we believe is the first quantitative examination of the co-occurrence of teachers' knowledge forms and PR&A stages, further research is required with more teachers working in different contexts before more definitive conclusions can be reached.

References

1. Shulman, L.S.: Knowledge and teaching: foundations of the new reform. Harvard Educ. Rev. **57**(1), 1–22 (1987)
2. Guerriero, S.: Teachers' pedagogical knowledge: what it is and how it functions. Educ. Res. Innov., 99–118 (2017)
3. Ogren, C.A.: Rethinking the "nontraditional" student from a historical perspective: state normal schools in the late nineteenth and early twentieth centuries. J. High. Educ. **74**(6), 640–664 (2003)
4. National Commission on Excellence in Education: A Nation at Risk. Washington D.C. (1983)
5. Tom, A.R., Valli, L.: Professional knowledge for teachers. In: Handbook of Research on Teacher Education, pp. 373–392 (1990)
6. Grimmett, P., MacKinnon, A.: Craft knowledge and the education of teachers. In: Grant, G. (ed.) Review of Research in Education. American Education Research Association, Washington DC (1992)
7. Nelson, B.: Teachers' special knowledge. Educ. Res. **21**(9), 32–33 (1992). https://doi.org/10.2307/1177020
8. Clandinin, D.J., Connelly, F.M.: Teachers' personal knowledge: what counts as 'personal' in studies of the personal. J. Curriculum Stud. **19**(6), 487–500 (1987)
9. Shulman, L.S.: Paradigms and research programs in the study of teaching. In: Wittrock, M. (ed.) Handbook of Research on Teaching, 3rd edn, pp. 3–36. Macmillan, New York (1986)

10. Shulman, L.S.: Those who understand: knowledge growth in teaching. Educ. Res. **15**(2), 4–14 (1986)
11. Shulman, L.S.: The wisdom of practice. In: Berliner, D. Rosenshine, B. (eds.) Talks to Teachers: A Festschrift for N.L. Gage, pp. 369–386. Random House, New York (1987)
12. Shulman, L.S.: Toward a pedagogy of cases. In: Case Methods in Teacher Education, pp. 1–30 (1992)
13. Ocak, C., Baran, E.: Observing the indicators of technological pedagogical content knowledge in science classrooms: video-based research. J. Res. Technol. Educ., 1–20 (2019)
14. Angeli, C., Valanides, N.: Preservice elementary teachers as information and communication technology designers: an instructional systems design model based on an expanded view of pedagogical content knowledge. J. Comput. Assist. Learn. **21**(4), 292–302 (2005). https://doi.org/10.1111/j.1365-2729.2005.00135.x
15. Koehler, M.J., Mishra, P., Yahya, K.: Content, pedagogy, and technology: testing a model of technology integration. Paper presented at the Annual Meeting of the American Educational Research Association, San Diego, CA, April 2004. http://www.matt-koehler.com/publications/Koehler_et_al_AERA_2004.pdf
16. Pierson, M.E.: Technology integration practice as a function of pedagogical expertise. J. Res. Comput. Educ. **33**(4), 413–430 (2001). https://doi.org/10.1080/08886504.2001.10782325
17. Harris, J., Phillips, M., Koehler, M., Rosenberg, J.: TPCK/TPACK research and development: past, present and future directions. Australas. J. Educ. Technol. **33**(3), i–viii (2017). https://doi.org/10.14742/ajet.3907
18. Fenstermacher, G.D.: Chapter 1: The knower and the known: the nature of knowledge in research on teaching. Rev. Res. Educ. **20**(1), 3–56 (1994)
19. Sachs, J.: Teacher professionalism: why are we still talking about it? Teach. Teach. **22**(4), 413–425 (2016)
20. Loughran, J., Mulhall, P., Berry, A.: In search of pedagogical content knowledge in science: developing ways of articulating and documenting professional practice. J. Res. Sci. Teach. **41**(4), 370–391 (2004)
21. Phillips, M.: Teachers' TPACK enactment in a Community of Practice, Monash University, Melbourne. eThesis (2014). http://arrow.monash.edu.au/1959.1/981787
22. Loughran, J., Keast, S., Cooper, R.: Pedagogical reasoning in teacher education. In: Loughran, J., Hamilton, M.L. (eds.) International Handbook of Teacher Education, pp. 387–421. Springer, Singapore (2016). https://doi.org/10.1007/978-981-10-0366-0_10
23. Smart, V.L.: Technological pedagogical reasoning: the development of teachers' pedagogical reasoning with technology over multiple career stages, Doctoral thesis, Griffith University, Queensland, Australia (2016). https://www120.secure.griffith.edu.au/rch/items/b658444f-8e00-4c61-9b95-a3b19c62d545/1/
24. Starkey, L.: Teachers' pedagogical reasoning and action in the digital age. Teach. Teach. Theor. Pract. **16**(2), 233–244 (2010). https://doi.org/10.1080/13540600903478433
25. Mishra, P., Koehler, M.J.: Technological pedagogical content knowledge: a framework for teacher knowledge. Teach. Coll. Rec. **108**(6), 1017–1054 (2006). https://doi.org/10.1111/j.1467-9620.2006.00684.x
26. Harris, J., Phillips, M.: If there's TPACK, is there technological pedagogical reasoning and action? In: Society for Information Technology & Teacher Education International Conference, pp. 2051–2061. Association for the Advancement of Computing in Education (AACE), March 2018
27. Shaffer, D.W., Collier, W., Ruis, A.R.: A tutorial on epistemic network analysis: analyzing the structure of connections in cognitive, social, and interaction data. J. Learn. Anal. **3**(3), 9–45 (2016)

28. Shaffer, D.W., et al.: Epistemic network analysis: a prototype for 21st-century assessment of learning. Int. J. Learn. Media **1**(2), 33–53 (2009). https://doi.org/10.1162/ijlm.2009.0013
29. Ferreira, R., Kovanović, V., Gašević, D., Rolim, V.: Towards combined network and text analytics of student discourse in online discussions. In: Penstein Rosé, C., et al. (eds.) Artificial Intelligence in Education, pp. 111–126 (2018). https://doi.org/10.1007/978-3-319-93843-1_9
30. Gašević, D., Joksimović, S., Eagan, B.R., Shaffer, D.W.: SENS: network analytics to combine social and cognitive perspectives of collaborative learning. Comput. Hum. Behav. **92**, 562–577 (2019). https://doi.org/10.1016/j.chb.2018.07.003
31. Shaffer, D.W.: Epistemic frames and islands of expertise: learning from infusion experiences. In: Proceedings of the 6th International Conference on Learning Sciences, pp. 473–480 (2004). http://dl.acm.org/citation.cfm?id=1149126.1149184
32. Nash, P., Shaffer, D.W.: Mentor modeling: the internalization of modeled professional thinking in an epistemic game. J. Comput. Assist. Learn. **27**(2), 173–189 (2011). https://doi.org/10.1111/j.1365-2729.2010.00385.x
33. Marquart, C.L., et al.: rENA: epistemic network analysis (Version 0.1.6.1) (2019). https://CRAN.R-project.org/package=rENA

Let's Listen to the Data: Sonification for Learning Analytics

Eric Sanchez[1(✉)] and Théophile Sanchez[2(✉)]

[1] University of Fribourg, Fribourg, Switzerland
eric.sanchez@unifr.ch
[2] Paris-Sud University, Orsay, France
theophile.sanchez@u-psud.fr

Abstract. This paper falls in the field of playing analytics. It deals with an empirical work dedicated to explore the potential of data sonification (i.e. the conversion of data into sound that reflects their objective properties or relations). Data sonification is proposed as an alternative to data visualization. We applied data sonification for the analysis of gameplays and players' strategies during a session dedicated to game-based learning. The data of our study (digital traces) was collected from 200 pre-service teachers who played Tamagocours, an online collaborative multiplayer game dedicated to learn the rules (i.e. copyright) that comply with the policies for the use of digital resources in an educational context. For one typical individual (parangon) for each of the 5 categories of players, the collected digital traces were converted into an audio format so that the actions that they performed become listenable. A specific software, SOnification of DAta for Learning Analytics (SODA4LA), was developed for this purpose. The first results show that different features of the data can be recognized from data listening. These results also enable for the identification of different parameters that should be taken into account for the sonification of diachronic data. We consider that this study open new perspectives for playing analytics. Thus we advocate for new research aiming at exploring the potential of data sonification for the analysis of complex and diachronic datasets in the field of educational sciences.

Keywords: Sonification · Playing analytics · Data visualisation · Game-based learning · Learning Analytics

1 Introduction

Rhythm and music are used as mnemonic means for teaching young students series of items such as days of the week or alphabet. Indeed, representing data through sounds offers the opportunity to benefit from human capacity to memorize or monitor complex temporal audio data. As a result, sonification of data has been recognized as an alternative to data visualization when visualization techniques are insufficient for comprehending certain features in the data or for the analysis of complex datasets. Sonification has already been applied to diverse scientific fields. In this paper we claim that sonification can be applied to playing analytics [11] i.e. the record and analysis of players' digital traces during a game session. We present an empirical work dedicated

© Springer Nature Switzerland AG 2019
B. Eagan et al. (Eds.): ICQE 2019, CCIS 1112, pp. 189–198, 2019.
https://doi.org/10.1007/978-3-030-33232-7_16

to explore the potential of data sonification for the analysis of players' behaviors. Indeed, one of the main challenges that data analysts face when they want to understand how a game is used consists in identifying the different gameplays performed by players. For this study, 200 pre-service teachers played Tamagocours, a collaborative multiplayer online game dedicated to learn the rules (i.e. copyright) that comply with the policies for the use of digital resources in an educational context. We first present a concise overview of the core concepts of sonification. We also describe the game, the data collected, the data processing and the different categories of players identified through a statistical analysis of the data. Third, we describe the results of the sonification of the data and the lessons learned from the work that we carried out with the data collected from the paragons (i.e. the typical individuals of the 5 different categories of players that identified with the statistical analysis of the data).

2   Traditional Approach for Playing Analytics

Data analysis tools such as KTBS4LA [4] Knime [2], Orange [6], WEKA [9] or R-Studio [15] can be used to conduct analyses on digital data, whether from an educational context or not. These analyses are generally statistical analyses, combined using graphical interfaces or scripts in dedicated languages enabling for data visualization. However, since the temporal aspect of data is essential to analyze the dynamics of knowledge in the context of learning, the time aspect of the data analyzed is not specifically taken into account by these tools. To take this temporal aspect into account, the work carried out by the Learning Analytics community focuses on timestamped traces representing learners' activity with the digital environment and theses traces enable for the visualization of learners' activity played with chronograms (e.g. [13]). Specific patterns might be identified in the traces collected from learners' activity (e.g. [13]). These patterns are key indicators of students' strategies. They are important pieces of information for the analysis of the learning process. Tools dedicated to identify these specific patterns have already been developed (e.g. [14]). However, they are not adapted to real time analysis and they do not make difference between patterns that have or have not information value. Thus we think that an alternative to data visualization is needed.

3   Sonification as an Alternative to Visualization

Sonification is defined as a systematic and reproducible transformation technique that can be used with different input data to produce sound that reflects objective properties or relations in the data [8]. Thus, sonification refers to audio-based data processing: data relations are transformed into perceived relations in an acoustic signal for the purposes of facilitating communication or interpretation [10]. Sonification is a recent research field but with already many ancient applications and sonification has been recognized to be a relevant alternative to data visualization. For examples, in a work published in 1961, Speeth demonstrated that auditory presentations of seismic data enable for the discrimination between earthquakes and bomblasts [14]. Different applications have been described. Kramer and al. [10] report successful sonifications such as the Geiger-counter, the discovery of a problem with Voyager 2 spacecraft due to high-speed collisions with electromagnetically charged particles. They also report sensory substitutions for visually impaired users of graphical information. More recently, sonification has been used as a mean for the analysis of complex datasets such as trading-data [16] and volcanoes activity [1].

Two main arguments are putted forward for the sonification of data: the limits of users' abilities to interpret visual information and the need to comprehend complex and big data. Indeed, research into auditory perception emphasizes the capacity of the auditory system. Different studies demonstrate that human beings are particularly sensitive to temporal characteristics or changes in sounds over time [7]. Complex and dynamic auditory patterns are generally well perceived and human auditory system organizes sound into perceptually meaningful elements [3]. Human hearing is also well designed to discriminate between periodic and aperiodic events and it enables for the detection of small changes in the frequency of continuous signals. Kramer's work also emphasizes that human hearing enables for discerning relationships or trends in data streams (Kramer 1994). Thus, sonification appears to be a powerful approach to study complex datasets.

For this preliminary study, we collected data from an empirical work dedicated to explore players's strategies during an online game-based learning session for pre-service teachers. We found that this data has the different characteristics listed above, such as diachronic and periodics events. The complexity and the volume of the information to proceed and analyze also lead us to consider that the data collected was good candidates for sonification.

4 A Game-Based Learning Case Study

4.1 Tamagocours, a Multiplayer Online Game

Tamagocours is a multiplayer online game dedicated to teach pre-service teachers the legal rules that comply with the policies for the use of digital resources within an educational context [12]. Tamagocours is a tamagochi-like game. It is based on a metaphor; a character (called Tamagocours) which needs to be fed with digital educational resources. A team is composed of 2 to 4 players randomly chosen by the system. Each player has to choose a can (which represents a given educational resource) on a shelf, according to its characteristics (eg. date of publication) and the format under which this resource will be used (eg. collective projection or online). The selected resources are stored in a fridge before the player feeds the character (the Tamagocours) with this resource. The players have access to the legislation about the copyright for the use of educational resources at any time. The players have also the possibility to chat with their teammates and to discuss about the compliance of one or other resource before feeding the Tamagocours. If a given resource complies with the copyright policies, it enables the character to stay healthy (the Tamagocours becomes green) and the player earns points. Otherwise, the Tamagocours gets sick (red colour) and dies if fed with too many inappropriate resources. The team can replay each level they have lost indefinitely until they reach the following level. The players have also access to the description of the rules of the game and the challenge to be addressed. A session encompasses 5 levels and lasts approximately one hour.

4.2   An Empirical Study

We carried out a study aiming at drawing a behavioral and epistemic models of the students (i.e. describing the strategies of the players/learners and evaluating to what extent they learn) [12]. The research methodology is Design-Based [5, 13]. It combines design and analysis within an iterative process carried out in ecological settings. This work implied various stakeholders: researchers, software developers, designers, students, tutors, pedagogical engineers, administrative representatives and a legal expert.

Our methodology is also based on recording and analyzing the digital traces of interaction produced by 200 pre-service teachers who played during a 60 min online session. For the analysis of the digital traces we used a statistical procedure based on an orthogonal transformation to convert a set of correlated variables into a set of linearly uncorrelated variables called principal components. This principal component analysis (PCA) enable for drawing 3 axis with significant eigenvalues. PCA was followed by a clustering method (hierarchical cluster analysis) to identify different classes of players depending on their strategies. As a result, the data collected enables to draw behavioral and epistemic models of the players/learners and to distinguish different gameplays. We also developed SOnification of DAta for Learning Analytics (SODA4LA), a software dedicated to the sonification of the data collected for individual typical of the different classes.

4.3   Five Categories of Players/Gameplays

One objective of our study consists in describing the different gameplays performed by the players. This has been done through the recording and the analysis of the actions performed by the students and the comparison of the typical individuals from each categories of players. We followed different steps for data processing and data analysis:

1. Digital traces are automatically collected when the students play. The different actions performed are recorded (id. of the player, click on a specific feature of the interface, chat with a teammate, success or failure...). Time stamped raw data is stored in a csv format file.
2. Raw data is coded according to a theoretical model of play [12]. In particular, the messages written by the students are tagged and allocated to different categories depending on their meaning in terms of how the students deal with the knowledge that they are expected to learn.
3. Aggregate data is produced from raw data. The data analysis process combines a principal component analysis followed by a clustering method. This method enables for identifying 5 categories of players according to the actions that they performed during the game session. Thus, these categories describe the different strategies followed by the players.

4. Sonification of coded data is automatically performed with SODA4LA, a software that we developed. A note is automatically allocated to each category of action performed by the player and time intervals between two actions are reduced. One audio file is produced for each player.

5. Data analysis is performed with MuseScore (c) a software enabling to listen to the files and to create a musical score. Data analysis is under the control of a theoretical model of play [12]. Hypothesis about students behavior are confirmed or disproved. In the following we will focus on the information conveyed by the audio files for different individuals that are recognized to be archetypal of 5 categories of players. These individuals are named parangons.

6. The conclusions drawn from data analysis are used by researchers to revise their theoretical model. Data is also used by computer scientists for the reengineering of the game. In the future, we plan to offer trainers and learners the opportunity to benefit from the analysis of this data as a mean for trainers to assess students or the opportunity to foster students' awareness (i.e. to reflect about the strategies that they perform during the game session).

In the following, we will examine the sonified data for the 5 categories of players:

- Students who belong to category 1 (p_1, 12,3%) are named *talkative*. They send many messages to theirs teammates during a play session. They are also active and perform a lot of actions.
- Category 2 (p_2, 20,2%) consists in students named *prudent*. They mainly feed the Tamagocours with paying attention to the characteristics of the resources that they use. Indeed, they consult the characteristics of the selected items before feeding the Tamagocours. They also often access to the legal documentation available in the game.
- Students from category 3 (p_3, 36%) are recognized to be *efficient*. They manage to have a good ratio of success in comparison to failures and they usually pay attention to the characteristics of the resources that they select before feeding the Tamagocours.
- Students from category 4 (p_4, 26,3%) are named *force feeders*. They frequently feed the character without paying attention to the characteristics of the resources that they use and the ratio success/failure is below average. It may seem at times as if the player is just using a trial and error approach
- Category 5 (p_5, 5,2%) consists in students named *experts*. They make few mistakes. They also pay attention to the characteristics of the resources and collaborate with their teammates. Indeed, they send many messages to express their opinion about the legal rules that have to be respected.

5   Listening to the Parangons

In this section we compare different methods for the data sonification and we discuss to what extend each method enable for the identification of specific patterns that are typical from the different identified categories. Thus, the different examples provided are parangons of the different categories of gameplays.

5.1   Coding the Actions with Sounds

Sonification consists of transforming data so that they will become listenable by a data analyst. For this exploratory study we decided to take into account only two dimensions: the type of action performed by a player and intervals between two action (the diachronic feature of the data).

Each type of action is replaced with a note (i.e. a sound with a specific frequency) from the C Dorian mode. The choices made are arbitrary and the following audio score indicates how the different actions performed by the players are matched with sounds. The different actions are the following (Fig. 1):

Fig. 1. Key for the coding of the data.

- "chat" (the player sends a message to his teammates): "C in octave 4"
- "showItemCupboard" (the player consults the characteristics of a given resource): "D in octave 4"
- "addToFridge" (the player stores a given resources in the fridge): "E-flat in octave 4"
- "feedTamago" (the player feeds the Tamagocours with a given resource): "F in octave 4"
- "help" (consulting information about the game): "G in octave 4"
- "helpLink" (consulting the legal library): "A in octave 4"
- "showItem" (the player checks the characteristic of a specific resource added by a teammate into the fridge): "B-flat in octave 4"
- "tuto" (the player consults a tutorial about how to play the game): "C in octave 5"
- "removeFromFridge": "D in octave 5"

Intervals between two actions have also to be coded. We tried different options: time between 2 notes is equal to the time between 2 actions (real_time), notes are played without taking into account the time between 2 actions (no_time) and time is reduced (compressed) with a log-transformation and rounded to eight possible pitches that are then associated to eight relative note values (duration) (1/4, 1/3, 1/2, 1, 2, 3, 4 and 5). The sonification of the data produces melodies with the same tune but with different tempos that illustrate the same gameplay. When the tempo is too low (real_time option), the musical dimension of the data tends to disappears and it becomes difficult to identify a 'music tune'. As a result, it seems to be important that the coding of the data enables for producing sounds that can be considered to be 'music'. We finally selected the third option for the sonification of data in order that the 'music' gives an idea of the tempo of the gameplay without too much silence between two notes.

This preliminary work shows that the main difficulty faced for the interpretation of the music in terms of gameplay (what is the player's strategy?) consists in giving meaning to the 'music' that is produced. For future work, an option might be to use sounds with more semantic content than notes. For example it is possible to replace a note with the sound made by the character when feed with an appropriate resource. Another option would be to offer data analysts the opportunity to choose how data is sonified. In this regard, an interface dedicated to select relevant parameters would be useful.

5.2 Identifying Similar Gameplays

The two examples below are audio files produced with the sonification of the data for 2 different parangons from the same category.

- p1_79171_compressed.wav
- p1_56107_compressed.wav
- p4_62127_compressed.wav

'Music' produced by players from the same category (p_1 talkative) has very similar features. By contrary they differ from the 'music' produced by a 'force feeder' (p4_62127). This observation tends to demonstrate that sonification enables to identify differences among different gameplays at a global level for analysis. It is also plausible that, with some experience and training, a data analyst becomes able to identify a specific gameplay with sonified data.

5.3 Identifying Specific Patterns

The five examples below are audio files produced with the sonification of the data produced by 5 different parangons.

- p1_79171_compressed.wav
- p2_61122_compressed.wav
- p3_63131_compressed.wav
- p4_62127_compressed.wav
- p5_4768_compressed.wav

Listening to the files produced by 5 different parangons confirms that the different gameplays are made 'visible' with music which varies among players. In addition, it is possible to identify some patterns that seem to be typical for the different parangons.

For example, a note (D) is played many times for the individual tagged p 1 (Fig. 2). This note corresponds to messages sent by the player.

Fig. 2. Sample of the musical score for p1 56107

Figure 3 shows a similar feature for p1 92206 (another talkative player).

Fig. 3. Sample of the musical score for p1 92206

There is also a clear difference between p2_61122 and p4_62127. For p2_61122, the music is constructed on a pattern of 3 notes whereas a pattern of 2 notes is dominant for player p4_62127. These patterns consist in the principal action that a player needs to perform (consulting the characteristics of a given resource in the shelf, putting this resource into the fridge and feeding the Tamagocours). The first action is missing for p4_62127, force feeders usually do not check the characteristics of a resource that the select before feeding the Tamagocours.

5.4   Identifying Gameplay Variations

Though that statistical analysis enables for allocating each player to a specific category, this analysis is based on the average number of variables (actions) for the whole session. Gameplay might change among time and it is important to identify when and how the player's behavior varies. The two following examples show that sonification enable for identifying such changes.

- p2_4984_compressed.wav
- p3_63131_compressed.wav

For player p2_4984, there is a clear difference of the melody at the end of the file. The melody becomes less monotonous. The data confirm that before 1:20 the majority of actions consist in checking the characteristics of the resources (ShowItemCupboard). After 1:20 he selects more resources (AddToFridge is more frequent) and feed more often the Tamaocours (FeedTamago). This change in the player behavior is well perceived with auditory information. A similar shift from one strategy to another is easily identified for player p3_63131. For this 'prudent' player, the shift occurs around 0:18 and correspond to an enrichment of the actions performed with more ChatAction and HelpLink.

6 Conclusion

It becomes apparent from this study that sonification enables for the identification of different features of our data: specific melodies linked with specific gameplay and variation of gameplay among time. Though that this preliminary work tends to demonstrate that sonification might be relevant for playing analytics, numerous challenges remain. First, there is a need for a specific software dedicated to match actions performed by players and sounds according to the need of the data-analysts. This tool should also provide with different options for the coding of the data. For example, the choice of different musical parameters (i.e. note pitch, value, tempo, timbre) should be under the control of data-analysts. In addition, it is plausible that sonified data are more easy to interpret (or at least to discriminate) if the produced audio file is close to music. In this regard, the musical competences of data-analysts might be important both for the coding of the data and for their interpretation. We need to imagine how to train researchers in charge of making meaning with sonified data. Another challenge lies in the fact that it is important to use sounds that have some semantic content. Within the game, some actions performed by the player such as success or failures for feeding the Tamagocours produce sounds that could be used for the sonification of data.

According to our knowledge, sonification has never been applied in educational research and playing analytics. However, it appears to be a powerful approach for studying complex systems such as learning setting. This empirical works aims to explore this idea. We think that new perspectives are now open. For example, it is extremely difficult to capture complex features such as peer-to-peer collaboration with technology enhanced learning systems. We hypothesize that data produced by different learners involved in collaborative learning will produce synchronized and harmonious musical phrases while data collected when learners act independently should produce cacophonic sounds.

New radical ideas are needed for playing analytics and learning analytics more broadly. We want to drastically reverse the traditional approach that is actually based on data visualization. Thus, we advocate for new researches aiming at exploring the potential of data sonification for the analysis of complex and diachronic datasets in the field of educational sciences. Indeed, we still need to build theories to guide decisions about how to sonify the data.

Supplementary Material

Scores, audio files, and midi files are available here: https://drive.google.com/open?id=1SLMg071XR7Mn14iIV7PdlkstHXGejRa4.

References

1. Avanzo, S., Barbera, R., De Mattia, F., La Rocca, G., Sorrentino, M., Vicinanza, D.: Data sonification of volcano seismograms and sound/timbre reconstruction of ancient musical instruments with grid infrastructures. Procedia Comput. Sci. 1(1), 397–406 (2010)
2. Berthold, M.R., et al.: KNIME-the konstanz information miner: version 2.0 and beyond. ACM SIGKDD Explor. Newsl. 11(1), 26–31 (2009)
3. Bregman, A.S.: Auditory Scene Analysis: The Perceptual Organization of Sound. The MIT Press, Cambridge (1990)
4. Casado, R., Guin, N., Champin, P.-A., Lefevre, M.: kTBS4LA: une plateforme d'analyse de traces fondée sur une modélisation sémantique des traces. In: Méthodologies et outils pour le recueil, l'analyse et la visualisation des traces d'interaction - ORPHEE-RDV, Font-Romeu, France, January 2017 (2017)
5. Design-Based Research Collective: Design-based research: an emerging paradigm for educational inquiry. Educ. Res. 32(1), 5–8 (2003)
6. Demšar, J., et al.: Orange: data mining toolbox in python. J. Mach. Learn. Res. 14(1), 2349–2353 (2013)
7. Handel, S.: Listening: An Introduction to the Perception of Auditory Events. The MIT Press, Cambridge (1993)
8. Hermann, T., Hunt, A., Neuhoff, J.G.: The Sonification Handbook. Logos Verlag, Berlin (2011)
9. Holmes, G., Donkin, A., Witten, I.H.: WEKA: a machine learning workbench, pp. 357–361 (1994). https://doi.org/10.1109/ANZIIS.1994.396988
10. Kramer, G., et al.: The sonification report: status of the field and research agenda. Report prepared for the national science foundation by members of the international community for auditory display. International Community for Auditory Display (ICAD), Santa Fe, NM (1999)
11. Sanchez, E., Mandran, N.: Exploring competition and collaboration behaviors in game-based learning with playing analytics. In: Lavoué, É., Drachsler, H., Verbert, K., Broisin, J., Pérez-Sanagustín, M. (eds.) EC-TEL 2017. LNCS, vol. 10474, pp. 467–472. Springer, Cham (2017). https://doi.org/10.1007/978-3-319-66610-5_44
12. Sanchez, E., Martinez-Emin, V., Mandran, N.: Jeu-game, jeu-play, vers une modélisation du jeu. Une étude empirique à partir des traces numériques d'interaction du jeu Tamagocours. Sciences et Technologies de l'Information et de la Communication pour l'Éducation et la Formation 22(1), 9–44 (2015)
13. Sanchez, E., Monod-Ansaldi, R., Vincent, C., Safadi-Katouzian, S.: A praxeological perspective for the design and implementation of a digital role-play game. Educ. Inf. Technol. 22(6), 2805–2824 (2017)
14. Speeth, S.D.: Seismometer sounds. J. Acoust. Soc. Am. 33(7), 909–916 (1961)
15. RStudio Team, et al.: Rstudio: Integrated Development for R. RStudio, Inc., Boston, 42:14 (2015). http://www.rstudio.com
16. Worrall, D.: Using sound to identify correlations in market data. In: Ystad, S., Aramaki, M., Kronland-Martinet, R., Jensen, K. (eds.) CMMR/ICAD-2009. LNCS, vol. 5954, pp. 202–218. Springer, Heidelberg (2010). https://doi.org/10.1007/978-3-642-12439-6_11

Examining the Impact of Virtual City Planning on High School Students' Identity Exploration

Mamta Shah[1], Aroutis Foster[2]($\boxtimes$) (iD), Hamideh Talafian[2] (iD),
and Amanda Barany[2] (iD)

[1] Elsevier Inc., Philadelphia, PA 19103, USA
[2] Drexel University, Philadelphia, PA 19104, USA
anf37@drexel.edu

Abstract. This paper is situated in an NSF CAREER project awarded to test and refine Projective Reflection (PR) as a theoretical and methodological framework for facilitating learning as identity exploration in virtual learning environments. PR structured the design, implementation, and refinement of *Virtual City Planning*, a play-based course that included identity exploration experiences mediated by a virtual learning environment (*Philadelphia Land Science*), and classroom experiences designed to augment the virtual learning experience. In this paper, Quantitative Ethnography techniques were applied to visualize and interpret changes at the group level (N = 20) for the first of three iterations of *Virtual City Planning*, as a result of exploring role-possible selves of an environmental scientist and urban planner. Changes were reflected in students' knowledge, interest and valuing, patterns of self-organization and self-control, and self-perceptions and self-definitions (KIVSSSS) in relation to the roles explored from the start of *Virtual City Planning* (starting self), during (exploring role-specific possible selves), and at the end of the play-based learning experience (new self).

Keywords: Identity exploration · Epistemic network analysis · Projective reflection · Virtual learning environments · Game-based learning

1 Virtual Learning Environments as Catalytic Contexts

Virtual learning environments such as digital games have designed potential to engage a learner's whole self (cognitive, affective, social, and motivational dimensions) in a dynamic environment, resulting in identity exploration - a catalytic transformation of game-players' knowledge of self in relation to the situated context [1, 2]. Some researchers have demonstrated how designed affordances of virtual learning environments can support personal identities and goals towards engagement in academic domains and professional careers [3, 4]. The domain of learning as identity exploration in virtual learning environments is still nascent, however, requiring new theories of change, evidence-based measurement, and design principles that can promote knowledge, identity processes, and career paths [5].

Few empirically tested theories currently exist to operationalize how learning relates to identity in gaming contexts. Shaffer's [6] Epistemic Frames theory supports

© Springer Nature Switzerland AG 2019
B. Eagan et al. (Eds.): ICQE 2019, CCIS 1112, pp. 199–210, 2019.
https://doi.org/10.1007/978-3-030-33232-7_17

the design of epistemic games as a process of learner enculturation in professional praxis ((I)dentity). Game scholars such as Chee [7] and Foster [3] stress intentionality, reflexivity, and a focus on the self ((i)dentity) as essential elements for examining identity exploration and change. Others have illustrated how guided interactions and peer perceptions shape trajectories of knowledge and interest development in certain academic domains and careers [8]. Though some examples exist [9, 10], theoretical gaps in the field persist, and few methodological approaches and methods have consequently emerged to (a) complement emerging theories and (b) guide assessment of learning as identity exploration in game-based interventions. Quantitative ethnography [11] offers one valuable method for exploring patterns of individual activity nested in situated discourse (i/Identity). Quantitative ethnographic techniques such as Epistemic Network Analysis (ENA) involve the quantification of qualitative data to generate visualizations that can represent the associations individuals establish across a network of constructs (i.e. identity exploration constructs). ENA has previously been used to characterize what players learn from gameplay in terms of knowledge, skills, values, and habits of mind [12], further illustrating the potential of this methodology for illustrating processes of identity exploration as they are enacted by learners using virtual learning environments in formal learning settings.

This paper addresses the identified research gaps and reports the results of a study that leveraged a robust theoretical and analytical framework to develop and assess the learning outcomes of a virtual learning environment and classroom curriculum that supported identity exploration and change. Projective Reflection, a theoretical and pedagogical model to conceptualize processes of identity exploration, structured the design and implementation of *Virtual City Planning (VCP)*, a play-based course that included identity exploration experiences mediated by a virtual learning environment (*Philadelphia Land Science*), and supportive classroom experiences. Given the affordances of the technique, Epistemic Network Analysis [11] was used to visualize and interpret changes in identity exploration trajectories for the student group as a result of exploring the role-possible selves of an environmental scientist and urban planner. The research question asked: *"What is the nature of high school students' identity exploration as a result of exploring the role-possible selves of an environmental scientist and urban planner in a play-based course?"*.

2   Theoretical Framework

Projective Reflection (PR) is a theory and methodology of learning that integrates a focus on content ((I)dentity anchored in a specific community of practice and enacted locally) and on the self ((i)dentity engaged in role-possible selves inspired by the community of practice reflecting an individual goal) in an integrated manner (i/Identity). PR defines learning as an intentional process of exploring role-possible selves in digital and non-digital play-based environments as a learner projects forward and reflects on who they are in relation to specific domains and careers (i.e. STEM professionals) [13]. Four theoretical constructs support exploration of identities through role-possible selves in PR to enable an integrated change in learners over time: (1) **Knowledge** (foundational, meta, humanistic) [14], (2) **Interests** (situated/perceptual, epistemic/personal))/**V**aluing (global, personal) [15], (3) patterns of **S**elf-organization/ **S**elf-control (co-regulation, socially-shared regulation, and self-regulation) [16], and (4) **S**elf-perceptions/**S**elf-definitions (self-concept, self-efficacy) [17] (**KIVSSSS**). Table 1 provides a more in-depth explanation of the constructs and sub-constructs as manifested by students in *Virtual City Planning*.

Table 1. Projective reflection construct definitions.

PR constructs	Definitions	Sample citations
Knowledge and game/technical literacy	Shifts in what a player knows about environmental science, urban planning, and urban planning systems from the beginning to the end of an intervention: • *Foundational knowledge*: awareness of complex and domain-specific content and processes that includes the ability to access information using digital technologies • *Meta-knowledge*: awareness of how to use foundational knowledge in relevant socially situated contexts • *Humanistic knowledge*: awareness of the self and one's situation in a broader social and global context	[14]
Interest and valuing	• Caring about environmental science and urban planning issues and viewing them as personally relevant or meaningful • Shifts in identification with environmental science • Viewing environmental science and urban planning as being relevant to the community or the world • Seeing the need for environmental science for self and for use beyond school contexts	[3, 15, 18]
Self-organization and self-control	Shifts in behavior, motivation, and cognition toward a goal: • *Self-regulated learning:* goal setting and goal-achievement conducted independently • *Co-regulated learning*: self-regulation processes supported by more knowledgeable real/virtual mentors • *Socially shared learning*: self-regulation is socially shared and defined in collaboration with peers	[16, 19, 20]
Self-perceptions and self-definitions	Shifts in how a participant sees himself/herself in relation to (environmental) science: • *Self-efficacy*: confidence in one's own ability to achieve goals and future roles • *Self-concept*: awareness of current aspects of self (i.e. skills, preferences, characteristics, abilities, etc.) • Specific roles one wants or expects to become in future	[21]

These constructs can be leveraged to scaffold and track changes across (a) the initial current self (Starting Self) that is established at the start of an intervention, (b) exploration of multiple role-possible selves (Exploring Possible Selves) measured repeatedly across an intervention, and (c) the New Self at the end of the intervention/experience [13] (See Fig. 1).

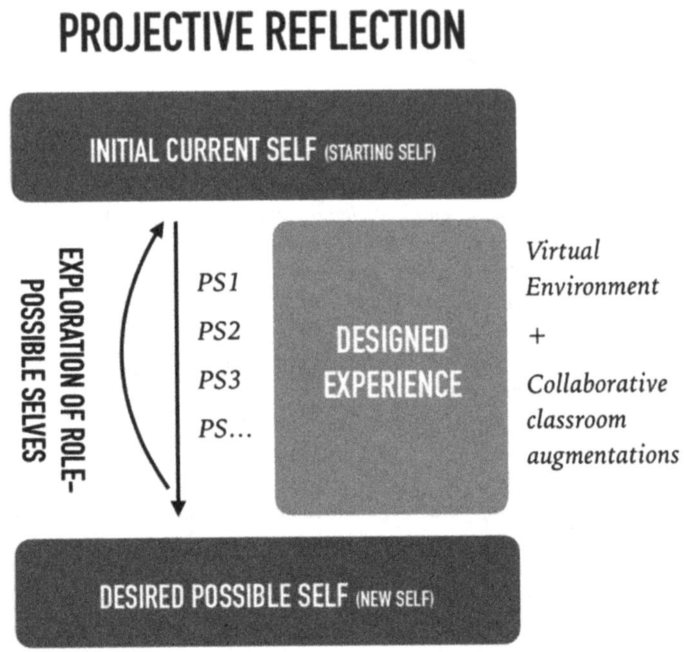

PROJECTIVE REFLECTION

INITIAL CURRENT SELF (STARTING SELF)

EXPLORATION OF ROLE-POSSIBLE SELVES

PS1
PS2
PS3
PS...

DESIGNED EXPERIENCE

Virtual Environment

+

Collaborative classroom augmentations

DESIRED POSSIBLE SELF (NEW SELF)

Fig. 1. The projective reflection framework for conceptualizing learning as identity change.

Throughout the virtual learning environment and supportive curriculum, learners are encouraged to engage in targeted and intentional reflection on aspects of self at repeated points throughout the situated designed experience (from Starting Self to New Self). For each student or cohort, the process identity exploration resulting in change is assessed chronologically based on how they explored a role-possible self as intentional changes in KIVSSSS - indicating the extent to which the identity exploration process was comprehensive or integrated.

3   Methods

This research was conducted as part of a 5-year (2014–2019) NSF CAREER project awarded to advance theory and research on promoting identity exploration and change in science using virtual learning environments through Projective Reflection [13]. Building on this broader agenda, *Virtual City Planning (VCP)*, a play-based course, was designed, developed, implemented, and refined using design-based research [22] to help freshman high school students explore and develop knowledge, interests and valuing, self-organization and self-control strategies, and self-perceptions and self-definitions related to urban planning and environmental science careers (identity exploration as defined by PR). This work reports findings from the first of three iterative *VCP* implementations (Session 1).

3.1 Data Collection and Procedures

VCP featured weekly use of both the virtual learning environment, *Philadelphia Land Science*, and supportive real-world augmentation in the classroom such as curricular activities that support reflection and discussion (see [23] for more information). *VCP* Session 1 was offered from September - November of 2016, across 9 weeks with 20 students. Data sources included in-game data (chats, written reflections, interactive map designs), pre-post surveys, classroom artifacts, and researcher observations (Table 2).

Table 2. Student demographics for *VCP* sessions 1–3.

VCP sessions	Sex	Race/Ethnicity	Total
VCP session 1	6 male 12 female 2 other/no response	4 Caucasian American 6 African American 5 Asian or Pacific-Islander 1 Hispanic or Latino/a 4 Multiple/other	20
VCP session 2	10 male 8 female 1 other/no response	8 Caucasian American 6 African American 0 Asian or Pacific-Islander 3 Hispanic or Latino/a 2 Multiple/other	19
VCP session 3	8 male 9 female 1 other/no response	4 Caucasian American 7 African American 0 Asian or Pacific-Islander 3 Hispanic or Latino/a 3 Multiple/other	18

In *VCP*, students roleplayed as interns in a fictitious urban planning firm that models how real-world professional settings are structured. Online and in-person mentors roleplaying as urban planners guided the group through the process of creating zoning proposals for Philadelphia. Starting Self data included student reflections and artifacts from weeks 1–2, in which students learned about their teams and the expected workflow, completed an intake interview (Likert and short-answer questions), and engaged in a focus group discussion. Exploring Possible Selves data included student reflections and artifacts from weeks 3–7, during which students researched the environmental and economic needs of the city and its stakeholders, and then collaboratively rezoned interactive models of the city to meet stakeholder needs (based on iterative virtual stakeholder feedback). New Self data included student reflections and artifacts from weeks 8–9, during which students worked towards finalizing a formal written proposal justifying their city zoning redesigns, and then completed an exit interview (Likert and short-answer questions).

3.2   Data Analysis

Student data was coded inductively and deductively to answer the research question. Quantitative Ethnography [11] was used to guide the data analysis procedures; researchers engaged in a deductive coding process for each student/case [24] using the qualitative data analysis software MAXQDA 2018. Lines of student data were coded as self-reflection on or demonstration of the four PR constructs and their sub-constructs, with agreement reached by two graduate-level coders. The qualitatively coded data was then quantified. Each line was coded for the occurrence (1) or non-occurrence (0) of the four constructs of PR that defined identity exploration (e.g. knowledge) and sub-constructs (e.g. foundational knowledge) to prepare the data for Epistemic Network Analysis (ENA).

We applied ENA [11] to look for patterns of identity change enacted across the start, during, and at the end of their participation in *VCP* using ENA1.5.2 Web Tool [25]. In this study, the association structure between the changes in students' KIVSSSS was modeled based on their co-occurrence in the specific implementation of *VCP* over time, by the three data points – starting self (SS), exploring role-possible selves (EPS), and new self (NS). As such, ENA offered a unique way to recognize the patterns of identity exploration at both the group and individual levels for engaging in Projective Reflection as a result of the play-based course. We referred back to interactions and activities coded in the data to close the interpretive loop and thus fully understand the phenomenon mirrored in the model for each student and the group at large. This last step was relevant for both individual and group findings as instrumental case studies [26], enabling the researchers to highlight the dominant issue for this paper - that is, nature of participants' identity exploration.

4   Results

To ascertain whether change across the three time periods was statistically significant, two sample t tests assuming unequal variance were conducted between the X and Y axes of Starting Self (SS), Exploring Possible Selves (EPS), and New Self (NS) data. Along each axis, a two-sample t test assuming unequal variance showed that these were not statistically significantly different at the alpha = 0.05 level (See Table 3).

While tests of statistically significant change in student data across the three time periods provides useful information regarding the degree of change enacted by the group, a lack of statistical significance does not necessarily indicate that students did not change over time. Processes of identity exploration as defined by Projective Reflection are most valuable when students can enact them in an integrated fashion; that is, when students can regularly connect Knowledge gains, emerging personal Interests and Values, the enactment of Self-organization and Self-control strategies, and specific Self-perceptions and Self-definitions in a domain (KIVSSSS). As such, an examination of the epistemic networks for each time period can help to illustrate what

Table 3. Paired sample t-test statistics.

Pair axes	Pairs	n	Mean	SD	t	df	Sig. (2-tailed)	Cohen's d
Pair 1 X-axis	SS	6	−0.27	0.69	0.95	7.24	0.37	0.47
	EPS	23	0.02	0.62				
Pair 1 Y-axis	SS	6	0.11	0.43	1.08	6.38	0.32	0.61
	EPS	23	0.09	0.30				
Pair 2 X-axis	EPS	23	0.02	0.62	0.60	10.96	0.56	0.24
	NS	7	0.17	0.55				
Pair 2 Y-axis	EPS	23	0.09	0.30	−1.17	6.99	0.28	0.71
	NS	7	−0.18	0.59				
Pair 3 X-axis	SS	6	−0.27	0.69	1.26	9.62	0.24	0.71
	NS	7	0.17	0.55				
Pair 3 Y-axis	SS	6	−0.11	0.43	−0.25	10.75	0.81	0.14
	NS	7	−0.18	0.59				

aspects of identity exploration (as defined by the four PR constructs) were discussed or enacted most by students at the beginning, middle, and end of an intervention.

Visualizations of data across the 3 time periods revealed differences in the strength of relationships between PR constructs (KIVSSSS), revealing shifting integration of identity exploration in participants over time that will be explored qualitatively below (See Figs. 2, 3 and 4).

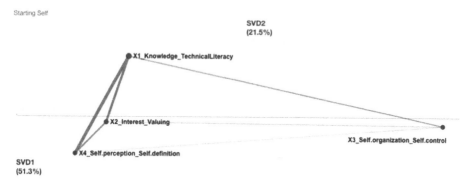

Fig. 2. Epistemic network for weeks 1–2 (Starting Self) for *VCP* Session 1 students.

Students in *VCP* Session 1 began the course with general foundational knowledge of environmental science and urban planning terms as a result of limited prior experience in these domains (e.g. Andrea [pseudonym] defined environmental science as "science that involves nature and the world that evolves around us"). Students reported frequent participation in online activities (e.g. typically engaging online or in gameplay for several hours a week). Students reported limited meta knowledge of the urban planning processes, but were able to recognize urban planning as relevant to

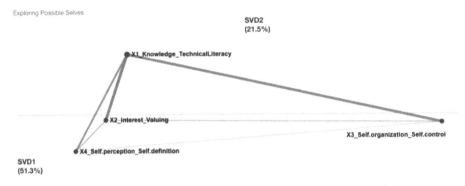

Fig. 3. Epistemic network for weeks 3–7 (Exploring Possible Selves) for *VCP* Session 1 students.

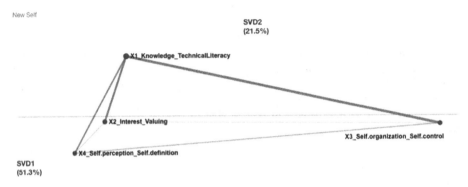

Fig. 4. Epistemic network for weeks 8–9 (New Self) for *VCP* Session 1 students.

themselves and sometimes to their community or city (e.g. Ciara wrote that building housing was important because "there are a lot of people in poverty and I think they should have the opportunity to get out"). While most students were able to recognize how recent activities could help them achieve desired future selves, the group began with varied levels of confidence in using science or engaging in urban planning activities (i.e. 5-point Likert scale responses ranging from Strongly Agree to Strongly Disagree when asked about confidence in their abilities). They demonstrated diverse interests and the desired future selves they wanted to strive towards (e.g. doctor, basketball player). Although most students had clear career goals and roles they hope to work for in future, only some saw connections with environmental science and their futures (e.g. Kimberly wrote, "I plan to use my RDA [urban planning company] skills from here and apply them to my work the next three years at [my school]."). Figure 2 above visualizes the groups' focus on connecting what they knew about environmental science and urban planning to what they wanted or expected to be, and to how they perceived themselves.

As *VCP* Session 1 progressed, students engaged in urban planning roles and enacted strategies for successful design with the help of in-game resources, peers, and in-person mentors. Exploration of urban planning was scaffolded through opportunities for design, discussion, and reflective questioning. During this period, students demonstrated an increase in detail and specificity in declarative knowledge of terms (e.g. Cynthia described her map design changes and how they affected environmental variables "improve on the low-income housing and also to improve on creating more wetland but not so much that it affected the run off"). This was reflected in the design decisions they made when rezoning the city the justifications they offered based on knowledge of issues important to their stakeholders. Some students thought of ways city design could meet their current needs and ways they could make changes (e.g. John wanted more housing and fewer commercial areas because "I personally don't enjoy crowded spaces. I feel like if there were less people in center city there would be less chaos"). Some students found the experience confusing or frustrating, while others deepened their interest in the experience. The extent to which students began caring for their roles was also an indication of the self-regulatory, co-regulatory, and socially shared regulatory actions in which they engaged. Some enacted strategies for success and engaged independently or with peers to achieve design goals, while others struggled with self-regulation and needed instructor support. Figure 3 visualizes how the group connected environmental science and urban planning knowledge to conceptualizations of their roles and the issues they found personally and globally relevant.

Session 1 of *VCP* relied heavily on the use of Philadelphia Land Science to support students' identity exploration. The intervention was impacted by internet connectivity issues, which interrupted the flow of events for some students. In weeks 8–9, students reported more detailed knowledge of urban planning terms and environmental and economic issues. They were able to describe urban planning processes and explain their importance to the city. They articulated specific interests related to personal values and relevance. Specifically, they recognized how city design can affect them and their community (e.g. Zola wrote, "urban development has greatly affected me. I live in an older part of the city where there are many large homes and open spaces. As the city has developed. Deeper city things have spread to my home area, and there are socially more chain shops and hotels"). Some expressed increased confidence in their roles and described ways they would continue to learn about issues or impact their communities (e.g. Rahim wrote "I learned that there is actually a lot of problems here in my very own city that I never even knew about. I actually am confident now because I learned everything about the map in this class"). Most students maintained their interest in the same desired future role/career they had shared at the start of *VCP*. However, most students concluded the experience affirming the relevance of urban planning to their lives and sometimes careers (e.g. Ali concluded the experience with a new desire to become a scientist and stated that "I want to develop good a relationship with science"). Figure 4 visualizes the groups' continued focus on connecting what they knew about environmental science and urban planning to their perception of the explored identity and the issues they found personally and globally relevant.

5 Conclusion, Discussions and Implications

Findings from Session 1 suggest that while students may not have enacted statistically significant changes in their processes of identity exploration over time (Starting Self to New Self), they were increasingly able to draw connections between the four Projective Reflection constructs. Students increasingly linked new knowledge of environmental science and urban planning to personal interests and valuing of relevant issues, perceptions of self and their desired future roles, and ultimately their identification and enactment of self-organization and self-control strategies. Session 1 of *VCP*, which leveraged affordances of a virtual learning environment (*Philadelphia Land Science*) and real-world curricular augmentations (i.e. roleplay and discussion) for promoting identity exploration, allowed for a more versatile and flexible learning experience that could meet personalized values, interests, and expertise to promote identity exploration and change in increasingly comprehensive and integrated ways.

Qualitative themes supported by ENA visualizations indicate that additional supports may be needed in play-based courses to support balanced student experiences (i.e. virtual and real-world activities) that can help players tailor interests and cope with difficulties or unforeseen issues. For students with limited digital literacy, stronger early scaffolding and mentorship may prove useful, while students who focus on some elements of PR over others may require other tailored supports. The development and enactment of the virtual experiences such as *Virtual City Planning* offers useful insights into unanswered questions in the field of identity exploration, such as how learning experiences may be optimally designed using a robust theoretical framework to support targeted identity exploration and change processes [27], and how trajectories of identity exploration and change pay progress when supported by virtual learning environments [13]. As practical and theoretical understandings of identity exploration trajectories emerge through this research, the capacity of learning practitioners to design early targeted supports for learners based on their starting-self characteristics increases.

In an era of research in which increasingly large and nuanced datasets are available from complex learning environments, robust theoretical frameworks are necessary to structure what counts as meaningful [28]. This paper provides an example of how Projective Reflection as a theoretical framework informed assessments of student learning using quantitative ethnographic techniques. Future studies of identity exploration in play-based learning environments will expand on the use of epistemic network analysis by leveraging a combination of ENA and social network analysis, two complementary methods that can shed more light on both individual/cognitive and social/collaborative learning (I/identity) [29].

References

1. Oyserman, D: Identity-based motivation. In: Pecher, D., Zeelenberg, R. (eds.) Emerging Trends in the Social and Behavioral Sciences: An Interdisciplinary, Searchable, and Linkable Resource, pp. 1–11 (2015)
2. Ryu, D., Jeong, J.: Two faces of today's learners: multiple identity formation. J. Educ. Comput. Res. **57**(6), 1351–1375 (2018)

3. Foster, A.: Games and motivation to learn science: personal identity, applicability, relevance and meaningfulness. J. Interact. Learn. Res. **19**(4), 597–614 (2008)
4. Squire, K.D., DeVane, B., Durga, S.: Designing centers of expertise for academic learning through video games. Theor. Pract. **47**(3), 240–251 (2008)
5. Honey, M.A., Hilton, M. (eds.): Learning Science Through Computer Games and Simulations. National Academies Press, Washington, DC (2011)
6. Shaffer, D.W.: Epistemic frames for epistemic games. Comput. Educ. **46**(3), 223–234 (2006)
7. Chee, Y.S.: Embodiment, embeddedness, and experience: game-based learning and the construction of identity. Res. Pract. Technol. Enhanced Learn. **2**(1), 3–30 (2007)
8. Fraser, J., Shane-Simpson, C., Asbell-Clarke, J.: Youth science identity, science learning, and gaming experiences. Comput. Hum. Behav. **41**, 523–532 (2014)
9. All, A., Castellar, E.P.N., Van Looy, J.: Assessing the effectiveness of digital game-based learning: best practices. Comput. Educ. **92**, 90–103 (2016)
10. Beier, M.E., Miller, L.M., Wang, S.: Science games and the development of scientific possible selves. Cult. Sci. Edu. **7**(4), 963–978 (2012)
11. Shaffer, D.W.: Quantitative Ethnography. Cathcart Press, Madison (2017)
12. Shaffer, D.W., Collier, W., Ruis, A.R.: A tutorial on epistemic network analysis: analyzing the structure of connections in cognitive, social, and interaction data. J. Learn. Anal. **3**(3), 9–45 (2016)
13. Foster, A.: CAREER: projective reflection: learning as identity exploration within games for science. National Science Foundation, Drexel University, Philadelphia, PA (2014)
14. Kereluik, K., Mishra, P., Fahnoe, C., Terry, L.: What knowledge is of most worth: teacher knowledge for 21st century learning. J. Digit. Learn. Teach. Educ. **29**(4), 127–140 (2013)
15. Wigfield, A., Eccles, J.S.: Expectancy-value theory of achievement motivation. Contemp. Educ. Psychol. **25**(1), 68–81 (2000)
16. Vygotsky, L.S.: Thought and Language. The MIT Press, Cambridge (1934/1986)
17. Kaplan, A., Flum, H.: Identity formation in educational settings: a critical focus for education in the 21st century. Contemp. Educ. Psychol. **37**(3), 171–175 (2012)
18. Hidi, S., Renninger, K.A.: The four-phase model of interest development. Educ. Psychol. **41**(2), 111–127 (2006)
19. Hadwin, A., Oshige, M.: Self-regulation, coregulation, and socially shared regulation: exploring perspectives of social in self-regulated learning theory. Teach. Coll. Rec. **113**(2), 240–264 (2011)
20. Zimmerman, B.J.: A social cognitive view of self-regulated academic learning. J. Educ. Psychol. **81**(3), 329–339 (1989)
21. Kaplan, A., Sinai, M., Flum, H.: Design-based interventions for promoting students' identity exploration within the school curriculum. In: Karabenick, S.A., Urdan, T.C. (eds.) Motivational Interventions: Advances in Motivation and Achievement, vol. 18, pp. 243–291. Emerald Group Publishing Limited, Bingley (2014)
22. Cobb, P., Confrey, J., DiSessa, A., Lehrer, R., Schauble, L.: Design experiments in educational research. Educ. Res. **32**(1), 9–13 (2003)
23. Foster, A., Shah, M.: The play curricular activity reflection and discussion model for game-based learning. J. Res. Technol. Educ. **47**(2), 71–88 (2015)
24. Krippendorff, K.: Content Analysis: An Introduction to Its Methodology. Sage Publications, Thousand Oaks (2004)
25. Marquart, C.L., Hinojosa, C., Swiecki, Z., Shaffer, D.W.: Epistemic Network Analysis [Software]. http://app.epistemicnetwork.org. Accessed 28 Aug 2019
26. Stake, R.E.: The Art of Case Study Research. Sage Publications, Thousand Oaks (1995)

27. DeVane, B.M.: Toward sociocultural design tools for digital learning environments: understanding identity in game-based learning communities. Doctoral dissertation, University of Wisconsin-Madison, Madison, WI (2010)
28. Wise, A.F., Shaffer, D.W.: Why theory matters more than ever in the age of big data. J. Learn. Anal. **2**(2), 5–13 (2015)
29. Gašević, D., Joksimović, S., Eagan, B.R., Shaffer, D.W.: SENS: network analytics to combine social and cognitive perspectives of collaborative learning. Comput. Hum. Behav. **92**, 562–577 (2019)

Multiple Uses for Procedural Simulators in Continuing Medical Education Contexts

Andrew R. Ruis[1(✉)] ⓘ, Alexandra A. Rosser[1], Jay N. Nathwani[1],
Megan V. Beems[1] ⓘ, Sarah A. Jung[1] ⓘ, and Carla M. Pugh[2] ⓘ

[1] University of Wisconsin–Madison, Madison, WI, USA
arruis@wisc.edu
[2] Stanford University, Palo Alto, CA, USA

Abstract. Simulators have been widely adopted to help surgical trainees learn procedural rules and acquire basic psychomotor skills, and research indicates that this learning transfers to clinical practice. However, few studies have explored the use of simulators to help more advanced learners improve their understanding of operative practices. To model how surgeons with different levels of experience use procedural simulators, we conducted a quantitative ethnographic analysis of small-group conversations in a continuing medical education short course on laparoscopic hernia repair. Our research shows that surgeons who had less experience with laparoscopic surgery tended to use the simulators to learn and rehearse the basic procedures, while more experienced surgeons used the simulators as a platform for exploring a range of hernia presentations and operative approaches based on their experiences. Thus simple, inexpensive simulators may be effective with both novice and more experienced learners.

Keywords: Surgery education · Procedural simulation · Continuing Medical Education (CME) · Quantitative ethnography · Epistemic Network Analysis (ENA) · Discourse analysis

1 Introduction

Procedural simulations—models of surgical cases that enable individuals or teams to implement operative techniques—have been widely adopted to help trainees learn procedural rules and acquire basic psychomotor skills, and research indicates that this learning transfers to clinical practice (see, e.g., [1]). However, as Madani and colleagues [2] argue, mere possession of knowledge or mastery of individual skills in isolation is not sufficient for basic competency, let alone mastery; rather, expert surgeons must be able to integrate these and other elements of operative practice to achieve optimal patient outcomes. Although procedural simulation has been studied extensively as a platform for developing basic knowledge and skills, little research has explored its use with more advanced learners [3]. This raises an important question: *Can procedural simulations help more advanced learners continue their professional development beyond learning and rehearsing basic procedural knowledge and skills?*

© Springer Nature Switzerland AG 2019
B. Eagan et al. (Eds.): ICQE 2019, CCIS 1112, pp. 211–222, 2019.
https://doi.org/10.1007/978-3-030-33232-7_18

This question is particularly pressing, as current approaches to surgical education in the United States do not adequately help new surgeons develop the competency to successfully implement operative procedures [4]. In two separate studies, educators [5] and fellows [6] alike expressed a lack of confidence in current educational approaches, particularly in minimally invasive surgery. Moreover, graduating general surgery residents are poorly prepared to operate independently: of the 121 procedures considered essential by the majority of program directors, the average resident had performed only 18 of them more than 10 times prior to graduation; for fully half of the procedures (63), the mode number of times completing the procedure was zero, indicating that graduating residents have never independently completed many essential operations [7].

While these findings indicate a clear problem with how general surgery residents are trained—or with the expectations for what can be learned in five years or surgical residency—they also have significant implications for subsequent training and professional development, of which *continuing medical education* (CME) is a significant component [8, 9]. CME was originally designed to help licensed, practicing physicians *maintain* competency, but it must increasingly help them *develop* it as well.

The goal of this study was to understand the use of procedural simulation in one CME short course. To do this, we conducted a quantitative ethnographic analysis [10] of small-group conversations in two separate implementations of a course on laparoscopic hernia repair held annually at a large surgical conference in the United States. The six-hour course consisted of a two-hour lecture and a four-hour practicum in which small groups of participants used basic, box-style simulators to learn or review various laparoscopic hernia repair techniques with an expert instructor. Our research shows that surgeons who had less experience with laparoscopic surgery tended to use the simulators to learn and rehearse the basic procedural steps and rules, and to work on identifying and managing common errors. That is, they used the simulated case as an opportunity for *procedural rehearsal*. More experienced surgeons, in contrast, used the sim-ulators as a platform for discussing and exploring a range of hernia presentations and operative approaches based on their real-world experiences. That is, they used the simulated case as an opportunity for *procedural analysis*. Our findings suggest that relatively simple, inexpensive simulators may be effective with both novice and more experienced learners.

2 Methods

2.1 Setting and Participants

This study included 58 surgeons (53 practicing surgeons and 5 general surgery residents) who participated in a one-day CME course on laparoscopic inguinal and ventral hernia repair at a large surgical conference in the United States. Data were collected during two implementations of the course held in two different years. The course involved an introductory lecture (2 h) and a practicum (4 h). The lecture covered the basic procedural steps and rules of minimally invasive ventral and inguinal hernia repairs. During the practicum, participants were assigned to groups of three based on their self-reported prior experience with laparoscopic surgery (this process is described in more detail below). Each group worked with one or two randomly assigned instructors—experts in minimally invasive hernia repair—who provided instruction using simple box-style simulators developed for training in laparoscopic hernia repair procedures (see Fig. 1) [11]. All procedures were taught as mesh repairs; cautery was discussed only if participants broached the topic, but cauterization tools cannot be used with the simulators. The participant groups completed two sessions during the practicum, learning a different laparoscopic hernia repair procedure with a different instructor in each session.

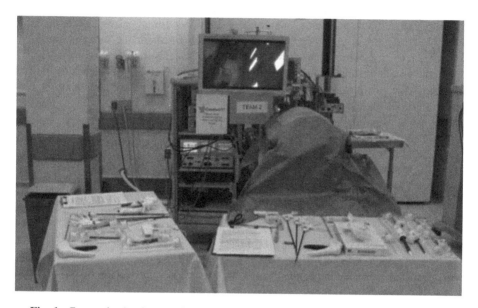

Fig. 1. Box-style simulator and laparoscopic tools used in the hernia repair practicum.

2.2   Data Collection

Before beginning the course, participants reported their experience performing six common laparoscopic procedures: cholecystectomy, appendectomy, colectomy, incisional hernia repair, totally extraperitoneal hernia repair (TEP), and transabdominal preperitoneal hernia repair (TAPP). Participants indicated their experience using a five-point Likert scale, where one is "beginner", three is "competent", and five is "master surgeon". Participants' mean laparoscopic surgery experience (i.e., the mean of their self-ratings on all six procedures) ranged from 1.83 to 4.67. Participants were assigned to groups based on their mean experience so that groups were composed of surgeons with similar levels of laparoscopic surgery experience. Four researchers directly observed each practicum, and all sessions were audio and video recorded. Audio was transcribed manually, and each transcription was subsequently verified by a second transcriber. Audio transcripts were then coded for analysis.

2.3   Coding Process

Based on ethnographic observations and conventional content analysis of the transcribed audio data [12], we defined six codes (see Table 1) to identify key topics and epistemic elements in participant conversations. Automated coding algorithms were developed to code the transcribed audio data.

To assess the reliability of the coding process, two independent raters coded a case-controlled random sample of 40 turns of talk for each code, and the automated coding algorithm coded each sample as well. Raters assigned a "1" to any turn of talk in which the code was present, and a "0" to those in which it was not. To calculate inter-rater reliability, we computed Cohen's kappa (κ) for each code for all pairwise combinations of raters. To determine whether the kappa values obtained for the samples could be generalized to the entire dataset, we computed Shaffer's rho (ρ) to estimate the expected Type I error rate of kappa given the sample size [10, 13]. For each of the six codes, the rate of agreement was statistically significant ($\alpha = 0.05$) for a minimum kappa threshold of 0.65 (see Table 1).

Table 1. Discourse codes and inter-rater reliability statistics.

Code	Description & Example	Human 1 vs. Human 2		Human 1 vs. Computer		Human 2 vs. Computer	
		κ^a	$\rho(0.65)$	κ^a	$\rho(0.65)$	κ^a	$\rho(0.65)$
Mesh Repair	Referencing mesh, tacking, or suturing "One of those tacks fell; you did not have control of the tacker."	1.00	<0.01	1.00	<0.01	0.97	0.01
General Anatomy	Referencing the anatomy of the abdomen "Spermatic vessel is more lateral to the vas."	1.00	<0.01	1.00	<0.01	0.96	<0.01
Pathological Anatomy	Referencing the anatomy of a hernia "We've got to make sure the hernia is on the midline too because sometimes the hernia isn't on the midline."	0.95	0.01	0.98	<0.01	1.00	<0.01
Requesting Advice	Asking what surgeons should do in a given situation "So what do you do if it doesn't tack? What do you do in the operating room?"	0.86	<0.01	0.86	0.04	0.75	0.04
Trouble-shooting	Managing or negotiating complications "This is the tough side. We need to go to the easy side. The easy side is over there."	0.85	<0.01	0.89	<0.01	0.80	0.03

(continued)

Table 1. (*continued*)

Code	Description & Example	Human 1 vs. Human 2		Human 1 vs. Computer		Human 2 vs. Computer	
		κ^{a}	$\rho(0.65)$	κ^{a}	$\rho(0.65)$	κ^{a}	$\rho(0.65)$
Real-World Case	Referencing real bodies, patients, or other cases "So that can be done with a suture passer and a suture … if the patient is thin enough that you can see the fascia."	0.80	0.02	0.95	0.01	0.73	<0.01

^a *All kappas are statistically significant for $\rho(0.65) < 0.05$.*

2.4 Epistemic Network Analysis

For the purposes of analysis, participants were divided into quartiles based on their mean laparoscopic surgery experience ratings. Participants with mean laparoscopic surgery experience scores less than or equal to 2.50 (lowest quartile, $n = 13$) were classified as *novices*, as their self-ratings indicated that they did not feel fully competent with common laparoscopic procedures. Participants with mean laparoscopic surgery experience scores greater than 3.00 (highest quartile, $n = 17$) were classified as *relative experts*, as their self-ratings indicated that they felt generally competent with common laparoscopic procedures. The participants in the second and third quartiles ($n = 23$) were classified as *intermediates*. Seven participants did not report their experience with minimally invasive surgery.

Epistemic network analysis (ENA) version 1.5.2 was used to analyze the conversations of novices, intermediates, and relative experts [10, 14–16]. We defined the units of analysis as all lines of data associated with a single participant (excluding instructors). The ENA algorithm uses a moving window to construct a network model for each line in the data, showing how codes in the current line are connected to codes that occur within the recent temporal context [17]. In this study, the window was defined as 5 utterances (each turn of talk plus the 4 previous turns) within a given practicum session. The resulting networks were aggregated for each unit of analysis in the model. In this model, networks were aggregated using a binary summation in which the networks for a given line reflect the presence or absence of the co-occurrence of each unique pair of codes. The networks in the ENA model were normalized for all units of analysis before they were subjected to a dimensional reduction, which accounts for the fact that different participants may have different numbers of coded utterances. For the dimensional reduction, we used a singular value decomposition (SVD), which produces orthogonal dimensions that maximize the variance explained by each dimension.

Networks were visualized using network graphs where nodes correspond to the codes, and edges reflect the relative frequency of co-occurrence, or connection, between two codes. The result is two coordinated representations for each unit of analysis: (1) an ENA score, or a point that represents the location of that unit's network in the projected space formed by the first two dimensions in the SVD, and (2) a weighted network graph projected into the same low-dimensional space. The positions of the network graph nodes are fixed, and the node positions are determined by an optimization routine that attempts to produce a high degree of correspondence between the ENA scores of the units and the corresponding network centroids. To do this, the optimization routine uses a least-squares approach that minimizes the sum of squared distances between the network centroids and the ENA scores on each dimension. Because of the co-registration of network graphs and projected space, the positions of the network graph nodes can be used to interpret the dimensions of the projected space and explain the positions of different units in the space. Our model has co-registration correlations (Pearson's and Spearman's r) of >0.93 on the first and second SVD dimensions. These measures indicate that there is a strong goodness of fit between the visualization and the original model.

3 Results

Figure 2 shows the ENA scores and difference graph for novices and relative experts. The novices appear primarily in the upper part (high y values) of the ENA space formed by the first two SVD dimensions, while the relative experts appear mostly in the lower part of the space (low y values). The first and second SVD dimensions account for 24.0% and 15.1% of the variance in patterns of connectivity, respectively.

To understand this difference between novices and relative experts, we plotted the difference graph for the two groups (see Fig. 2, bottom). For the novices, the most distinguishing connections were from *Requesting Advice* to *General Anatomy*, *Mesh Repair*, and *Troubleshooting*. That is, the novices focused mostly on the procedural aspects of the simulated case: asking questions about basic anatomy, about the procedural steps and rules, and about managing errors or complications. The relative experts, in contrast, made proportionately stronger connections to *Real-World Case* and *Pathological Anatomy*. Like the novices, the relative experts discussed the overall anatomy relevant to any abdominal procedure (*General Anatomy*), but they focused more on the anatomy specific to the hernia (*Pathological Anatomy*). Moreover, the relative experts were more likely than novices to discuss and ask questions about the procedure and the anatomy in the context of real-world cases or scenarios. These patterns of conversation suggest that relative experts used the simulators less as an opportunity to learn the surgical procedure or practice implementing the procedural steps, and more as an opportunity to discuss specific hernia repair issues that arise in actual cases. In other words, the novices used the simulators in the traditional sense—to learn the operative rules of the modeled hernia repair and rehearse key skills and techniques—while the relative experts treated each simulated case as a specific instantiation of a broader class of operative problem, using it to explore how expert surgeons adapt to different clinical presentations.

These differences are evident in the qualitative data as well. For example, consider the following excerpt from a conversation among novices.

Line 1 Novice 1:   *So, the principle, is the mesh going to cover everything?*
Line 2 Instructor: *That's right, the mesh going to cover everything. Alright, so let's look over here again. Let's see, are your vas and vessels separated?*

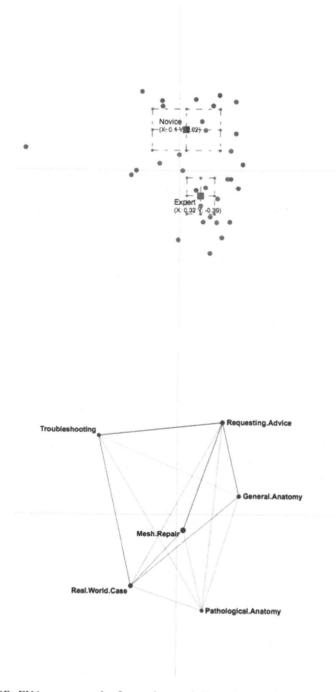

Fig. 2. TOP: ENA scores on the first and second dimensions (points) for novices (red) and relative experts (blue) with means (squares) and 95% confidence intervals (dashed boxes). BOTTOM: Difference graph of the novice and relative expert mean networks (Color figure online).

Line 3 Novice 2:   *This is the vas here?*
Line 4 Instructor: *That's the vas.*
Line 5 Novice 1:   *You are still in the wrong plane.*
Line 6 Instructor: *Yeah, you have to cut the white [fascia].*
Line 7 Novice 1:   *I thought we did cut the white.*
Line 8 Novice 2:   *I thought we did. That's what I was originally going into. Was going to practice with the balloon. Now we need the trocar in below it*

This excerpt begins with a participant asking a question about the mesh (Line 1). The instructor replies by orienting the participant to the basic anatomy associated with the procedure (Lines 2–4). A second participant identifies an error with the procedure (Line 5) and works with the others to find a solution (Lines 6–8). In this brief excerpt, the novices focus on a specific procedural step—placing the mesh—which requires understanding the anatomy, properly orienting the endoscope, and managing errors as they occur. In addition, novice conversations involve more reiteration and requests for basic information. As this example shows, novices tended to use the simulated case as an opportunity to learn and rehearse the basic procedural steps and fundamental skills of laparoscopic hernia repair.

In contrast, the relative experts tended to draw more on their own experience with laparoscopic hernia repair, as the following excerpt illustrates.

Line 1 Expert 1:   *I do a lot of melanoma patients that I have had with superficial internal dissection. ... In theory, their complication is a pulled femoral hernia. Would that just be sufficient to cover up the hole?*
Line 2 Instructor: *If I've over-dissected the space ... then I'll fixate... with a tack that covers medially and a really one high lateral tack.*
Line 3 Expert 1:   *So what do you do there? I mean, do you just kind of move the mesh down lower to make sure it's covered?*
Line 4 Instructor: *So, so if you've got a femoral, you should dissect that space out and cover it*

This excerpt begins with the participant asking a question about specific hernia pathology based on prior cases (Line 1). The instructor then responds by outlining strategies for tacking the mesh (Line 2). The participant asks a follow up question about mesh placement (Line 3), and the instructor addresses how to achieve coverage. In comparison to the novices, the relative experts were less focused on the procedure in front of them and more on using it as a platform to talk—and think—about real clinical scenarios, often ones that they experience in practice. Their conversations extended beyond discussion of basic procedural steps or skills; in Line 3, for example, even though the participant is asking about a procedural step (mesh placement), he does so in the context of a more complex clinical presentation that he sees in his practice.

4 Discussion

This study examined how surgeons with different levels of experience in laparoscopic surgery used basic procedural simulators in a CME short course. While surgeons with less experience used the simulators to learn and rehearse the procedural rules and basic laparoscopic skills, as evidenced by the strong connections they made from *Requesting Advice* to *General Anatomy*, *Mesh Repair*, and *Troubleshooting*, more experienced surgeons used the simulators to explore various hernia presentations and best practices for addressing them, as evidenced by the strong connections they made to *Real-World Case* and *Pathological Anatomy*. In other words, novices engaged in *procedural rehearsal*—seeking to develop basic knowledge and skills—while the relative experts engaged in *procedural analysis*—using the simulated case as a platform with which to explore different hernia presentations or operative challenges they experience in practice.

This suggests that there are two primary functions that procedural simulators can serve. The first, and the one most commonly acknowledged, is that they enable surgeons and surgical trainees to learn and rehearse a specific operative procedure or the associated skills and best practices in a setting that involves no risk to patients, that provides useful feedback, and that is low-cost and logistically feasible. The novices in this study used the simulators primarily in this way. The second function the procedural simulators serve is to facilitate a guided version of what Schön calls *reflection in action* [18]. Reflection-in-action takes place as experts in a domain (a) identify similarities between novel problems and past problems, (b) adapt the solutions from those past problems based on their understanding of the current problem, and then (c) evaluate the results of applying the adapted solution to the problem at hand. The relative experts in this study used the simulators as a platform for engaging in reflection-in-action with the guidance of their peers and a more experienced instructor. That is, they drew on past experiences to pose questions and construct scenarios, and then used the simulated case to work through how to solve those problems with the guidance of an even more experienced surgeon.

Facilitation of reflection in this way is an important affordance of procedural simulation that requires further study [3]. While numerous interventions and training protocols have been used to promote reflective practice and prepare clinicians to be self-directed, lifelong learners (see, e.g., [19–21]), reflective practice was found to be negatively correlated with physician age and experience [22]. This suggests that reflection on practice may decline as surgeons progress in their careers. CME courses grounded in procedural simulation could provide an effective mechanism for promoting on-going reflection among practicing surgeons.

Of course, this study has several limitations. First, the sample size is small, and we studied only one CME course that uses only one type of procedural simulator. Thus, our conclusions cannot be generalized without further research on learning across a range of CME course and simulation models. However, small-group practica are very common in CME contexts, and our research suggests that grouping participants by expertise may facilitate more productive learning interactions. This is critical, as many

CME courses are quite short, and many physicians have limited time for ongoing training. Future research should explore the affordances and limitations of this approach.

Second, this study modeled learning processes by analyzing participant conversations, but there was no outcome measure. Thus, we were unable to assess the extent to which the course helped participants develop knowledge, skills, or other competencies. This is a broader problem with CME as it is implemented in the United States. Few CME courses assess learning, and it is difficult to control for the effects of CME courses in longitudinal studies that document changes in clinical skill or practice. While there is limited evidence that simulation is more effective in CME contexts than traditional approaches to clinical education [23], the basic CME model is not particularly effective for changing physician behavior or improving patient health outcomes [24]. Considerably more research is needed on CME in particular and, more generally, on how physicians develop and maintain expertise over the course of their careers [9, 25]. We argue that quantitative ethnography provides a method and a set of tools for exploring such learning processes, and prior research has shown that it could provide an effective means of assessing operative competency as well [26].

Lastly, this study did not explore the role that instructors played in guiding participant conversations. While instructor discourse was included in the analysis, the instructors themselves were not included as units. Research on this same data has found, for example, that instructors were significantly more likely to answer questions with anecdotes when responding to relative experts and with prohibitions (what not to do) when responding to novices [27]. In future research, we will explore in more detail the relationship between teaching practices and learning processes in order to understand better the effects of instructional strategies on learning with procedural simulators.

Despite these limitations, our findings suggest that inexpensive, basic procedural simulators, such as the box-style simulators used by participants in this study, can help both novices and more experienced surgeons improve their understanding of operative practices by facilitating both rehearsive and reflective practice.

Acknowledgements. This work was funded in part by the National Science Foundation (DRL-1661036, DRL-1713110), the Wisconsin Alumni Research Foundation, and the Office of the Vice Chancellor for Research and Graduate Education at the University of Wisconsin-Madison. A. R. Ruis was supported by a University of Wisconsin-American College of Surgeons Education Research Fellowship. The opinions, findings, and conclusions do not reflect the views of the funding agencies, cooperating institutions, or other individuals.

References

1. Dawe, S.R., Windsor, J.A., Broeders, J.A., Cregan, P.C., Hewett, P.J., Maddern, G.J.: A systematic review of surgical skills transfer after simulation-based training: laparoscopic cholecystectomy and endoscopy. Ann. Surg. **259**, 236–248 (2014)
2. Madani, A., et al.: What are the principles that guide behaviors in the operating room? creating a framework to define and measure performance. Ann. Surg. **265**, 255–267 (2017)

3. Sullivan, S.A., Ruis, A.R., Pugh, C.M.: Procedural simulations and reflective practice: meeting the need. J. Laparoendosc. Adv. Surg. Tech. **27**, 455–458 (2017)
4. Mattar, S.G., et al.: General surgery residency inadequately prepares trainees for fellowship: results of a survey of fellowship program directors. Ann. Surg. **258**, 440–449 (2013)
5. Subhas, G., Mittal, V.K.: Minimally invasive training during surgical residency. Am. Surg. **77**, 902–906 (2011)
6. Osman, H., Parikh, J., Patel, S., Jeyarajah, D.R.: Are general surgery residents adequately prepared for hepatopancreatobiliary fellowships? a questionnaire-based study. HPB **17**, 265–271 (2015)
7. Bell, R.H.: Operative experience of residents in us general surgery programs: a gap between expectation and experience. Ann. Surg. **249**, 719–724 (2009)
8. Iyasere, C.A., Baggett, M., Romano, J., Jena, A., Mills, G., Hunt, D.P.: Beyond continuing medical education: clinical coaching as a tool for ongoing professional development. Acad. Med. **91**, 1647–1650 (2016)
9. Sachdeva, A.K.: The new paradigm of continuing education in surgery. Arch. Surg. **140**, 264–269 (2005)
10. Shaffer, D.W.: Quantitative Ethnography. Cathcart Press, Madison (2017)
11. Pugh, C., Plachta, S., Auyang, E., Pryor, A., Hungness, E.: Outcome measures for surgical simulators: Is the focus on technical skills the best approach? Surgery **147**, 646–654 (2010)
12. Hsieh, H.-F., Shannon, S.E.: Three approaches to qualitative content analysis. Qual. Health Res. **15**, 1277–1288 (2005)
13. Eagan, B.R., Rogers, B., Pozen, R., Marquart, C., Shaffer, D.W.: rhoR: Rho for inter rater reliability (2016)
14. Marquart, C., Hinojosa, C.L., Swiecki, Z., Eagan, B.R., Shaffer, D.W.: Epistemic network analysis (2019)
15. Shaffer, D.W., Collier, W., Ruis, A.R.: A tutorial on epistemic network analysis: analyzing the structure of connections in cognitive, social, and interaction data. J. Learn. Anal. **3**, 9–45 (2016)
16. Shaffer, D.W., Ruis, A.R.: Epistemic network analysis: a worked example of theory-based learning analytics. In: Lang, C., Siemens, G., Wise, A.F., and Gasevic, D. (eds.) Handbook of Learning Analytics, pp. 175–187. Society for Learning Analytics Research (2017)
17. Siebert-Evenstone, A.L., Irgens, G.A., Collier, W., Swiecki, Z., Ruis, A.R., Shaffer, D.W.: In search of conversational grain size: modelling semantic structure using moving stanza windows. J. Learn. Anal. **4**, 123–139 (2017)
18. Schön, D.A.: The Reflective Practitioner: How Professionals Think in Action. Basic Books, New York (1983)
19. McGlinn, E.P., Chung, K.C.: A pause for reflection: Incorporating reflection into surgical training. Ann. Plast. Surg. **73**, 117 (2014)
20. Jordan, M.E., McDaniel Jr., R.R.: Managing uncertainty during collaborative problem solving in elementary school teams: the role of peer influence in robotics engineering activity. J. Learn. Sci. **23**, 490–536 (2014)
21. Husebø, S.E., O'Regan, S., Nestel, D.: Reflective practice and its role in simulation. Clin. Simul. Nurs. **11**, 368–375 (2015)
22. Mamede, S., Schmidt, H.G.: Correlates of reflective practice in medicine. Adv. Health Sci. Educ. **10**, 327–337 (2005)
23. McGaghie, W.C., Issenberg, S.B., Cohen, M.E.R., Barsuk, J.H., Wayne, D.B.: Does simulation-based medical education with deliberate practice yield better results than traditional clinical education? a meta-analytic comparative review of the evidence. Acad. Med. **86**, 706–711 (2011)

24. Davis, D., O'Brien, M.A.T., Freemantle, N., Wolf, F.M., Mazmanian, P., Taylor-Vaisey, A.: Impact of formal continuing medical education: do conferences, workshops, rounds, and other traditional continuing education activities change physician behavior or health care outcomes? J. Am. Med. Assoc. **282**, 867–874 (1999)
25. Sachdeva, A.K., Blair, P.G., Lupi, L.K.: Education and training to address specific needs during the career progression of surgeons. Surg. Clin. North Am. **96**, 115–128 (2016)
26. Ruis, A.R., Rosser, A.A., Quandt-Walle, C., Nathwani, J.N., Shaffer, D.W., Pugh, C.M.: The hands and head of a surgeon: modeling operative competency with multimodal epistemic network analysis. Am. J. Surg. **216**, 835–840 (2018)
27. Godfrey, M., Rosser, A.A., Pugh, C.M., Shaffer, D.W., Sachdeva, A.K., Jung, S.A.: Teaching practicing surgeons what not to do: an analysis of instruction fluidity during a simulation-based continuing medical education course. Surgery **165**, 1082–1087 (2019)

Cause and Because: Using Epistemic Network Analysis to Model Causality in the Next Generation Science Standards

Amanda Siebert-Evenstone[1(✉)] and David Williamson Shaffer[1,2]

[1] University of Wisconsin – Madison, Madison, WI, USA
alevenstone@wisc.edu
[2] Aalborg University, Copenhagen, Denmark

Abstract. The Next Generation Science Standards propose an integrated and holistic view of science education that teaches science through three-dimensional learning. In this vision of science, content and practices are interconnected and inseparable. While the NGSS has influenced K-12 education standards in 40 states, there has not been a systematic analysis of the standards themselves. In this study, we investigate three-dimensional learning in order to identify new insights into underlying relationships between science concepts as well as make comparisons between different science disciplines. We used Epistemic Network Analysis to measure and models the structure of connections among crosscutting concepts and practices within and across disciplines. Results show systematic differences between how Physical and Life Sciences use and describe cause and effect relationships in which Physical Sciences predominantly focuses on the generation of causal relationships while Life Sciences focuses on the explanation of causal relationships.

Keywords: Epistemic network analysis · Next generation science standards · Three-dimensional learning

1 Introduction

Following the turn in science education toward teaching science as a practice [1], the Next Generation Science Standards (NGSS) [2, 3] constructed a practice-based vision for science education in the United States. The NGSS propose an integrated and holistic view of science education that organizes science into *three-dimensional learning*: a coherent combination of disciplinary core ideas, crosscutting concepts, and science and engineering practices. In this vision of science, content and practices are interconnected and inseparable. As such, this document provides an important artifact of what scientists and science educators deem valuable and core to the pursuits of this discipline and how students could learn how to think like scientists.

While there are many articles, books, and websites that provide resources for teacher implementation, there has been less research on the implications and rhetoric of the standards themselves. In this study, we investigate overarching claims about the

© Springer Nature Switzerland AG 2019
B. Eagan et al. (Eds.): ICQE 2019, CCIS 1112, pp. 223–233, 2019.
https://doi.org/10.1007/978-3-030-33232-7_19

interconnected nature of the NGSS, specifically what are the relationships among the three dimensions of science learning.

2 Theory

With the goal of improving K-12 science education, the Next Generation Science Standards were developed through a collaboration between the National Research Council (NRC), the National Science Teachers Association, the American Association for Advancement of Science, and Achieve, Inc. [2, 3]. This project united science experts, researchers, and educators to create a new vision for science education and consequently a new set of education standards to be followed in K-12 classrooms. These standards proposed and organized important and overarching themes in science into what the standards call 3-dimensional learning including

> Dimension 1: *Science and Engineering Practices*, which are the skills and knowledge scientists and engineers employ;
> Dimension 2: *Crosscutting Concepts*, which are the common themes and unifying ideas across the disciplines; and
> Dimension 3: *Disciplinary Core Ideas*, which are specific and fundamental concepts and contexts necessary for understanding the discipline.

In this vision of science, content and practices are interconnected and inseparable. Instead of learning content and then applying it, the NGSS proposed an integrated and holistic view of science education. As such, this document provides an important artifact of what scientists and science educators deem valuable and core to the pursuits of this discipline and how students could learn how to think like scientists.

While the NGSS has influenced K-12 education standards in 40 states, there has not been a systematic analysis of the standards themselves. Recent work has analyzed components of the standards, such as genetics content [4], sustainability [5], or a single crosscutting concept (i.e. scale, proportion, and quantity) [6]. One reason there may have been few systematic analyses is that the publicly available version of the standards is an unwieldy and dense set of tables within a lengthy document.

To dive deeper into this conception of science thinking I use David Shaffer's [7] epistemic frame theory to describe the pattern of associations among skills, knowledge, and other cognitive elements that characterize groups of people who share similar ways of framing, investigating, and solving complex problems. More specifically, epistemic frame theory considers the ways in which certain groups of people think and suggests that in specific communities there is a set of systematic patterns of relationships among skills, knowledge, identity, values, and epistemology that form the epistemic frame for that community.

Importantly, epistemic frame theory shifts the focus of learning from accumulating isolated pieces of knowledge to focusing on the structure of connections among them. Similarly, diSessa [8] argued that deep understanding results from linking basic disciplinary concepts within a theoretical framework. For example, diSessa describes how novices have "knowledge-in-pieces", whereas experts have a deep and systematic understanding of how these disciplinary concepts are connected. Other learning scientists have similarly conceptualized learning as developing patterns of connections between concepts [9, 10].

Therefore, in order to model what it means to adopt the epistemic frame of a scientist, or more simply what it means to think like a scientist, we need a way to analyze the relationships among elements in that domain. One way to measure the relationships among elements in an epistemic frame is by using epistemic network analysis (ENA), a tool designed to analyze the structure of connections by identifying the co-occurrence of domain elements in a particular community of practice [11]. The resulting models can be visualized as networks in which the nodes in the model are the codes and the lines connecting the nodes represent the co-occurrence of two codes. Thus, I can quantify and visualize the structure of connections between science practices and crosscutting concepts making it possible to characterize important connections for each science discipline.

In this study, I investigate how modeling and measuring the connections between practices and concepts can identify new insights into underlying relationships between science concepts as well as make comparisons between different science disciplines.

3   Methods

3.1   Data Source

The NGSS provide a set of 208 K-12 science standards organized across three science disciplines (Earth and Space Sciences, Life Sciences, and Physical Sciences) as well as sections addressing Engineering and Technical Sciences. Each discipline has standards that are arranged by Performance Expectations (PEs) that constitute what should be learned by students by the end of that grade level.

3.2   Segmentation and Coding

In this analysis, each line of data represents a single chunk of written content from the NGSS. For example, in the performance expectation for MS-PS4-1, the standards outline 6 pieces of information and each unique piece of text was segmented into a different row. Based on the structure of the NGSS layout, each performance expectation was further segmented by the specific science and engineering practice (SEP, Table 1) and crosscutting concept (CCC, Table 2) that was identified.

For example, MS-PS4-1 asks students to describe the amplitude of waves using mathematics and computational thinking (SEP) and identify patterns (CCC).

3.3   Epistemic Network Analysis

To analyze the connections with the NGSS, I used Epistemic Network Analysis (ENA) [11, 14], which models the structure of connections among NGSS code elements. ENA measures connections by quantifying the co-occurrence of practice and concept within a defined conversation. In this case, a conversation is a collection of lines of data such that lines within a conversation are assumed to be closely related.

Table 1. List of science and engineering practice codes adapted from National Science Teaching Association (NSTA) [12].

Science and Engineering Practice (SEP)	Definition
Asking questions and defining problems	A practice of science is to ask and refine questions that lead to descriptions and explanations of how the natural and designed world works and which can be empirically tested.
Developing and using models	A practice of both science and engineering is to use and construct models as helpful tools for representing ideas and explanations.
Planning and carrying out investigations	Scientists and engineers plan and carry out investigations in the field or laboratory, working collaboratively as well as individually. Their investigations are systematic and require clarifying what counts as data and identifying variables or parameters.
Analyzing and interpreting data	Scientific investigations produce data that must be analyzed in order to derive meaning.
Using mathematics and computational thinking	In both science and engineering, mathematics and computation are fundamental tools for representing physical variables and their relationships.
Constructing explanations and designing solutions	The products of science are explanations and the products of engineering are solutions.
Engaging in argument from evidence	Argumentation is the process by which explanations and solutions are reached.
Obtaining, evaluating, and communicating information	Scientists and engineers must be able to communicate clearly and persuasively the ideas and methods they generate. Critiquing and communicating ideas individually and in groups is a critical professional activity.

For the NGSS, I defined the conversation as a single PE. For example, across MS-PS4 there are 3 total PEs. Each separate expectation has an associated SEP and CCC which would be considered in the same conversation because these two codes specifically relate to one another based on their PE. MS-PS4-1, above, would have a connection between mathematic and computational thinking to the crosscutting concept of patterns However, MS-PS4-1 would not be considered related to MS-PS4-2 because the SEP and CCC relate to a different topic.

ENA constructs a network model for each unit of analysis, showing how the codes within a conversation are connected to one another. The resulting models can be visualized as network graphs where the nodes correspond to the codes and edges reflect the relative frequency of the connection between two codes. Thus, we can quantify and visualize the structure of connections among SEP and CCC, making it possible to characterize three-dimensional learning ideas within each discipline.

Table 2. List of Science and Engineering Practice codes adapted from the NSTA [13].

Crosscutting Concepts (CCC)	Definition
Patterns	Observed patterns of forms and events guide organization and classification, and they prompt questions about relationships and the factors that influence them.
Cause and effect	Events have causes, sometimes simple, sometimes multi-faceted. A major activity of science is investigating and explaining causal relationships and the mechanisms by which they are mediated. Such mechanisms can then be tested across given contexts and used to predict and explain events in new contexts.
Scale, proportion, quantity	In considering phenomena, it is critical to recognize what is relevant at different measures of size, time, and energy and to recognize how changes in scale, proportion, or quantity affect a system's structure or performance.
Systems and system models	Defining the system under study-specifying its boundaries and making explicit a model of that system-provides tools for understanding and testing ideas that are applicable throughout science and engineering.
Energy and matter	Flows, cycles, and conservation. Tracking fluxes of energy and matter into, out of, and within systems helps one understand the systems' possibilities and limitations.
Structure and function	The way in which an object or living thing is shaped and its substructure determine many of its properties and functions.
Stability and change	For natural and built systems alike, conditions of stability and determinants of rates of change or evolution of a system are critical elements of study.

4 Results

I used ENA to measure and model connections between practices and concepts for each discipline. For this ENA model, connections were counted for each performance expectation and accumulated across disciplinary core ideas and grades to model each of the four disciplines. In this paper, I specifically compare Physical Sciences and Life Sciences.

In Fig. 1, the network graph for Physical Sciences identifies many connections across the PEs and shows a few main connections, including Cause and Effect to Explanations, Cause and Effect to Argument, and Analyze to Patterns. On the other hand, Physical sciences (Fig. 2) shows the most connections between Cause and Effect to Investigations, Explanations to Energy, and Energy to Models.

One way to consider the differences in connections is to choose a common node and then analyze the similarities and differences in how each discipline connects to this idea. In the next section, we focus on a single concept and analyze the difference in connections to this idea.

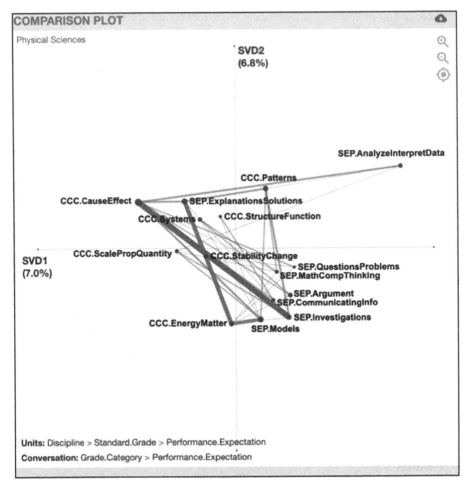

Fig. 1. Network for physical sciences (purple). Thicker lines represent more frequent connections, thinner lines represent less frequent connections (Color figure online).

4.1 Cause and Effect

Within the crosscutting concepts, Cause and Effect is the most prominent concept within all three disciplines occurring a total of 56 times. In Life Sciences, Cause and Effect is included in 31% of PEs while in Physical Sciences this concept is included in 33% of all PEs. In their networks, both Life and Physical sciences make many connections between concepts and practices and in both sets of standards, there are many connections to Cause and Effect (seen by a larger diameter node and thick lines connecting to that node).

Another way to consider the differences between disciplines is to construct a difference graph (Fig. 2). The difference graph subtracts the edge weights of the mean networks of each unit visualizing the differences in weights. Connections represented

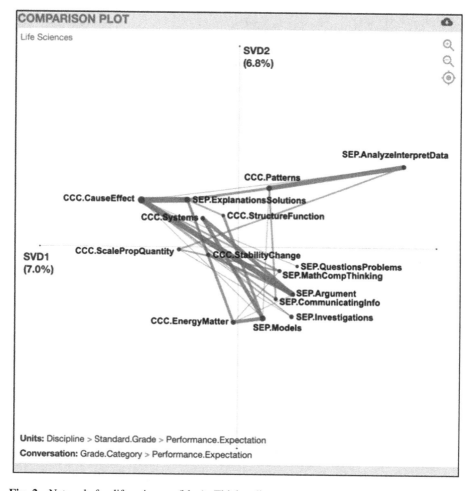

Fig. 2. Network for life sciences (blue). Thicker lines represent more frequent connections, thinner lines represent less frequent connections (Color figure online).

by purple lines were stronger among PS standards, while connections in blue occurred proportionally more often among LS standards (Fig. 3).

As the difference network shows, Life Sciences made proportionally more connections between Cause and Effect to Constructing Explanations and Designing Solutions as well as to Engaging in Argument from Evidence. On the other hand, Physical sciences were more likely to connect this idea of causality with Analyzing and Interpreting Data as well as Planning and Carrying out Investigations. These differences indicate different treatments of the ways cause and effect are used and treated in this representation of the disciplines. Life Sciences were more likely to link causality with ways to explain causal relationships while Physical Science expectations were more likely to propose the investigation and analysis of causal relationships.

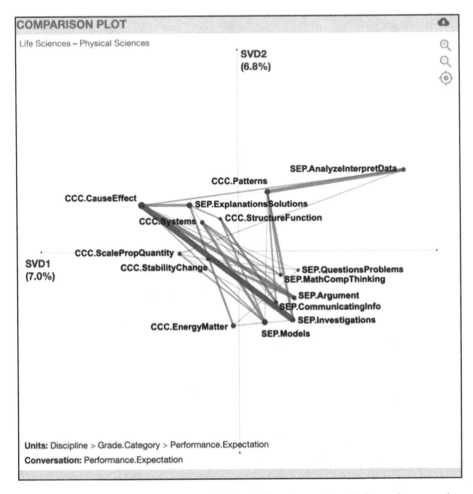

Fig. 3. Difference network between physical and life sciences, in which purple connections occur more frequently in the Physical Sciences and blue connections occur more frequently in the Life sciences (Color figure online).

This difference is also apparent in how Cause and Effect is described in the crosscutting concepts section of the performance expectations:

"Phenomena may have more than one cause, and some cause and effect relationships in systems can only be described using probability." (MS-LS1-4)

In this Life Sciences text, causality has an emphasis on the ways and potentially the only ways a certain relationship can be described which is related to both explanation and argumentation. On the other hand, Physical Sciences expectations about causality are described in a different way:

"Simple tests can be designed to gather evidence to support or refute student ideas about causes." (1-PS4-3)

In this Physical Sciences text, there is an emphasis on using "simple tests" which is related to planning and carrying out investigations.

These are two descriptions for Cause and Effect, but there were also many cause and effect descriptions that were listed across the disciplines, including: "Events have causes that generate observable patterns" which is listed in Earth and Space Sciences, Life Sciences, and Physical Sciences. Another example is that "Empirical evidence is required to differentiate between cause and correlation and make claims about specific causes and effects" which can be found in both Life Sciences and Earth and Space sciences. There are even three versions of one Cause and Effect text that vary when the idea ends, "Cause and effect relationships are routinely identified [.; and used to explain change; tested, and used to explain change]". This idea and the different versions occur across the three disciplines.

However, while each of these disciplines share common language about causality, the two incline texts above were unique to their field. No other discipline listed that "phenomena may have more than one cause, and some cause and effect relationships in systems can only be described using probability." Likewise, no other discipline listed that "simple tests can be designed to gather evidence to support or refute student ideas about causes." This dichotomy highlights the systematic differences with which common ideas like causality are described differently and may serve different roles across the disciplines.

5 Discussion

Preliminary evidence suggests that there are interesting and systematic differences between science disciplines. Both Life and Physical Sciences make many connections between concepts and practices and in both sets of standards, there are many connections to Cause and Effect. In comparison, Life Sciences makes more connections between causality and explanations as well as to argumentation. On the other hand, Physical Sciences more often connects this idea of causality with investigations and analysis.

Science has been defined in terms of empirically deriving causal explanations [15], and within the NGSS, they state, "A major activity of science is investigating and explaining causal relationships and the mechanisms by which they are mediated" (Appendix G, p. 1) [2]. In each of these two definitions, there are two main components, (1) generation and (2) explanation of causal relationships. Both disciplines in this analysis address these ideas, however, Physical Sciences predominantly focuses on generation while Life Sciences focuses on explanation of causal relationships. This is not to say that the physical science standards do not address explanation and argumentation or that life science standards omit investigations, rather, that in terms of causality they make more connections to the practices of investigating and analyzing data.

One goal of crosscutting concepts was to help students use ideas like cause and effect to relate and understand core ideas across disciplines. But the separation of ideas identified in this analysis may foster misconceptions for how students learn the goals of science. If the purpose of science is both investigating *and* explaining causality and students are taught *either* investigating *or* explaining these relationships students may incorrectly associate a discipline with only one component of causality. Moreover, there are important differences that should be explored in how teachers, students, and curricula unite concepts to create coherent learning experiences for students [16].

These systematic differences also have implications for how teachers and curriculum designers may create lessons and assess learning. While the NGSS and NRC *Framework* were created as guidelines for science instruction, they also serve as starting points and references for how to teach science. As such it is important to know which connections were important within and across disciplines, which can be used to compare with curricula and discussions to identify which connections occur in either, neither, or both the standards and real-world examples. Further, and more importantly, this epistemic network analysis created a metric space that can be used to measure and differentiate science thinking in other datasets. Therefore, I can create a metric space based on what the NGSS proposed and then project coded data about three-dimensional learning from real-world implementations into that space.

Of course, this analysis was limited to a comparison of two disciplines. Future analyses will further explore three-dimensional learning for earth and space sciences as well as engineering and technical sciences. Another limitation is that this analysis focuses solely on the standards themselves and does not look at science education in real classrooms. Future work will investigate how the practices and concepts are connected during actual student discussions about science and compare connection-making in the real-world with connections represented in the NGSS.

This work empirically identifies the underlying structure of the NGSS and provides both a way to compare science ideas within the standards as well as compare real-world data using the structure of the standards.

Acknowledgements. This work was funded in part by the National Science Foundation (DRL-1661036, DRL-1713110), the Wisconsin Alumni Research Foundation, and the Office of the Vice Chancellor for Research and Graduate Education at the University of Wisconsin-Madison. The opinions, findings, and conclusions do not reflect the views of the funding agencies, cooperating institutions, or other individuals.

References

1. Ford, M.J., Forman, E.A.: Redefining disciplinary learning in classroom contexts. Rev. Res. Educ. **30**, 1–32 (2006). https://doi.org/10.3102/0091732X030001001
2. NGSS Lead States: Next Generation Science Standards: For States, by States (Appendix F – Science and Engineering Practices). Achieve, Inc. behalf twenty-six states partners that Collab. NGSS, pp. 1–103 (2013). https://doi.org/10.17226/18290
3. National Research Council: A Framework for K-12 Science Education: Practices, Crosscutting Concepts, and Core Ideas. National Academies Press, Washington, DC (2012). https://doi.org/10.17226/13165
4. Lontok, K.S., Zhang, H., Dougherty, M.J.: Assessing the genetics content in the next generation science standards. PLoS One **10**, 1–16 (2015). https://doi.org/10.1371/journal.pone.0132742

5. Feinstein, N.W., Kirchgasler, K.L.: Sustainability in science education? how the next generation science standards approach sustainability, and why it matters. Sci. Educ. **99**, 121–144 (2015). https://doi.org/10.1002/sce.21137

6. Chesnutt, K., et al.: Next generation crosscutting themes: factors that contribute to students' understandings of size and scale. J. Res. Sci. Teach. **55**(6), 876–900 (2018). https://doi.org/10.1002/tea.21443

7. Shaffer, D.W.: How Computer Games Help Children Learn. Palgrave Macmillan, New York (2006). https://doi.org/10.1057/9780230601994

8. DiSessa, A.: Knowledge in pieces. Constr. Comput. age. 49–70 (1988). https://doi.org/10.1159/000342945

9. Bransford, J.D., Brown, A.L., Cocking, R.R.: How people learn: brain, mind, experience, and school. National Academies Press, Washington D.C. (1999). https://doi.org/10.17226/9853

10. Chi, M.T.H., Feltovich, P.J., Glaser, R.: Categorization and representation of physics problems by experts and novices. Cogn. Sci. **5**, 121–152 (1981). https://doi.org/10.1207/s15516709cog0502_2

11. Shaffer, D.W.: Quantitative Ethnography. Cathcart Press, Madison (2017)

12. National Science Teaching Association: Science and Engineering Practices (2014)

13. National Science Teaching Association: Crosscutting concepts (2014)

14. Shaffer, D.W., Ruis, A.R.: Epistemic network analysis: a worked example of theory-based learning analytics. In: Handbook of Learning Analytics Education data Mining (2017). in press. https://doi.org/10.18608/hla17.015

15. Berland, L.K., Schwarz, C.V., Krist, C., Kenyon, L., Lo, A.S., Reiser, B.J.: Epistemologies in practice: making scientific practices meaningful for students. J. Res. Sci. Teach. **53**, 1082–1112 (2016). https://doi.org/10.1002/tea.21257

16. Reiser, B.J., Mcgill, T.A.W.: Coherence from the students' perspective: why the vision of the framework for K-12 Science Requires More than Simply " Combining " Three Dimensions of Science Learning, 1 (2017)

Student Teachers' Discourse During Puppetry-Based Microteaching

Takehiro Wakimoto[1]([✉]) [iD], Hiroshi Sasaki[2], Ryoya Hirayama[3],
Toshio Mochizuki[3] [iD], Brendan Eagan[4], Natsumi Yuki[3],
Hideo Funaoi[5], Yoshihiko Kubota[6], Hideyuki Suzuki[7],
and Hiroshi Kato[8]

[1] Yokohama National University, 79-1 Tokiwadai, Hodogaya-ku,
Yokohama, Kanagawa 240-8501, Japan
educeboard@mochi-lab.net
[2] Kyoto University, 54 Kawaharacho, Syogoin, Sakyo-ku, Kyoto,
Kyoto 606-8507, Japan
[3] Senshu University, 2-1-1 Higashi-mita, Tama-ku,
Kawasaki, Kanagawa 214-8580, Japan
[4] University of Wisconsin-Madison, Madison, WI 53706, USA
[5] Soka University, 1-236 Tangi-machi, Hachioji, Tokyo 192-8577, Japan
[6] Tamagawa University, 6-1-1 Tamagawagakuen,
Machida, Tokyo 194-8610, Japan
[7] Ibaraki University, 2-1-1 Bunkyo, Mito, Ibaraki 310-8512, Japan
[8] The Open University of Japan, 2-11 Wakaba, Mihama-ku,
Chiba, Chiba 261-8586, Japan

Abstract. This study investigates how puppetry-based tabletop microteaching systems can contribute to student teacher training compared with normal microteaching. The study analyzes student teachers' discourse using a puppetry-based microteaching system called "EduceBoard" introduced to a university class. The analysis included an epistemic network analysis to identify the specific features that influence changes and clarify particular discourse patterns that were found and a qualitative analysis of the discourse data. Results indicate that the puppetry-based microteaching and improvisational dialogs that it elicited enhanced student teachers' practical insights and gave them the opportunity to develop their students' learning and run the class smoothly.

Keywords: Microteaching · Teacher education · Puppetry

1 Introduction

Nurturing students to explore things in a meaningful way, discover problems, reflect on their opinions, and engage in problem-solving alongside their peers is essential for preparing for 21st century society. To prepare for the conversations that will develop these skills in pupils, teachers need to imagine children's various voices, reactions, and questions to such issues [1].

© Springer Nature Switzerland AG 2019
B. Eagan et al. (Eds.): ICQE 2019, CCIS 1112, pp. 234–244, 2019.
https://doi.org/10.1007/978-3-030-33232-7_20

Microteaching is a method of implementing such dialogic pedagogy in teaching. To practice microteaching as a part of teacher training, student teachers are usually introduced into the roles of teachers and students. Playing the role of the teacher allows student teachers to evaluate and improve their classes and teaching skills, whereas playing the role of the pupil gives them a greater understanding of the pupils' psychology in the teaching process [2]. However, previous studies indicate that student teachers roleplaying as students can lead to excessive self-consciousness [3] and evaluation anxiety [4]. Concerns have also been raised about evaluation methods that lead to psychological resistance or inhibition, which can result in over-adaptation [5].

To make the roleplay in microteaching realistic, a puppetry roleplay learning system called "EduceBoard" was created, and past empirical research on the system found that a wide range of students' voices are elicited through puppetry-based microteaching [6]. However, this study did not focus on how these dialogic exchanges occurred and what such changes meant for prospective teachers in teacher education programs. Thus, the present study reveals how the discourse patterns within the EduceBoard puppetry-based microteaching differ from normal microteaching to identify the former's specific effects.

2 Methods

2.1 Target Class and Participants

The practical evaluation included 36 Japanese undergraduates studying to acquire an elementary school teacher's license. The target class was conducted twice, both of which were 3 h and 30 min long. The students were instructed to prepare a teaching plan and materials for their microteaching. The participants were divided into 12 groups of 3 and each participant took turns being in charge of teaching (10 min of microteaching), while the remaining two participants participated as children. Table 1 shows the outline of the target class. In the first and third sessions of the microteaching practice, the participants roleplaying as a teacher taught in front of the whiteboard and pupils studied in front of the teacher in a self-performed format, whereas in the second microteaching practice, all of the students performed with puppets on the EduceBoard system as a group (Fig. 1). In the puppetry session, each participant playing pupils' roles manipulated two pupils' puppets simultaneously. The microteaching was recorded either by a camera or the EduceBoard system.

2.2 Assessment

A coding scheme based on Fujie [7] was created for the present study, the scheme classified classroom dialog structures into two types of utterances: formal utterances (teacher-formal TF, student-formal SF) and informal utterances (teacher-informal TI, student-informal SI; Table 2). If an utterance was considered as a mixture of formal and informal utterances, the coders coded both categories. If utterances from the participants roleplaying as pupils were judged as not having any meaning (such as only "ah" or "umm"), those utterances were coded as other kinds of utterances. The first and

Table 1. Outline of target class.

Time	Activities		
20 (min)	Guidance for announcing the outline of the class.		
10	1st session: Self-performance	Self-performed micro-teaching role-play	Student A taught Students B & C. (B & C played pupils' roles)
10		Reflections on their performances while watching the recorded video independently	
20		Mutual feedback discussion while watching video together	
10		Writing short essays independently regarding what they learned in this session	
10	Break		
10	2nd session: Puppetry on EduceBoard	Tutorial of EduceBoard system	
10		Puppetry microteaching role-play	Student B taught Students A & C. (A & C played pupils' roles)
10		Reflections on their performances while watching the animation on the Web application independently.	
20		Mutual feedback discussion while watching the animation together.	
10		Writing short essays independently regarding what they learned in this session	
10	Break		
10	3rd session: Self-performance	Self-performed micro-teaching role-play	Student C taught Students A & B. (A & B played pupils' roles)
10		Reflections on their performances while watching the recorded video independently	
20		Mutual feedback discussion while watching video together	
10		Writing short essays independently regarding what they learned in this session	
10	Reflections	Discussion of what they learned during the three sessions	

fourth authors independently coded each microteaching utterance in accordance with the coding scheme (Cohen's kappa coefficient = .874). Table 3 shows quantified utterances classified into each relevant category and utterances coded for more than one category.

To conduct an epistemic network analysis (ENA; Shaffer [8]), ENA 1.5.2 (Marquart et al. [9]) was used to quantitatively analyze the discourse pattern in the microteachings and identify the characteristics of possible changes. ENA is a quantitative ethnographic method used for modeling the structure of speech, which can quantify and model the co-occurrence of codes in a conversation, provide a relevant visualization for each unit of analysis throughout the data, and create co-occurrence weighted networks. In this study, the structure of the utterance chain was quantitatively analyzed to study the changes that occurred in all three samples of microteaching and analyze the qualitative content using the feature values. Four codes were used in the ENA: TF, TI, SF, and SI. The number of the microteaching session and the conversation number and follow numbers used for the dataset for each group were included in the analysis. When differences were identified in the ENA, the strength of the co-

Normal microteaching

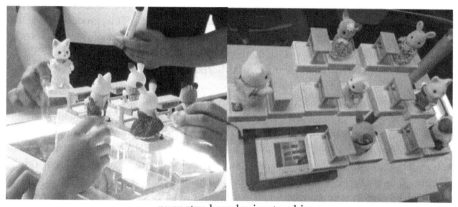

puppetry-based microteaching

Fig. 1. Normal microteaching and puppetry-based microteaching.

Table 2. Definition of utterances in roleplay microteaching [7].

Utterances	Definition
Teacher-Formal (TF)	A teacher's utterance that follows his/her lesson plan or is academic related
Teacher-Informal (TI)	A teacher's utterance based on his or her individual experience and reaction to the students
Student-Formal (SF)	A student's utterance that follows the teacher's instructions or is academic related
Student-Informal (SI)	A student's utterance based on his or her individual experience and intention (not academic)

Table 3. Total number of categorized sentences according to utterances in the discourse.

	1st (Self-performance)	2nd (Puppetry)	3rd (Self-performance)
Teacher-Formal (TF)	992	1494	1153
Teacher-Informal (TI)	111	529	276
Double-coded utterances (included in TF & TI)	45	206	112
Student-formal	597	603	731
Student-informal	223	342	409
Double-coded utterances (included in SF & SI)	61	44	109

occurrence connection between the individual codes was mentioned as a feature of the discourse network to be further examined. The package rENA was used to calculate and compare the intensity of individual co-occurrences of any two codes (i.e., between the first and second data points and the second and third). This process made the isolation of any connection between any two codes for any individual, group, or set of groups possible. Comparing individual connection strengths made it possible to determine whether statistically significant differences existed between two groups for any given connection in an ENA network, even when no statistical differences exist between groups on the ENA dimensions. In addition, this approach can be used to compare individual connection strengths to quantify the extent to which different connections contribute to statistical differences between groups along an ENA dimension. This approach is particularly useful when an ENA model has many connections and is hard to interpret visually and in studies where researchers focus on the differences in theoretic connections, connections detected in qualitative analysis, or those that emerged as important when conducting an ENA. Therefore, this approach was used to identify a number of factors affecting the discourse network of microteaching by identifying the statistical differences among the three microteaching sessions. Furthermore, the discourse from the puppetry microteaching session that have significant differences in the co-occurrence connection strength analyses were qualitatively analyzed.

3   Results and Discussion

3.1   Results of the ENA

The ENA results are summarized in Fig. 2, which shows the first (self-performance), second (puppetry), and third (self-performance) ENA networks. The findings show that the TF–SF connection was the strongest and some of the other connections seem to have differences when comparing the different microteaching sessions. Figure 3 shows the differences between the first and second discourse networks and the second and third. Strong co-occurrence relationships was found for the second time in TF–TI, TF–SI, and TI–SI connections, whereas the TF–SF connection was stronger for the first and

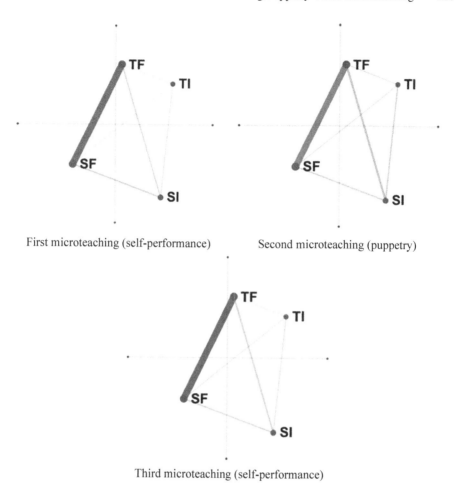

First microteaching (self-performance)   Second microteaching (puppetry)

Third microteaching (self-performance)

Fig. 2. First (self-performance), second (puppetry), and third (self-performance) microteaching discourse networks according to the ENA.

third times. The TF–SF connection consists of a normative IRE sequence [10]. These differences indicate possible significant changes in TF–SI, TF–SI, and TI–SI due to the puppetry microteaching.

Figure 3 shows that the mean of the plotted points of the discourse network fluctuate in each of the three rounds. When the differences between the first, second, and third rounds were measured with ENA, a significant difference was found along the X-axis between the first and second rounds (Mann–Whitney's $U = 931.00$, $p = .000$, $r = .48$) and the second and third rounds (Mann–Whitney's $U = 456.00$, $p = .03$, $r = .30$). This finding suggests that a significant discourse structure fluctuation occurred, especially during the puppetry microteaching session. As shown in the figures, the second microteaching session elicited a co-occurrence network with more connections to informal utterances. To investigate how such a significant fluctuation

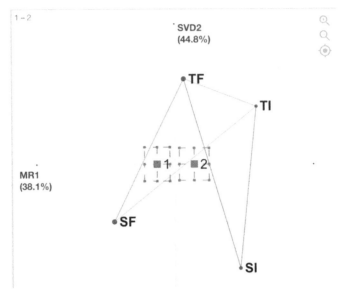

Differences between the 1st (self-performance) and 2nd (puppetry) discourse networks

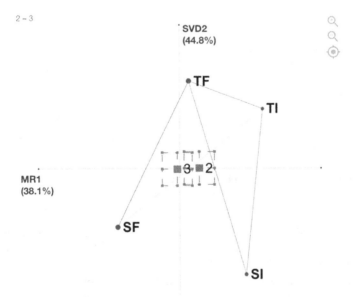

Differences between the second (puppetry) and third (self-performance) discourse networks

Fig. 3. Differences observed in the ENA between the first (self-performance) and second (puppetry) discourse networks and the second (puppetry) and third (self-performance) discourse networks.

was generated, a Freedman's test was conducted using rENA on the strength of the co-occurrence connections between codes for each session. The results revealed a significant difference among the three microteaching sessions. In particular, the strength of the co-occurrence connections of TF–TI (χ^2 (2) = 6.167, p = .046, Cramer's V = .72) and a small yet significant difference in the co-occurrence of TI–SF (χ^2 (2) = 5.167, p = .076, Cramer's V = .66) was observed. Although the TF–TI co-occurrence network was higher in the second session due to the increase of double codes, the high co-occurrence of TI–SF still needed to be investigated by focusing on the puppetry microteaching discourse to identify the kind of specific discourse prompted in this context.

3.2   Content Analysis of TI-SF

We used ENA webtool to analyze the puppetry-based microteaching discourse containing the co-occurrences of TI–SF. The categories chosen for the analysis were generated from the dialog in a bottom-up manner, and the syntactic break (e.g., the end of the sentence) was chosen as the end of the conversational turn. Six categories were chosen, which will be discussed at length in the following subsections.

3.2.1   Confirming the Children's Understanding in a Group

This category included scenes where participants roleplaying as teachers asked questions using the puppets to confirm pupils' understanding. The participants roleplaying as students played various children similar to a real classroom setting. The following is an excerpt from an arithmetic class, which discussed how to find the sum of four rectangular angles (Teacher is teacher role student, and Child is pupils role student.). The student playing the teacher role asked child questions such as "who understands?" or "Do you understand?". Then based on their reactions, the student playing the teacher role told the children that the figure is a rectangle or that the sum of the angles can be obtained when the four angles are combined only by hints:

Child 1: Yes (SF)
Teacher: And who understands that [pointing to a rectangle] angle? (TF)
Teacher: Is it a rectangle? (TF & TI)
Child 2: Angle … (SI)
Teacher: How much would it be if the four were combined? (TF)
Teacher: Do you understand? (TF & TI)
Child 1: There are four 90 s … I guess I should multiply. (SF)

3.2.2   Instructions for Children in a Group

This category included scenes where the teacher deals with a group of students with different learning levels and capacities. The example below is an excerpt from an arithmetic class where students struggle with addition and their calculations were wrong. Accordingly, the teacher improvised by instructing the student to use their hand:

Teacher: So, with your left hand? Because it is "2"? (TF & TI)
Teacher: Let's do the same as me everyone. (TF & TI)
Teacher: Raise your hand. (TF & TI)
Child 3: Well. (SF)

In the puppetry-based microteaching, there were children with various degrees of understandings, so the student playing the teacher role needed to watch his/her pupils and to teach them in an improvised manner.

3.2.3   One-to-One Interactions Regarding a Child's Understanding

This category included teachers' one-to-one interactions with children who are struggling to understand in a group setting. The following excerpt is from the arithmetic class that discusses the area of a circle:

> Child 4: I do not know too. (SI)
> Teacher: Hmmm, what do you not understand? (TI)
> Teacher: Well, you could draw a circle, right? (TF & TI)
> Child 5: I could draw a circle. (SF)
> Teacher: Could you draw a circle? (TF & TI)
> Teacher: Oh, fan-shaped … Do you know where the fan-shapes are? (TF & TI)
> Child 5: Here, here, here, here, and here? (SF)

The student playing the teacher role did not only teach the whole of the class as described in Sects. 3.2.1. and 3.2.2, but also responded for each pupil's understanding in an improvised manner.

3.2.4   One-to-One Interactions Regarding Class Content

This category included teachers' one-to-one interactions with students about specific topics as part of a class discussion. These comments included an in-depth exploration of the child's remarks, which is usually in the form of the teacher asking the child more questions and replying in an improvised manner to the utterance of each child. The following excerpt is from a Japanese language class where the teacher asks more questions regarding the topic of "wisdom":

> Child 6: Well. (SF)
> Child 6: I, grandma's, grandma's wisdom, I have read. (SF & SI)
> Teacher: Grandma's wisdom? (TF & TI)
> Teacher: Oh. (TF & TI)
> Teacher: Grandma's wisdom … (TF & TI)
> Teacher: Do you understand what that means? (TF)
> Child 6: Well, maybe not, Grandma has a lot of wisdom because she has been alive so long. (SF)
> Child 6: So, when I burned myself she applied aloe to my wounds. (SF)
> Child 7: Oh, there is aloe; there is aloe. (SI)
> Child 7: I am a growing aloe. (SI)
> Child 8: Oh, it's amazing. (TI)

Thus, in order to deepen the children's remarks, the student playing the teacher role drew remarks from the children by asking them more questions. The student playing the teacher role was required to accept the utterance of each child and to speak in an improvised manner that leads to learning. In the above-mentioned excerpt, the student playing the teacher role asked them to say what the grandmother's wisdom specifically means, and to share it with the class, so that the class theme of "what wisdom is" would be addressed. As mentioned above, the student playing the teacher role was able to experience enriching the class-wide conversation by talking with each child in an improvised manner.

3.2.5 One-to-One Interactions Regarding Discipline

This category included scenes where children required extra guidance regarding discipline. The following excerpt is from a Japanese language class where one child was instructed to listen more attentively to another child:

> Teacher: I'm sorry Hana-chan, would you like me to say it one more time? (TF & TI)
> Teacher: Well then. (TF & TI)
> Teacher: I'm listening to Sho-kun attentively. (TF & TI)
> Child9: Well. (SF)
> Child10: Mameta was able to go to the Buddha's place, but after all [inaudible]… I think that he is a coward because he did not change his mind that he relied on his grandfather. (SF)

Thus, in the puppetry-based microteaching, student teachers playing pupil roles played various types of pupils who do not necessarily listen to their teacher's instructions. The student teacher playing a teacher role is required to respond to such situations in an improvised manner. In the above situation, the student teacher playing the teacher role noticed that there would be pupils who did not listen to other children's remarks and could experience giving instructions to such problematic pupils.

3.2.6 Summary

In the puppetry-based microteaching, all five of the categories discussed above required the participants roleplaying as teachers to improvise their responses to participants roleplaying as students. In a real classrooms, teachers constantly attempt to deepen their students' learning and conduct their classes smoothly. Teacher training can reflect this reality and improve teacher professionalism and confidence. As Sato [11] concluded, teachers' professional competencies are made up of practical insights requiring reflection, including (1) improvisational thinking, (2) situational thinking, (3) multidimensional thinking, (4) contextualized thinking, and (5) reflective thinking for frameworks. In puppetry-based microteaching where participants roleplay as students and teachers, classes do not proceed according to the prepared teaching plan as they do in normal microteaching training. This context allows for random events and examples of learning to occur. It also allows student teachers to practice responding to changing situations, respond in improvised ways, and get more involved. The EduceBoard system provides a more authentic, less scripted simulation of classroom involvement for teachers. Having these experiences as part of teacher training can thus contribute to the improvement of the student teacher's professionalism.

4 Conclusion and Future Issues

This study aimed to analyze the discourse patterns used in a puppetry-based microteaching system (EduceBoard [6]) as part of a university class for student teachers. We hoped to clarify how puppetry-based microteaching changes the learning discourse and how this influences students who wish to become teachers. The results show that the EduceBoard system used for microteaching elicits a great variety of voices that are more realistic to a classroom setting compared with normal microteaching training. As part of the exercise, student teachers roleplaying as students were engaged in improvised dialogs with student teachers roleplaying as teachers. This process gave student teachers the opportunity to reflect on microteaching and develop practical insights.

Further analyses of the results still need to be conducted, including an analysis of the students' reflections after participating in microteaching training.

Acknowledgements. This work was supported in part by JSPS KAKENHI Grants-in-Aids for Scientific Research (B) (Nos. JP26282060, JP26282045, JP26282058, JP15H02937, & JP17H02001) from the Japan Society for the Promotion of Science, as well as the National Science Foundation (DRL-1661036, DRL-1713110), the Wisconsin Alumni Research Foundation, and the Office of the Vice Chancellor for Research and Graduate Education at the University of Wisconsin-Madison. The opinions, findings, and conclusions do not reflect the views of the funding agencies, cooperating institutions, or other individuals.

References

1. Bakhtin, M.: Discourse in the novel. In: Holquist, M. (ed.) The Dialogic Imagination, pp. 259–422. University of Texas, Austin (1981)
2. Sakamoto, T.: Meaning of teaching simulation. Mod. Educ. Technol. **47**, 5–11 (1978)
3. Ladrousse, G.P.: Role Play. Oxford University Press, Oxford (1989)
4. Cottrell, N., Wack, D., Sekerak, G.: Rittle, R: social facilitation of dominant responses by the presence of an audience and the mere presence of others. J. Pers. Soc. Psychol. **9**(3), 245–250 (1968)
5. Kira, S., Sato, S., Yoshida, M.: A study on microteaching technique in teaching training. Kumamoto University Department-of-Education Bulletin: Human sciences, 29, 221–236 (1980). in Japanese
6. Sasaki, H., et al.: Development of a tangible learning system that supports role-play simulation and reflection by playing puppet shows. In: Kurosu, M. (ed.) HCI 2017. LNCS, vol. 10272, pp. 364–376. Springer, Cham (2017). https://doi.org/10.1007/978-3-319-58077-7_29
7. Fujie, Y.: Children's in-class participation mixing academic and personal material: teacher's instructional response. Japan. J. Educ. Psychol. **48**, 21–31 (2000). in Japanese
8. Shaffer, D.W.: Quantitative Ethnography. Cathcart Press, Madison (2017)
9. Marquart, C.L., Hinojosa, C., Swiecki, Z., Eagan, B., Shaffer, D.W.: Epistemic network analysis (version 1.5.2) [software]. http://app.epistemicnetwork.org. Accessed 16 Jul 2019
10. Mehan, H.: Learning Lessons: Social Organization in the Classroom, Cambridge. Harvard University Press, MA (1979)
11. Sato, M.: Pedagogy as practical inquiry. Japan. J. Educ. Res. **63**(3), 275–278 (1996)

Using Epistemic Network Analysis to Explore Outcomes of Care Transitions

Abigail R. Wooldridge(✉) 🆔 and RuthAnn Haefli

University of Illinois at Urbana-Champaign, Urbana, IL 21801, USA
arwool@illinois.edu

Abstract. Care transitions are important to patient safety, but we lack consensus on what outcomes of transitions to evaluate. We interviewed 28 physicians and nurses who participate in transitions of adult and pediatric trauma patients from the operating room to the intensive care unit. The handoff (i.e., communication about patient information) in the pediatric care transition was done together as a team while the other handoff was separated by profession. In this study, we identify nine care transition outcomes: (1) communication sufficiency, completeness and accuracy, (2) handoff timing, (3) patient outcomes, (4) change in workload, (5) individual situation awareness, (6) team situation awareness, (7) organization awareness, (8) team experience and (9) timing of feedback. These outcomes could be positive and negative (i.e., good or bad). This study also investigates relationships between outcomes in the two groups using epistemic network analysis (ENA). While we found the no difference between the outcomes in the team and separate handoff when comparing frequency counts, relationships between outcomes did differ when using ENA. Interviewees with the team handoff described more relationships between care team level outcomes – team situation awareness and team experience – and other outcomes, while interviewees with the separate handoffs focused on the relationship between communication and patient outcome. Future work should investigate differences in relationships between positive and negative valences of the outcomes.

Keywords: Care transition outcomes · Handoffs · Epistemic network analysis

1 Introduction

Care transitions, i.e., the exchange of information, authority and responsibility for a patient between two or more healthcare professionals [1], are exceedingly important to patient safety. While care transitions provide opportunities to detect and correct errors [2], they are also opportunities for information loss and delays in care [3]. Because of this importance, and more than 7000 instances of patient harm in the United States between 2009 and 2013 [4], much work has focused on improving various care transitions. This paper focuses on care transitions between hospital units.

Research on care transitions between hospital units has focused on identifying key information to communicate between healthcare professionals and developing interventions to ensure that information is transferred, such as checklists and mnemonics [5].

© Springer Nature Switzerland AG 2019
B. Eagan et al. (Eds.): ICQE 2019, CCIS 1112, pp. 245–256, 2019.
https://doi.org/10.1007/978-3-030-33232-7_21

An important component of developing these interventions is the ability to assess their impact on the care transitions, which requires a way to measure outcomes. However, there is a notable lack of consensus around outcomes of care transitions; in fact, a systematic review of care transitions between hospital units identified 82 unique measures from 29 reviewed studies [6]. Measures included information transfer, satisfaction of participating healthcare professionals, clinical patient outcomes and compliance with care transition protocols. These various outcomes of care transitions may not be meaningful to the healthcare professionals who participate in the care transitions; in these studies, outcome measures were defined by the researchers, who were not all practicing healthcare professionals themselves. In other words, while those measures may be *etic*, they may not be *emic* [7].

The distinction of emic and etic can be very important. Emic focuses on the perspective and interpretation of members of the group, such as healthcare professionals; etic focuses on the perspective and interpretation of those outside of the group, such as researchers or scientists [7]. Similarly, in ergonomics – the authors' area – there is a strong tradition of focusing on the *activity* of workers (i.e., healthcare professionals), rather than the *task* prescribed by others [8]; to do this, we must obtain some level of understanding of the perspective of the healthcare professionals themselves. In other words, we strive for an emic understanding the work and outcomes under investigation, which in this study is care transitions. Therefore, the key premise of this study is that emic measures, i.e., measures based on the perspective and interpretation of healthcare professionals participating in care transitions, will be more useful in assessing interventions to improve care transitions and/or impacting healthcare professional acceptance of those interventions.

1.1 Background on Care Transitions

Previous research has focused on care transitions between hospital units of trauma patients [9–11], as trauma is the leading cause of death in children and young adults in the United States [12]. Wooldridge [11] compared care transitions of pediatric and adult trauma patients from the operating room (OR) to the pediatric intensive care unit (PICU) or intensive care unit (ICU), respectively, using SEIPS-based process modeling [13]. Both care transition processes studied involved coordination and communication work while the patient was still in the operating room, physical transport between the two units, handoff communication between healthcare professionals in the OR and the ICU to transfer information about past, ongoing and future patient care and follow up after the handoff [11]. Work in each phase was distributed across three physical environments: the OR, patient transport (movement between the units) and the (P)ICU.

However, the way the handoff communication between the two units was organized was very different in the two transitions. In one transition, the handoff was done as a team, with surgeons, anesthesiologists, nurses and physicians-in-training from the OR physically going to the PICU with the patient and communicating as a team at the patient's bedside to the physicians, nurses and physicians-in-training from the PICU; we will refer to this as the *team transition*. In the other transition, the handoff communication was separate by profession (i.e., nurse to nurse and physician to physician) and could occur via telephone, in-person at bedside or in-person away from the patient;

we will refer to this as the *separate transition*. The team and separate handoffs are two very different ways of organizing work in the care transition – in fact, this difference was the impetus for the original study – and could influence what participating healthcare professional identify as outcomes.

This paper identifies care transition outcomes described by healthcare professionals who participate in two care transitions. This paper also compares the frequency each outcome was mentioned healthcare professionals in each care transition, i.e., treats the outcomes as independent when comparing the two transition. Lastly, we explore the relationships between care transition outcomes using epistemic network analysis (ENA), which is a powerful tool to explore relationships between coded elements that have previously been treated as independent [14]. Specifically, our objective is to compare the relationships between outcomes described by participants in the team transition with outcomes described by the participants in the separate transition.

2 Methods

This research is part of a larger project focused on designing health information technology to support care transitions of trauma patients, in particular comparing handoffs organized by team or separately by profession. We obtained approval for this study from the IRB at the University of Wisconsin-Madison and the secondary data analysis from the IRB at the University of Illinois at Urbana-Champaign.

2.1 Setting and Sample

We conducted this at a level 1 trauma center in the Midwestern United States. The team transition was from the pediatric OR (8 operating suites) to the PICU (21 beds). The separate transition was from the adult OR (27 operating suites) to an adult ICU (24 beds). We collected interview data between 2016 and 2018 with 28 healthcare professionals (15 from the team transition and 13 from the separate transition). Our sample was ten nurses, three intensivists, six anesthesiologists, two anesthetists and seven surgeons. We determined sample size by monitoring for saturation [11].

2.2 Data Collection

Our interviews were semi-structured, focusing on soliciting descriptions of the entire care transition, examples of good transitions, examples of bad transitions and opportunities for improvement. The interview guide is available at https://cqpi.wisc.edu/research/health-care-and-patient-safety-seips/teamwork-and-care-transitions-in-pediatric-trauma/. Each interview took on average about 50 min.

2.3 Data Analysis

Qualitative Data Analysis. The interview transcripts were cleaned to remove identifying information. The interview transcripts were analyzed using an inductive thematic analysis method inspired by grounded theory [15], iterating between reviewing the interview data to allow outcomes to emerge from the data and reviewing care transitions and human factors literature to connect emic concepts with etic terminology, thus developing outcomes with both emic and etic meaning. Two researchers read one transcript, generating paper-based notes, and then discussed coding challenges to refine the coding scheme in a consensus-based process. We repeated this process for a second interview.

To enhance the rigor of the quantitative analysis of qualitative data analysis, a triangulation of analysts [16] was employed. Two researchers coded four interviews with the nine care transition outcomes separately; they met and reviewed their coding to discuss differences and refine outcome definitions. The two researchers then coded two more interviews separately to evaluate inter-rater reliability to strengthen the internal validity of the research. Cohen's κ was calculated for all outcomes; all values were above the acceptable value of 0.8 and indicate that the interpretation and coding of interview data are reliable. Both authors reviewed and coded the remaining transcripts; any disagreements were resolved by discussion.

Comparison of Outcomes. Following the qualitative data analysis, the frequency of each code application was counted for both team and separate transitions. Chi-Squared tests were conducted using Microsoft Excel© to determine if the outcomes were dependent on the care transition (team or separate).

ENA Analysis. For the ENA analysis, one researcher segmented the interview data by hand following the interview guide structure. Each segment was one question from the interview guide, the interviewee's initial response and any follow-up, probing questions and responses.

We used ENA to compare the relationships between the nine outcomes described by the two groups, i.e., team transition and separate transition. To conduct this analysis, we used the ENA web tool (version 1.5.2) [17]. ENA uses an algorithm that constructs a matrix for each segment of data that shows connections between the coded outcomes (i.e., the recent temporal context was defined as 1 line and the codes included in the model are the nine outcomes seen in Table 1). All matrices for each interviewee are then aggregated to represent all of the connections made. The adjacency vectors of all of the individual interviewees are then normalized and subjected to a means rotation to account for different interviewees having different amounts of coded lines. Finally, an average network for each group – team transition and separate transition – are developed. See Shaffer et al. [18] for a more detailed explanation of the mathematics and information about the resulting networks. For this paper, we compared the average network for each group – team transition and separate transition – by examining the centroids and differences between the average network of both groups. To test for differences, we applied a two-sample t-test assuming unequal variance to the centroids of the average team transition and separate transition networks.

Table 1. Care transition outcomes identified by participating healthcare professionals.

Outcome	Definition	Team handoff frequency count	Separate handoff frequency count	ENA abbreviation
Communication sufficiency, completeness and accuracy	Whether or not sufficient, accurate information was communicated [19] for healthcare professionals involved in the handoff, i.e., the receiver is/is not able to care for the patient with information provided	39	69	O.CommSCA
Handoff timing	Whether or not there was a delay to starting the handoff, which was/was not appropriate length (duration) with/without repeating information and/or distractions or interruptions	20	29	O.HandoffTiming
Patient outcomes	Whether or not the handoff impacts patient safety and quality of care, e.g., harmed with lost/inaccurate information potentially causing patient safety issues, medical errors or improved with errors caught, decisions revisited, etc	7	23	O.PtOutcome
Change in workload	Whether or not the handoff impacts healthcare professionals' workload, for example causing an increase to seek additional information via phone calls, chart review, etc	4	13	O.Workload
Individual situation awareness	Whether or not each healthcare professionals in the handoff perceives information elements in the handoff, comprehends the meaning of these elements and projects the status of these elements in the near future [20]; in other words, whether or not healthcare professionals can contextualize the patient and the information they received	8	18	O.SA

(continued)

Table 1. (*continued*)

Outcome	Definition	Team handoff frequency count	Separate handoff frequency count	ENA abbreviation
Team situation awareness	Whether or not each healthcare professional on the care team is on the same page. At the team level, this is the aggregate of unique and shared SA; there should be some minimum overlap	18	29	O.TSA
Organization awareness	Whether or not each healthcare professional is aware of how their role fits in the organization [21]	4	5	O.OrgAware
Team experience	Whether or not the handoff impacts how experienced care teams are in terms of familiarity, length of time working together and how frequently they work together	10	9	O.TeamExper
Timing of feedback	Whether or not sending healthcare professionals immediately know if handoffs were successful, because they may not perceive a problem until much later	2	2	O.DelayedFB

3   Results

3.1   Emic Care Transition Outcomes

Nine handoff outcomes were identified in this study: (a) communication sufficiency, completeness and accuracy, (b) handoff timing, (c) patient outcomes, (d) change in workload, (e) individual situation awareness, (f) team situation awareness, (g) organization awareness, (h) team experience and (i) timing of feedback. See Table 1 in the for the full definition of each outcome.

The most common care transition outcome was communication clarity, accuracy and sufficiency, which was often linked with patient outcomes, such as by the adult critical care fellow:

> *"I felt that the care was most complete and safest [when]... the fellow, the resident, and the nurse and the RT all received the same story and sign-out and directions from me directly. ... [N]othing would be missed in that way" (adult fellow).*

Related to clear, accurate and sufficient communication was that not only was the information communicated, but that the healthcare professional(s) in the (P)ICU were able to *"contextualize the patient"* (adult surgery chief resident). In other words, their individual situation awareness was good [20, 22].

One outcome of the transition was timing, leading to delay that prevented participants from completing other work, or avoiding delays. An adult anesthesia resident described this: *"[I]t was a nice transition because I had, the surgeons were there in the room when I got there, the receiving surgical team. The nurse was there. The attending was there."*

The transition also impacted the workload of healthcare professionals. In particular, a poor transition could increase their workload and result in effort spent searching for information, as described by a PICU fellow: *"[T]here's a lot of phone tag that happens until everyone is back on the same page again."*

Care transitions also impacted team situation awareness, i.e., whether or not the care team members are on the same page about the patient after the handoff [23, 24]. It is important to note that team situation awareness is related to the care team – everyone involved in caring for the patient – and therefore applies to both care transitions, no matter the organization of the handoff. This was described by the pediatric critical care fellow:

> *"[When handoffs are performed over the phone,]... I'm supposed to be able to ask questions for everybody. But it's hard for me to think of all those questions because I'm not in that role ...[It] really gets back to the point where if you get everyone on the same page, it makes [the handoff] easier"* (pediatric fellow).

Participants described how participating in care transitions in the (P)ICU could positively impact organization awareness, i.e., awareness of the impact of their actions upstream and downstream [21]. For example, participating in the care transition in the PICU *"helps with the flow because our PICU beds, like when we're on high census, they're full"* (pediatric OR nurse). Being in the PICU lets the surgical team notice if beds are available for future patients, and could lead to changes in the surgical schedule.

Interestingly, participants noted that each care transition was also an opportunity to increase team experience. Team experience refers to the how experienced the care team is working together. Team experience therefore also applies to the experience of the care team working together in the care transitions with separate handoffs. For example, a pediatric anesthesia resident noted that *"the same anesthesia attendings and the same surgeons are dropping people off day [in] and day out,"* and went on to say that this increase in familiarity made subsequent transitions easier.

The participants also noted that sometimes feedback about the care transition was delayed, meaning that if it went well or poorly, they might not hear until the next day, or later. For example, a surgery resident described a case where she transitioned the patient to the PICU overnight and felt that the transition went well from a communication perspective. But, the next day, *"based on like the questions I had got asked the next morning by both the nurses and the residents, there clearly was a lack of understanding"* (pediatric surgery resident).

Table 1 summarizes the frequency of the outcomes by care transition types. The distribution of outcome frequency counts was not dependent on by the type of care transition (team or separate; $\chi^2(8) = 7.02$, p = 0.534).

3.2   ENA Results

Figure 1 shows the resulting ENA plots. The resulting ENA model had co-registration correlations of 0.95 (Pearson) and 0.96 (Spearman) for the first dimension and co-registration correlations of 0.95 (Pearson) and 0.94 (Spearman) for the second. These measures indicate that there is a strong goodness of fit between the two dimensional visualization and the original model. Outcomes that are at the individual level fall in the negative x-axis direction, while outcomes at the team level fall in the positive x-axis direction. Further, sufficient, complete and accurate communication is nearly at the center of the network, likely because most other outcomes are related (i.e., connected) to it.

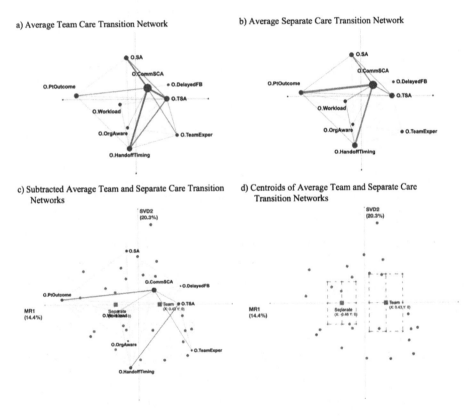

Fig. 1. ENA plots.

Figure 1a shows the average network for the interviewees participating in the team transition, while Fig. 1b shows the average network for interviewees participating in the separate transition. Node size indicates relative frequency of the coded outcome while edge darkness indicates strength (i.e., frequency) of co-occurrence or relationship. We can compare these networks visually, but Fig. 1c shows the network resulting when the average separate transition network is subtracted from the average team transition network. When we compare these networks, first we see that there are stronger relationships involving the team level outcomes, i.e., team situation awareness and team experience, and handoff timing in the team transition group. On the other hand, the separate transition group shows a stronger relationship between patient outcome and complete, sufficient and accurate communication. When we examine the centroids of the average network, shown in Fig. 1d with the 95% confidence interval, we see that there is a significant difference in the location of the centroids along the x axis ($p < 0.01$), but not the y axis ($p = 1.00$).

4   Discussion

In this study, we investigated care transition outcomes described by healthcare professionals participating in two care transition processes – one in which the handoff was done as a team and one in which the handoff was separated by profession. We identified emic outcomes by first reviewing interview transcripts to identify themes related to outcomes of the care transition (i.e., what happened after the process was completed) and then iterating between reviewing the care transitions and human factors bodies of literature and the interview data to relate etic descriptions from literature to the emic ideas from the interview [11]. Notably, only two of these nine emic outcomes were related to the measures identified in the systematic review: the communication being sufficient, accurate and complete and patient outcomes.

When comparing the frequency that each outcome was mentioned by healthcare professionals in the two different care transitions, we found that there was not a significant difference in the distribution of frequency counts between the two care transitions. In other words, including a team or separate handoff did not influence the outcomes that were discussed by healthcare professionals participating in the transition. However, this approach treated the nine outcomes as independent and isolated; in reality, we suspected that the nine outcomes were related. For example, it follows logically that a patient outcome would be related to sufficient, accurate and complete communication. Therefore, we used ENA – with its focus on examining relationships – to examine the relationship between care transition outcomes described by healthcare professionals. This analysis demonstrated that complete, sufficient and accurate communication is related to all of the other outcomes; in other words, we cannot assume they are unrelated.

Further, ENA demonstrated that the relationships between the outcomes varied by team or separate transition process. Interviewees from the team transition made stronger connections with care team level outcomes and handoff timing, while interviewees from the separate transition made stronger connections between complete, sufficient and accurate communication and patient outcome. This is interesting for multiple reasons. First, these findings could indicate that the team handoff is more effective at the team-outcome level. The outcomes identified in this study did not describe valences – they could be positive (e.g., communication could be sufficient, accurate and complete; team situation awareness could be high, etc.) and they could be negative (e.g., communication could be insufficient, inaccurate and/or incomplete; team situation awareness could be low, etc.). The valences of outcomes described in each care transition could vary; e.g., one might expect more positive care team outcomes in the team transition but more positive handoff timing in the separate transition. Confirmation would require a future analysis examining whether the data resulting in these connections described positive or negative outcomes (i.e., coding an additional level to describe polarity or valence of the outcomes in each excerpt). Second, it could indicate that the most appropriate outcome measure(s) to study depends on the design of the process that is being analyzed. This may not be a surprise to pragmatic researchers who select research methods based on research questions; however, this may not be intuitive to the clinical researchers subscribe to different research philosophies, and thus could have important implications for future work.

Nonetheless, this study does have limitations. The sample is limited to one participating site, which does limit generalizability. This analysis compares transitions in two different units in the same hospital, which are different contexts and could influence some of the differences found. However, all of each unit's transitions were organized in the same way (i.e., a comparison of within one unit was not feasible). A more pressing limitation is that these outcomes do not include the perspective of the patient and/or their family, who are becoming readily accepted as members of the care team. Finally, during data collection another adult ICU at the hospital implemented a team handoff from the adult OR, and some of our interviewees participated in them and reflected on this during the interviews. While we did not include those reflections in the dataset for this paper, it could nonetheless influence our findings.

5 Conclusion

This study identified nine care transition outcomes described by healthcare professionals who participate in two different OR to ICU care transition processes. Those emic outcomes could guide the evaluation of future care transition outcomes. The frequency that each outcome was mentioned did not differ between the two care transitions. ENA showed that not only are the outcomes related, but the relationships are different in the two care transitions. Interviewees from a team transition described more relationships between team-level outcomes such as team situation awareness and team experience and other outcomes, while interviewees from the separate transition focused on the relationship between completed, accurate and sufficient communication and patient outcomes. This difference indicates that some outcomes may be more or less relevant, depending on how the care transition is organized; the current study could help guide the selection of outcome measures that have importance to healthcare professionals participating in care transitions. Future work should investigate differences in relationships between positive and negative valences of the outcomes, which may give further insight into the benefit of including a team handoff rather than separate handoffs in a care transition process.

Acknowledgements. Funding for this research was provided by the Agency for Healthcare Research and Quality (AHRQ) [Grant No. R01-HS023837] (PI: Ayse P. Gurses, Johns Hopkins University; PI of UW-Madison subcontract: Pascale Carayon) and the Research Experience for Undergraduates program of the Department of Industrial and Enterprise Systems Engineering at the University of Illinois at Urbana-Champaign. This work was funded in part by the National Science Foundation (DRL-1661036, DRL-1713110), the Wisconsin Alumni Research Foundation, and the Office of the Vice Chancellor for Research and Graduate Education at the University of Wisconsin-Madison. The content is solely the responsibility of the authors and does not necessarily represent the official views of the funding agencies. We thank the study participants, as our research would not be possible without them.

References

1. Abraham, J., Kannampallil, T., Patel, V.L.: A systematic review of the literature on the evaluation of handoff tools: implications for research and practice. J. Am. Med. Inform. Assoc.: JAMIA **21**(1), 154–162 (2014)
2. Perry, S.J.: Transitions in care: studying safety in emergency department signovers. Focus Patient Saf. **7**(2), 1–3 (2004)
3. Arora, V.M., Manjarrez, E., Dressler, D.D., Basaviah, P., Halasyamani, L., Kripalani, S.: Hospitalist handoffs: a systematic review and task force recommendations. J. Hosp. Med. **4**(7), 433–440 (2009)
4. The Joint Commission: Inadequate hand-off communication. Sentinel Event Alert **58**, 1–6 (2017)
5. Abraham, J., Ihianle, I., Burton, S.: Exploring information seeking behaviors in inter-unit clinician handoffs. In: International Symposium on Human Factors and Ergonomics in Health Care, pp. 226–231. SAGE Publications Sage India, New Delhi, India (2017)
6. Robertson, E.R., Morgan, L., Bird, S., Catchpole, K., McCulloch, P.: Interventions employed to improve intrahospital handover: a systematic review. BMJ Qual. Saf. **23**(7), 600–607 (2014)
7. Shaffer, D.W.: Quantitative Ethnography. Madison, Indianapolis (2017)
8. Daniellou, F.: The French-speaking ergonomists' approach to work activity: cross-influences of field intervention and conceptual models. Theor. Issues Ergon. Sci. **6**(5), 409–427 (2005)
9. Wooldridge, A.R., et al.: Complexity of the pediatric trauma care process: implications for multi-level awareness. Cogn. Technol. Work **21**, 397–416 (2018)
10. Wooldridge, A.R., et al.: Work System Barriers and Facilitators in Inpatient Care Transitions of Pediatric Trauma Patients (Submitted)
11. Wooldridge, A.R.: Team Cognition Distributed in Spatio-Temporal Processes: A Macroergonomic Approach to Trauma Care. ProQuest, University of Wisconsin, Madison (2018)
12. Injury Prevention and Control: Data & Statistics (WISQARS) (2015). www.cdc.gov/injury/wisqars. Accessed 28 Feb 2019
13. Wooldridge, A.R., Carayon, P., Hundt, A.S., Hoonakker, P.L.T.: SEIPS-based process modeling in primary care. Appl. Ergon. **60**, 240–254 (2017)
14. Swiecki, Z., Ruis, A.R., Farrell, C., Shaffer, D.W.: Assessing individual contributions to collaborative problem solving: a network analysis approach. Comput. Hum. Behav. (2019)

15. Robson, C.: Real World Research, 3rd edn. Wiley, Chincester (2011)
16. Devers, K.J.: How will we know "good" qualitative research when we see it? Beginning the dialogue in health services research. Health Serv. Res. **34**(5), 1153–1188 (1999)
17. Marquart, C.L., Hinojosa, C., Swiecki, Z., Eagan, B., Shaffer, D.W.: Epistemic Network Analysis (Version 1.5.2) (2018). 1.5.2 ed
18. Shaffer, D.W., Collier, W., Ruis, A.R.: A tutorial on epistemic network analysis: analyzing the structure of connections in cognitive, social, and interaction data. J. Learn. Anal. **3**(3), 9–45 (2016)
19. Lingard, L., et al.: Communication failures in the operating room: an observational classification of recurrent types and effects. BMJ Qual. Saf. **13**(5), 330–334 (2004)
20. Endsley, M.R.: Toward a theory of situation awareness in dynamic systems. Hum. Factors: J. Hum. Factors Ergon. Soc. **37**(1), 32–64 (1995)
21. Schultz, K., Carayon, P., Hundt, A.S., Springman, S.R.: Care transitions in the outpatient surgery preoperative process: facilitators and obstacles to information flow and their consequences. Cogn. Technol. Work **9**(4), 219–231 (2007)
22. Endsley, M.R.: Proceedings of the National Aerospace and Electronics Conference. IEEE, New York (1988)
23. Prince, C., Salas, E.: Team situation awareness, errors, and crew resource management: research integration for training guidance. In: Situation Awareness Analysis and Measurement, pp. 325–347 (2000)
24. Endsley, M.R., Jones, W.M.: Situation Awareness Information Dominance & Information Warfare (1997)

Exploring the Development of Reflection Among Pre-service Teachers in Online Collaborative Writing: An Epistemic Network Analysis

Yuhe Yi[✉], Xiaoxu Lu, and Jing Leng

Department of Educational Information Technology,
East China Normal University, Shanghai, China
313699389@qq.com

Abstract. Facilitating reflection of pre-service teacher is becoming a more and more important topic in teacher education. There are a number of social media tools which can support teacher professional development. It also enables us to examine the development of individuals' reflective process and group dynamics. In this study, 50 pre-service teachers were involved to write scripts collaboratively using wikis and they were encouraged to reflect upon their written texts and script-writing strategies during the online collaborative writing process. In particular, epistemic network analysis is adopted in order to characterize learners' reflection dynamics during the two phases of collaborative script writing. The research results show that the characteristics of reflection type in different phases are different. Also, teachers tend to reflect on the content and methods of the group in the first phase; while in the second phase, they tend to reflect on the group methods and personal gains. Using content analysis and epistemic network analysis, this paper characterize the development of reflection during collaborative writing activities and provides reference for the cultivation of reflection among pre-service teachers.

Keywords: Teacher reflection · Teacher professional development · Collaborative writing · Epistemic network analysis

1 Introduction

As the practitioners of education and teaching, teachers should become active learners to improve their professional ability so as to meet teaching and learning needs in the age of information technology. Reflection is the link between individual's recent experience and past experience, and the process of individual's critical reflection on his own behavior and thoughts (Colton and Sparks-Langer 1993). It is based on past experience, a more profound internal psychological state. Therefore, it is important to understand how to cultivate pre-service teachers' reflection (Good and Whang 2002). The establishment of online communities provides a way to cultivate teachers' reflection and promote their professional development. However, one's reflection, as a mental process inside the individual's mind, is regarded as implicit and invisible. While engaging pre-

© Springer Nature Switzerland AG 2019
B. Eagan et al. (Eds.): ICQE 2019, CCIS 1112, pp. 257–266, 2019.
https://doi.org/10.1007/978-3-030-33232-7_22

service teachers in online collaborative activities, it is important to foster their reflection upon script writing and collaborative work. In order to characterize the dynamics of reflective process, this study analyzes students' reflection journals in the process of collaborative learning.

At present, researches on reflection mainly focus on the analysis of different reflection level, the exploration of relevant cultivation modes and concepts as well as the exploration of affecting factors (Blomberg et al. 2014). However, the content of reflection journals involves multiple aspects, and each aspect may involve different types of reflection. It is necessary to find out how to dynamically construct the development of reflection and whether collaborative writing promote a more complex and interrelated mental schema among pre-service teachers during an extended period of online collaboration. In this study, epistemic network analysis (ENA) is used to explore the characteristics of individual or group cognitive framework by quantifying qualitative data. Due to its dynamic and coupling characteristics, ENA plays an important role in deep data mining, dynamic assessment of learners' ability development and improvement. It has also been favored by researchers in related fields (Shaffer 2017). The aim of this study is to use ENA method to analyze and compare the development of reflection among pre-service teachers in different phases of online collaborative writing activities.

2   Literature Review

2.1   The Cultivation of Pre-service Teachers' Reflection

The cultivation of teachers' reflection is generally considered to be an important part of their professional development. In a way, Teachers' reflection can predict the degree to which a teacher can teach students well, which is considered as evidence of teachers' effective teaching skills (Blomberg et al. 2014). Without reflection, practical experience is only a quantitative change rather than a qualitative change of experience, and teacher professional growth will not occur. And through teaching reflection, teachers can correct teaching problems and improve teaching skills (van Es and Sherin 2008). The significance of reflection for pre-service teachers is to enable them to reflect on different experiences, integrate theories with their own practical experience, form their own understanding of education and teaching beliefs, as well as put them into practice. However, pre-service teachers' beliefs and awareness of effective teaching can directly affect their attention to classroom teaching (Huang and Li 2012). At the same time, the effective reflection of pre-service teachers in their own teaching practice is an important part for them to grow and mature from novice teachers to expert teachers as soon as possible. Effective reflection can enable pre-service teachers to review and study their own teaching practice in the future teaching process. Therefore, it is necessary to pay attention to the cultivation of pre-service teachers' reflection and explore the development of their reflection.

The development of information technology has opened up new ideas for the development of teacher education. The use of technology in teaching not only promotes teachers' teaching reflection as a teaching method, but also changes and enriches the content of teachers' teaching reflection (Mckinney 1998). For example, Krutka et al. (2014) used the social networking website Edmodo to enable teachers to conduct collaborative learning and collaborative reflection. Potter (2001) explored the development of teachers' critical reflection in the process of online collaborative research activities. Research shows that teachers' reflection is promoted during the process of collaborative learning, and at the same time, the background experience and opinions of peers have important contributions to the improvement of teachers' teaching skills and teachers' professional development.

2.2   The Epistemic Network Analysis

Ethnography is considered as an effective method to develop teachers' reflection, so ethnography can be used as a means to explore teachers' reflection (Beyer 1984). Traditional ethnography is time-consuming and difficult to be analyzed on a large scale. To solve this bottleneck, Shaffer proposed a method of "qualitative ethnography", the most important of which is epistemic network analysis. Epistemic network analysis (ENA) is not only an evidence-centered tool for quantitative analysis of textual discourse, but also a technique for modeling network topics of professional competence (Shaffer 2017). ENA has three core concepts: code, units, and sections. The code represents a set of conceptual elements, the purpose of ENA is to understand the relationships between these elements, not the individual ideas in the discourse (Shaffer and Graesser 2010). Besides, Analysis units represent ENA objects, such as activity phases or group divisions. Meanwhile, section represents the scope in which the code appears together. ENA's core idea is to establish a network reflecting the connection between different ability codes in the whole dialogue process, on the basis of the co-occurrence times of each ability codes in the context of the dialogue.

ENA has been successfully applied to the analysis of teachers' collaborative learning and scientific reasoning, and has demonstrated its outstanding characteristics (Csanadi et al. 2018). Based on this, this study will adopt the epistemic network analysis to analyze the reflection journals written by pre-service teachers during the process of collaborative learning, and explore the development process and mode of teachers' reflection in this process by comparing teachers' reflection in different phases of activities.

2.3   Research Questions

1. What are the characteristics in teachers' reflection journals while engaging them in online collaborative script writing?
2. Is there any difference in the epistemic networks of teachers' reflection between different phases of the online activity phases? If yes, what are the different epistemic characteristics of teachers in the two phases?

3 Methodology

3.1 Research Context

In this study, 50 sophomore students (23 male students and 27 female students) who took interactive courseware development course in a university were selected as experimental subjects. All of them have strong information technology literacy, and can skillfully operate various software platforms, besides, they will become an information technology teacher after graduation. Therefore, they were defined as pre-service teachers in our study. The students are randomly assigned in different groups with each containing 4 or 5 pre-service teachers. As a result, the whole class was divided into 12 groups. Before collaborative learning, the course teacher assigned the task of script writing, and required them to complete the writing task within their groups, and the activity lasted for three months.

3.2 Design

The whole collaborative scripting activity can be divided into two phases. In the first phase, the teacher group carries out online and offline discussions respectively to determine the theme and framework of script and the division of labor. After each group member completed the division of tasks, the team carried out collaborative modification to form the prototype script. The course teachers gave feedback on each group's prototype script scheme and evaluated the individual teachers' performance in the collaboration process. Then, each teacher made reflection and summary, and wrote the corresponding reflection journals towards their own understanding of the script, personal writing, group collaboration, the quality of the script, personal gains and shortcomings, etc.

The second phase is mainly to improve the prototype script. Based on the discussion results within the class and within the group, the teacher group further improved the writing of script on the Wiki platform. Each teacher participated in the collaborative modification process and finally formed the final script of the group. After course teachers' feedback, Each pre-service teacher wrote the corresponding reflection journals according to their performance in the second phase of collaborative learning activities. And the content of reflective journals mainly includes "individual performance ", "team coordination", "suggestions for script improvement" as well as "personal gains".

3.3 Coding Scheme

The data obtained in this study are mainly 100 reflection journals written by the pre-service teachers, and 50 for each phase. Taking the characteristics of collaborative script writing and reflection journals into account, we chose the reflection coding framework which is proposed by Hatton and Smith (1995) as a coding scheme to characterize different types of reflection in the reflection journal. The specific description of this framework is shown in Table 1.

Table 1. The different types in student reflections

Different kinds of reflection	Description	Coding
Descriptive writing	Not reflective at all, but merely reports events or literature	R0
Descriptive reflection	Does attempt to provide reasons based often on personal judgment or behavior	R1
Dialogic reflection	A form of discourse with one's self, an exploration of possible reasons	R2
Critical reflection	Involving reasons giving for decisions or events which takes account of the broader historical, social, and/or political contexts	R3

According to the content in student reflections, five main aspects could be identified: reflection on individual script, reflection on personal methods, reflection on group script, reflection on group method, and summative reflection. The explanation and description of each aspect are shown in Table 2.

Table 2. The different aspects in student reflections

Reflection aspects	Description	Coding
Individual content	Describing the content of individual script writing	CC1
Individual method	Explaining the methods used by himself or herself in script writing	CC2
Group content	Describing the content of the group script	CC3
Group method	Explaining the writing method of group script	CC4
Summary reflection	Analysis of its experience and harvest in the process of activities, or reflection on its shortcomings	CC5

3.4   Data Collection and Analysis

The coding of the reflection journals is done by two assistants of the course. Due to the length of teachers' reflection journals, a series of symbols representing the end of a sentence such as period, question mark, exclamation mark and so on were used as the interval points of the meaning unit during the coding in this study, and the content between each adjacent two symbols was used as a basic unit of analysis. Prior to the formal coding, the two assistants negotiated and confirmed the content of the coding framework to ensure that their understanding of it was consistent. At the same time, they randomly selected 40% of the original corpus for pre-coding, and used SPSS software to analyze the coding results, and found that the Kappa coefficient was greater than 0.7 (Kappa = 0.87), which indicating that the results of their coding were basically consistent and scientific. The two then went on to code the rest of the reflection journals.

4 Results

4.1 What Are the Characteristics in Teachers' Reflective Journals While Engaging Them in Online Collaborative Script Writing?

For the first phase, the most common reflection type was descriptive writing (R0, 56.63%). The frequency of critical reflection was the lowest (R3, 3.75%). In terms of aspects of reflection content, reflection in the first phase focused on the group content (CC3, 38.61%). In the second phase, pre-service teachers have the highest frequency in the descriptive reflective type (R1, 48.78%). The type with lowest frequency was critical reflection (R3, 10.42%). The proportion of dialogue reflection (R2, 10.42%) and critical reflection (R3, 7.32%) in the second phase was significantly higher than that in the first phase. For reflection content, the highest proportion of reflection aspect was group method (CC4, 32.88%) (Table 3).

Table 3. Categories and frequency distributions of teachers' refection domains

	Reflection types				Reflection aspects				
	R0	R1	R2	R3	CC1	CC2	CC3	CC4	CC5
Phase one	56.63%	34.35%	5.27%	3.75%	12.24%	5.78%	38.61%	22.45%	20.92%
Phase two	33.03%	48.78%	10.42%	7.32%	8.88%	12.44%	19.96%	32.88%	26.22%
Total	46.63%	40.65%	7.51%	5.30%	10.79%	8.67%	30.34%	20.97%	20.22%

4.2 What Are the Differences Between the Epistemic Network Characteristics of Teachers in Different Activity Phases?

The epistemic network maps of each group in the first and second phase were drawn and the results showed in Fig. 1. It is shown that there are significant differences in the reflection ability of each group in the first and second phases (the first phase M = 2.27, the second phase M = −2.27, t = −7.44, P = 0.00 < 0.05, Cohen's d = 4.29).

In order to further analyze the differences between the two phases of epistemic network structure, the overall average epistemic network of the two phases is shown in Fig. 2. At the same time, the epistemic network connection coefficients of the first and second phases are shown in Table 4. The values in the table indicate the weight of the number of times each connection appears in the reflection journals. As a result, for the first phase of epistemic network, there are more connections between R0-CC3 and R0-CC4. For the second phase of epistemic network, there are more connections between R1-CC4 and R1-CC5, which indicates that they appear more frequently in pairs of pre-service teachers' reflection journals. By subtracting Fig. 2a, b and c can clearly show the difference of average epistemic network between pre-service teachers' reflection journals in the first and second phases. As can be seen from Fig. 2c, teachers' reflection in the second phase focuses more on the connection between elements in the left area, while in the first phase focuses more on the right area, which also confirms the conclusion that there are significant differences between the two phases.

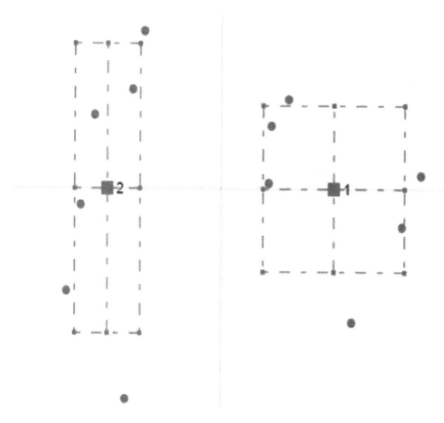

Fig. 1. Epistemic networks of the first phase (red color) and the second phase (blue color). (Color figure online)

Table 4. Connection coefficients of the two phases (Phase I & Phase II)

Connection	Phase I	Phase II	Connection	Phase I	Phase II	Connection	Phase I	Phase II
R0-CC1	1.96	1.34	R1-CC3	2.54	2.25	R2-CC5	0.64	0.45
R0-CC2	0.56	0.81	R1-CC4	1.34	3.31	R3-CC1	0	0
R0-CC3	4.75	1.58	R1-CC5	2.13	3.25	R3-CC2	0	0
R0-CC4	3.01	3.05	R2-CC1	0.03	0.15	R3-CC3	0	0
R0-CC5	0.56	0.66	R2-CC2	0.10	0.36	R3-CC4	0	0.20
R1-CC1	0.34	0.56	R2-CC3	0.16	0.56	R3-CC5	0.76	1.46
R1-CC2	0.49	1.60	R2-CC4	0.10	0.86			

Note. The connection between R0 and CC1 represents the co-occurrence of the R0 and CC1 in a stanza.

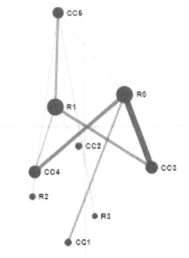

Figure 2a. Phase I

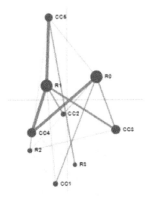

Figure 2b. Phase II

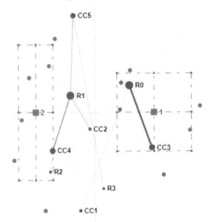

Figure 2c. The difference network between the two phases

Fig. 2. The epistemic networks of the two phases (Phase I & Phase II)

5 Discussion

Based on the online collaborative environment, this study carried out the pre-service teacher script collaborative writing activities. And research results show that the characteristics of reflection type in different phases are different. Teachers can think dialectically in a wider range of fields. Meanwhile, in the first phase, teachers tend to reflect on the content and methods of the group directly, while in the second phase, they tend to reflect on the group methods and personal gains.

In the order of the first phase and the second phase, pre-service teachers carried out collaborative activities on the online platform. In the second phase, the reflection level of teachers is higher than that of the first phase, which indicates to some extent that with the progress of the activity, a series of activity designs can effectively promote the development of teachers' reflection and enhance their reflection consciousness. Through providing specific learning feedback, pre-service teachers can deepen their understanding of the script and promote their in-depth reflection on the content.

During the activity, the reflection of teachers gradually developed. And the teachers' group discussion has strengthened the dialogue between each other to a certain extent. In the process of communicating with others, teachers have accumulated experience, which provides a reference for later individual reflection. At the beginning, the teacher simply elaborated the group content and method, and later began to deeply explore the behavioral reasons behind the method, and on this basis, reflected on its gains and shortcomings. In the process of reviewing and reflecting, the teacher continuously deepens the understanding of the script content, expands and extends the script, and improves the ability of self-reflection. Teachers' reflection is no longer a single content theory, but came to a broader field. By summarizing its experience and methods, it provides reference value for the future study.

Follow-up study will further expand the breadth and explore the epistemic network analysis in teachers' reflection development the possibility of related research. We will use the methods of multi-modal data modeling and weighted network, to construct the relation map between teachers' reflective content and their reflection, which can provide a reference for later pre-service teachers' reflection ability training.

References

Beyer, L.E.: Field experience, ideology, and the development of critical reflectivity. J. Teach. Educ. **35**(3), 36–41 (1984)

Blomberg, G., Sherin, M.G., Alexander, R., et al.: Understanding video as a tool for teacher education: investigating instructional strategies to promote reflection. Instr. Sci. **42**(3), 443–463 (2014)

Colton, A.B., Sparks-Langer, G.M.: A conceptual framework to guide the development of teacher reflection and decision making. J. Teach. Educ. **44**(1), 45–54 (1993)

Csanadi, A., Eagan, B., Kollar, I., Shaffer, D.W., Fischer, F.: When coding-and-counting is not enough: using epistemic network analysis (ENA) to analyze verbal data in CSCL research. Int. J. Comput.-Support. Collaborative Learn. **13**, 419–438 (2018) van Es, E.A., Sherin, M.G.: Mathematics teachers' "learning to notice" in the context of a video club. Teach. Teach. Educ. **24**(2), 244–276 (2008)

Good, J.M., Whang, P.A.: Encouraging reflection in preservice teachers through response journals. Teach. Educ. **37**(4), 254–267 (2002)

Hatton, N., Smith, D.: Reflection in teacher-education - towards definition and implementation. Teach. Teach. Educ. **11**(1), 33–49 (1995)

Krutka, D.G., Bergman, D.J., Flores, R., et al.: Microblogging about teaching: Nurturingpar-ticipatory cultures through collaborative online reflection with pre-service teachers. Teach. Teach. Educ. **40**, 83–93 (2014)

Mckinney, M.: Preservice teachers' electronic portfolios: integrating technology, self-assessment, and reflection. Teach. Educ. Q. **25**(1), 85–103 (1998)

Potter, G.: Facilitating critical reflection on practice through collaborative research. Aust. Educ. Res. **28**(3), 117–139 (2001)

Huang, R., Li, Y.: What matters most: a comparison of expert and novice teachers' noticing of mathematics classroom events. Sch. Sci. Math. **112**(7), 420–432 (2012)

Shaffer, D.W.: Quantitative Ethnography. Cathcart Press, Madison (2017)

Shaffer, D.W., Graesser, A.: Using quantitative model of participation in a community of practice to direct automated mentoring in an ill-formed domain. In: Intelligent Tutoring Systems (ITS), Pittsburgh, PA (2010)

Epistemic Network Analysis for Semi-structured Interviews and Other Continuous Narratives: Challenges and Insights

Szilvia Zörgő[1(✉)] and Gjalt-Jorn Ygram Peters[2]

[1] Institute of Behavioural Sciences, Semmelweis University, Budapest, Hungary
zorgoszilvia@gmail.com
[2] Faculty of Psychology and Education Science, Open University,
Heerlen, The Netherlands
gjalt-jorn@behaviorchange.eu

Abstract. Applying Quantitative Ethnography (QE) techniques to continuous narratives in an inquiry where manual segmentation with a multitude of codes is preferred poses several challenges. In order to address these issues, we developed the Reproducible Open Coding Kit – convention, open source software, and interface – that eases manual coding, enables researchers to reproduce the coding process, compare results, and collaborate. The ROCK can also be employed to prepare data for Epistemic Network Analysis software. Our paper elaborates the challenges we encountered and the insights we gained while conducting a research project on decision-making regarding therapy choice among patients in Budapest, Hungary. Our aim is to broaden the usage of QE, while facilitating Open Science principles and transparency.

Keywords: Epistemic Network Analysis · Semi-structured interviews · Methodology

1 Introduction

Epistemic Network Analysis (ENA) has been applied in modelling a wide range of data [1], but continuous narratives, such as semi-structured interviews, are sparse within these worked examples. Continuous narratives are distinguished from discontinuous narratives by the lack of naturally occurring possibilities for segmentation. Due to the fact that ENA is a nascent application, the lack of clear segmentation possibilities needs to be addressed when dealing with continuous narratives in order for ENA to be more broadly utilized. Gaining insight into questions of segmentation is especially important if the project warrants manual segmentation and coding of data.

This paper provides an overview of our project, some of the challenges we encountered and how we addressed these. As a part of weighing options and making decisions regarding these challenges, we developed the Reproducible Open Coding Kit (ROCK), which is a protocol and open source software to aid the preparation of data for ENA. The ROCK guides the researcher in managing their raw data, performing manual coding and segmentation, as well as aggregating coded data for ENA software.

© Springer Nature Switzerland AG 2019
B. Eagan et al. (Eds.): ICQE 2019, CCIS 1112, pp. 267–277, 2019.
https://doi.org/10.1007/978-3-030-33232-7_23

2 Research Topic and Relevance

The manner in which individuals encounter, filter, and process information is vital in decision-making processes; an exemplary domain of said processes is how patients choose a therapy to treat their illness. Decisions regarding healthcare are multifactorial and occur throughout the patient journey.

In Western pluralistic healthcare systems, a frequent distinction is made between conventional and non-conventional medicine, the latter comprised of various modalities referred to as complementary and alternative medicine (CAM). Complementary signifies treatments used in tandem with, alternative connotes treatments employed instead of biomedicine. Most studies agree that CAM use is increasing throughout the Western world [2–4], and depending on the scrutinized illnesses and modalities, ranges between 40–83% in the United States [5] and reaches 86% in Europe [6].

Characterizing CAM users poses a challenge. Many authors resort to categorization as a descriptive tool, and "the average CAM user" has been described as a middle-aged, wealthy, well-educated, Caucasian female most likely suffering from cancer [7, 8]. In addition to problems stemming from introducing such artificial discontinuities [9], there may be a variety of sampling biases [10] and undervalued aspects involved. Clinical factors, such as the type and prognosis of illness, greatly influence choice of therapy [11, 12]. Furthermore, CAM is often only employed for managing the side-effects of biomedical treatment [13], thus directing attention to the possible differences between motivations in complementary and alternative usage.

The scrutiny of CAM use is clinically significant, as, from a biomedical point of view, it often indicates patient non-compliance and non-adherence. Discontinuing the recommended biomedical treatment may occur in an a priori (refusal to undergo treatment) or a posteriori (discontinuing the treatment) manner [14]. Refusal to undergo biomedical treatment or its discontinuation concerning life-threatening illness is often difficult to understand from a physician's perspective. The above phenomena may pose a threat to patient safety and denote tensions in doctor-patient communication, thus warranting thorough analysis.

CAM users are frequently conceptualized as a static, homogeneous group, yet decisions regarding choice of therapy take place throughout the patient journey and are influenced by e.g. novel symptoms, changes in the patient's overall condition, information from the social network of the patient, and the patient's interpretation of these. Because we wanted to explore factors influencing decision-making in a context where little is known of relevant cognitive processes and behavior, we chose qualitative methods. Yet we wanted an analytical system that enables us to handle large amounts of data and capture the systemic nature of many variables. A method suited for this is ENA, which allows for the visualization of a large number of variables in a single system. In order to apply ENA, we borrowed terms and concepts from Quantitative Ethnography (QE) that provide for segmentation and data preparation. We also created novel additions to this framework to suit the treatment of our data and continuous narratives in general.

3   Data Collection

Our inquiry presupposes that patient cognitive and behavioral patterns can be accessed via semi-structured interviews and their scrutiny. The study takes place in Budapest, Hungary, as a continuation of a previous exploratory research project dealing with the same general topic [15]. Data collection began in February 2019 and is on-going; at the time of submission, 30 interviews have been conducted with patients, at least another 18 are planned.

Subjects were recruited via convenience sampling, but adhering to a predetermined quota (see point 4: "Sampling"). Interviews were conducted by four researchers: the principle researcher and three assistants. Assistants were trained in qualitative methods (especially interview techniques), the employed interview structure and code system. Interviews lasted 60 min on average, ranging from 40 to 120 min; they were sound-recorded and transcribed verbatim. Table 1 shows the main topics of the semi-structured interview, its subtopics, and the number of codes pertaining to these.

Table 1. Areas of the semi-structured interview and related codes.

Realm of inquiry	Topic of inquiry	Subtopics	Question load	Code load
Epistemology	Information	Sources of health-related information, appraisal of information	6 questions + probes	Parent codes (N = 1) Child codes (N = 2) Grandchild codes (N = 16)
Ontology	Explanatory Model	Concepts of illness and health; metaphors of illness and health	4 questions + probes	Parent codes (N = 1) Child codes (N = 2) Grandchild codes (N = 23)
Behavior	Patient journey	Choices of therapy, evaluation of therapeutic efficacy, dispositions	5 questions + probes	Parent codes (N = 1) Child codes (N = 3) Grandchild codes (N = 13)

For each interview we registered the following: interview date, interviewer ID, interviewee ID, interviewee sex, age, and level of education, diagnosis type (D1–4), specific illness, comorbidities, illness onset, time of diagnosis, and therapy choice (treatment type concerning primary diagnosis: biomedicine only, complementary use of non-conventional medicine, alternative use of non-conventional medicine). For CAM users we also registered type of CAM use (product and/or practitioner, for details see "Sampling" below), attendance in CAM-related courses, disclosure of CAM use to conventional physician and the employed CAM modalities (inductively coded). For users of solely biomedicine, reason for rejecting CAM was deductively coded.

In addition, at the end of each interview, a brief survey containing six questions was administered. Four of these consisted of 3-point Likert scales and inquired about how important "natural" and "holistic" therapies were for the patient, how important it was for their chosen therapy to be "evidenced-based", and whether they make a conscious effort to avoid all pharmaceuticals. Two were open-ended questions and asked participants to define what "natural" and "holistic" meant to them, as these expressions may have vastly different meanings for individuals.

4 Sampling

Our sampling was based on the primary mode of investigation found within the relevant literature, that is, distinguishing between groups of biomedicine versus CAM users. We utilized non-proportional quota sampling to maximize sample heterogeneity, stratifying on therapy choice (Biomedical and CAM), primary diagnosis (D1, D2, D3, D4), and sex (males and females). Inclusion criteria were the following: 18 years of age or above; having received a diagnosis of D1–D4; and resident of Budapest, Hungary. Our nosological groups were: D1 – Diabetes (I, II, pre-diabetes), D2 – Musculoskeletal diseases, D3 – Digestive illnesses (excluding "sensitivities" and "intolerance"), D4 – Nervous system diseases. Inclusion based on primary diagnosis was defined by the patient retaining a diagnosis of a specific illness (within the given nosology of D1–4) made by a conventional doctor based on biomedical test results.

4.1   Stratum: Therapy Choice

Challenges
Although CAM research scrutinizes biomedicine use vis-à-vis non-conventional medicine use, this distinction implies a false dichotomy. This is sometimes acknowledged by authors who distinguish between three subgroups (biomedicine, complementary, alternative), arriving at the conclusion that these are based on varying patient motivations [16]. Yet such research projects are scarce and provide little theoretical guidance for selecting stratification criteria. Thus, finding a way to create sampling guidelines and code therapy choice proved to be an important factor in our research design.

Insights
Aside from our sampling strata, we wanted to take advantage of the ability to explore our data by specifying and comparing a variety of group categorizations post-hoc based on participants' characteristics (i.e. conditional exchangeability). In order to achieve this, we coded patients for sampling groups, but also coded them for other, more detailed choice of therapy: complementary use (CM) or alternative use (AM). Both of these had two subcategories: CM/AM through a practitioner or acquisition of products. All of these subgroups may signify distinct cognitive patterns and demarcate different potential subgroups within the sample (Table 2).

Table 2. Examples for registered participant characteristics enabling conditional exchangeability.

Category	Label	Possible values
Sampling group	GroupID	B or CAM
Diagnosis	Diag	D1 or D2 or D3 or D4
Specific illness	Illness	Inductive
Treatment type	TreatType	Biomedicine only (B)/Complementary (CM)/Alternative (AM)
CAM use	CAMuse	Complementary product (CMprod); Complementary practitioner (CMprac); Alternative product (AMprod); Alternative practitioner (AMprac)

4.2   Stratum: Diagnosis

Challenges

Type of illness connotes an important stratum in our sampling because the illness experience, lay theories of illness causation, and the available biomedical cures all interact with it closely and in turn, affect therapy choice. Patients were included based on a primary diagnosis belonging to one of our nosological groups (D1–4). However, patients rarely suffer only from their primary diagnosis, and as such, often discuss or refer to comorbidities as well. Since cognitive and behavioral patterns are likely to vary for different afflictions, it is important to code this information.

Insights

We decided that diagnoses needed to constitute a fundamental part of coding so that in later phases of the project, with enough data, we can create distinctive networks based on illnesses and see what kinds of cognitive and behavioral patterns arise. An exhaustive nosological coding tree is unwieldy. Thus, for each participant, we recorded a primary diagnosis (D1–4) and coded the specific illness and comorbidities inductively, classifying them only afterwards. Therefore, a participant received a primary diagnosis code (e.g. D4) and a specific illness (e.g. epilepsy) that was registered at the time of the interview and also coded in the interview transcript. Comorbidities were coded in the transcript (e.g. myoma, migraine) as well, and additionally, symptoms not referring to a specific illness were also coded inductively, as they may be significant in understanding the securitized cognitive processes.

Albeit illnesses signify the only inductive coding in our study, they follow the general structure of all other codes (discussed below). A separate code was designated for narrative segments referring to illness in general.

5   Operationalization

5.1   Coding

Our deductive coding system is based on a qualitative project conducted between Jan. 2015 and June 2017; participant observation was carried out at four sites of Traditional

Chinese Medicine (TCM) and 105 patients were involved. Furthermore, semi-structured interviews were conducted with patients and practitioners of TCM (N = 20). The collected data was analyzed with Interpretative Phenomenological Analysis with the aid of Atlas.ti 6.0, and a code system was developed inductively. Our present code system is founded on this previous one; it is comprised of three levels of abstraction, containing 52 low-level codes in total.

Challenges

The software used for Epistemic Network Analysis requires data in a spreadsheet-like format. However, manually coding transcripts using spreadsheets is not feasible, let alone with such a large number of codes. Completing the binary coding for each utterance would have meant constant sideways scrolling and an inability to view the entire code set; this endeavor would have led to unreliable coding and results. Furthermore, we wanted to code the narratives manually via hermeneutic analysis, which necessitates a full view of each utterance, their proximal context and all applied and applicable codes.

Insights

We decided to involve four researchers in the coding process. Three raters were checked for inter-rater reliability vis-à-vis the principal investigator. The coding tasks were divided up amongst the four raters based on the parent codes plus the inductive coding for illnesses. The three research assistants were trained in coding and specialized in detecting their designated parent code and subcodes.

5.2 Segmentation

Sentences are the smallest unit of segmentation in our project, they constitute the QE term "utterances". The verbatim transcription of spoken speech comes with inherent subjectivities; as a sentence in speech may persist across vast reaches, the transcriber makes many judgement calls in punctuation. Albeit coding occurs on the level of utterances, co-occurrences are computed based on a higher level of segmentation, the "stanza". Continuous narratives introduce the challenge of defining stanzas as an analytical unit. Stanza size crucially determines analysis results, thus addressing this issue is vital.

Challenges

Segmentation based on question-response was not an option, as related utterances may persist over several "turns of talk" between interviewer and interviewee. Furthermore, narratives may abruptly digress or terminate. We briefly considered numbering stanzas so that they are linked to each other topic-wise, i.e. interviewees frequently begin a topic, digress, and then return to it again (uptake on a large scale). If so, would stanzas need to receive the same identifier then? We came to the conclusion that determining this would lead to very subjective choices and that we should allow the coding to make this judgement; if stanzas should be linked, they will be linked through code co-occurrence.

Insights

ENA networks are based on adjacency matrices, where co-occurrence is literal co-occurrence of codes within the same analytical unit (i.e. stanza). Stanza size reflects how much content the researchers consider indicative of psychological proximity. Researchers who are only interested in tightly connected concepts may prefer shorter stanzas, e.g. stanzas may overlap completely with utterances. However, if the research topic concerns broader, more complex issues, researchers may want to define larger stanza sizes to study constructs with higher cognitive distance. Our working definition of a stanza is: a set of one or more utterances that occur in close proximity and discuss the same topic (i.e. recent temporal context).

Due to the fact that stanzas are so pivotal to the overall network, but our working definition of them is subjective, we decided that discourse segmentation should be performed by three autonomous raters: the principal investigator, a research assistant, and someone not involved in the research project. The latter's unfamiliarity with the questions under scrutiny and employed codes lends a different treatment of the narratives themselves. Albeit coding necessitates specialized knowledge on the part of the rater, the key to optimal discourse segmentation in continuous narratives may be a "naïveté" regarding the subject. Because we want to test this hypothesis, we will retain segmentation from the three raters and run analyses for all to see which set provides the more accurate model; in later phases we might only use one of them or collapse them.

6 The Reproducible Open Coding Kit (ROCK)

To summarize our circumstances: we wanted to be able to code and segment our data manually with a multitude of codes and three versions of segmentation. We wanted to employ both deductive and inductive coding in the hermeneutic analysis of continuous narratives. We did not want to work with spreadsheets because accomplishing this would have been problematic. Finally, the large number of participant characteristics (to be used for subgroup exchangeability) required a simpler way of adding this information, side-stepping binary coding. For these reasons, we developed the Reproducible Open Coding Kit (ROCK).

The ROCK is simultaneously a convention for specifying sources of qualitative data and the coding of those sources; it is a Free/Libre Open Source Software R package [17]. As such, the ROCK facilitates sharing qualitative coded data, enabling other researchers to reproduce the coding process, compare results, and collaborate by expanding the coding system. The ROCK facilitates adhering to Open Science principles in research, as well as enables transparency and minimizes research waste.

In ROCK terminology, a source is a document that is coded. All sources are stored as plain-text files and consist of one or more utterances, which receive unique utterance identifiers (UIDs) that enable merging or comparing coding by different raters. Utterances are separated by line breaks, and to code an utterance, the relevant code can be appended to the utterance's line in the source file, where each code is delimited by two square opening brackets and two square closing brackets (Table 3).

Table 3. Basic ROCK terminology and related functions

Term	Examples and specifics	Function
Source	document that is coded (plain-text file)	delimiting sources of information (e.g. interviews) and labeling them
CID	`[[cid=alice]]` case identifier used to specify the data provider (e.g. interviewee)	identifying each participant separately
UID	`[[uid=73ntnx8n]]` unique identifier for each utterance	Identifying each utterance separately (which allows, for example, aggregating information from various raters for each utterance)
Deductive code format	`[[example_code1]]`	providing a universal way of coding utterances deductively
Inductive code format	`[[examples>child1>code1]]` hierarchical marker included in code label	providing a universal way of specifying hierarchical relationships between codes while coding utterances inductively
Segmentation (section breaks)	`<<stanza-delimiter>>` several types of section breaks can be employed of which "stanza" is one	Parsing the narrative into groups of utterances

Attributes	`---`
	`rock_attributes:`
	`  -`
	`    caseId: 1`
	`    gender: female`
	`    age: 50s`
	`---`
		Providing additional information about cases, such as metadata about data collection or characteristics of participants.

The ROCK supports both deductive and inductive coding. Inductive codes use the "greater-than" sign (>) as hierarchical markers to signify parent-child relationships, e.g.: `[[examples>code1]]`. The rock R package can read all codes and reconstruct the inductive code tree.

In addition to coding functionality, the ROCK allows specification of dynamic sets of identifiers, e.g.: a case identifier (CID). CIDs can be used to specify the data provider of sets of utterances, for example by specifying which individual said what. The utterances identified by a given case identifier can then easily be supplemented with the attributes specified for that case (e.g. metadata or demographic variables). Finally, the ROCK allows specifying sections in each source. Sections can be automatically counted and numbered, providing easy means to segment sources (e.g.: into stanzas) using a variety of segmentations in tandem.

To perform coding and segmentation, we developed iROCK, a simple online interface consisting of a file that combines HTML, CSS, and javascript to provide a rudimentary graphical user interface. Because iROCK is a standalone file, it does not need to be hosted on a server, which means that no data processing agreements are required (as per the GDPR). The iROCK interface allows raters to upload a source, a list of codes, and segmentation identifiers; coding consists of dragging and dropping codes upon utterances at the end of their line (see Fig. 1 for illustration). Once coding is finished, the coded sources can be saved. The rock R package then reads these and constructs a dataframe in R, which is then ready for further processing by the rENA R package or for export to a comma separated values file that the ENA web application can use.

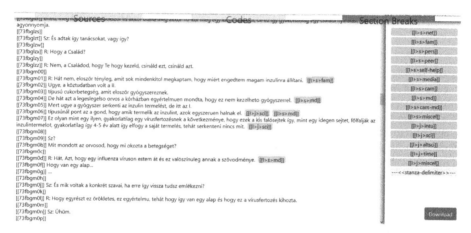

Fig. 1. The iROCK interface. The top ribbon displays controls for uploading sources, codes, and section breaks. The panel on the right contains the uploaded codes that are appended to the chosen utterance on the left. The narrative is parsed according to utterances, each receives a unique identifier. (The coded fragment is merely illustration.)

Our Application of ROCK

First, all interview transcripts were stored as plain-text files with the .rock extension. Because all interviews were individual interviews, the first line of every source was labelled with a CID that linked that source's utterances to that case. Second, the sources were "cleaned" by the rock R package (e.g. indicating utterances by inserting newline characters at appropriate places), and the cleaned versions were stored in a different directory. Third, every utterance received a unique UID, and the resulting versions were stored in yet another directory. This process documents each step, enabling inspection of intermediate steps if required at any point. Codes (3 lists of subcodes based on their parent) and 3 ways of denoting segmentation (professional, middle, and "naïve") were created and placed into separate plain-text files, with the attributes for each source listed in a separate plain-text file. Raters import sources, codes, segmentation identifiers into iROCK, perform coding, then download their document into a designated folder. The `rock` R package collapses all information within the folders based on UID and can then construct R dataframes (Fig. 2).

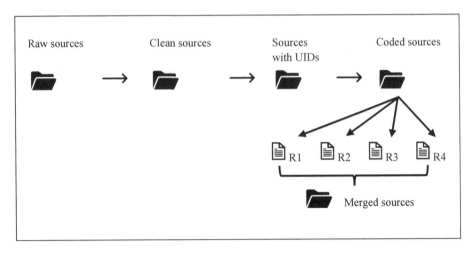

Fig. 2. This flowchart shows the process of cleaning and merging sources. The first folder (Raw sources) contains the transcripts in plain text documents. The second folder (Clean sources) contains the narratives parsed according to one utterance per line. The third folder (Sources with UIDs) contains the parsed narratives where each line receives a unique identifier. These sources are uploaded to iROCK where coding and segmentation are performed. The last folder (Coded sources) contains the parsed, coded narratives downloaded from iROCK. The coded sources from our four researchers (R1–4) are then merged based on UIDs, thus creating master documents with sources containing all codes for each utterance. Merged sources can then be converted by the ROCK to qualitative data tables and uploaded to the ENA interface.

7   Closing Remarks

We hope the disclosure of our challenges and means of addressing them offers insight into QE techniques applied to continuous narratives. Furthermore, the aim of this pilot study is to operationalize the ROCK and make it available for use by other researchers as well.

Acknowledgements. The authors would like to acknowledge the support of ÚNKP-18-3-III New National Excellence Program of the Ministry of Human Capacities, Hungary. We would also like to thank Brendan Eagan for his valuable insights throughout the planning and implementation phases of this research. Lastly, we are grateful to research assistants Anna Geröly, Anna Jeney, and Krisztina Veres for their rigorous work.

References

1. Shaffer, D.: Epistemic network analysis: understanding learning by using big data for thick description. In: Fischer, F., Hmelo-Silver, C.E., Goldman, S.R., Reimann, P. (eds.) International Handbook of the Learning Sciences, pp. 520–531. Routledge, New York (2018)

2. Frass, M., Strassl, R., Friehs, H., Müllner, M., Kundi, M., Kaye, A.: Use and acceptance of complementary and alternative medicine among the general population and medical personnel: a systematic review. Ochsner J **12**, 45–56 (2012)
3. Thomas, K., Nicholl, J., Coleman, P.: Use and expenditure on complementary medicine in England: a population based survey. Complement. Ther. Med. **9**, 2–11 (2001)
4. Tindle, H., Davis, R., Phillips, R., Eisenberg, D.: Trends in use of complementary and alternative medicine by US adults: 1997-2002. Altern. Ther. Health Med. **11**, 42–49 (2005)
5. Arthur, K., Belliard, J., Hardin, S., Knecht, K., Chen, C., Montgomery, S.: Practices, attitudes, and beliefs associated with complementary and alternative medicine (CAM) use among cancer patients. Integr. Cancer Ther. **11**, 232–242 (2012)
6. Eardley, S., et al.: A systematic literature review of complementary and alternative medicine prevalence in EU. Forsch Komplementmed. **19**(Suppl 2), 18–28 (2012)
7. Harris, P.: Prevalence of complementary and alternative medicine (CAM) use by the general population: a systematic review and update. Int. J. Clin. Pract. **66**, 924–939 (2012)
8. Ernst, E.: Prevalence of use of complementary/alternative medicine: a systematic review. Bull. World Health Organ. **78**, 258–266 (2000)
9. Altman, D.: Problems in dichotomizing continuous variables. Am. J. Epidemiol. **15**, 442–445 (1994)
10. Henrich, J., Heine, S., Norenzayan, A.: The weirdest people in the world? Behav. Brain Sci. **33**, 61–83 (2010)
11. Faith, J., Thorburn, S., Tippens, K.: Examining CAM use disclosure using the behavioral model of health services use. Complement. Ther. Med. **21**, 501–508 (2013)
12. Thomson, P., Jones, J., Browne, M., Leslie, S.: Psychosocial factors that predict why people use complementary and alternative medicine and continue with its use: a population based study. Complement. Ther. Clin. Pract. **20**, 302–310 (2014)
13. Stratton, T., McGivern-Snofsky, J.: Toward a sociological understanding of complementary and alternative medicine use. J. Altern. Complement. Med. **14**, 777–783 (2008)
14. Zörgő, S., Olivas Hernández, O.: Patient journeys of nonintegration in hungary: a qualitative study of possible reasons for considering medical modalities as mutually exclusive. Integr. Cancer Ther. **17**, 1270–1284 (2018)
15. Zörgő, S., Purebl, G., Zana, Á.: A qualitative study of culturally embedded factors in complementary and alternative medicine use. BMC Complement. Altern. Med. **18**, 25 (2018)
16. Astin, J.: Why patients use alternative medicine: results of a national study. JAMA **279**, 1548–1553 (1998)
17. Peters, G., Zörgő, S.: Introduction to the Reproducible Open Coding Kit (ROCK) (2019). Psyarxiv. doi:https://doi.org/10.31234/osf.io/stcx9

Short Papers

Quantitative Multimodal Interaction Analysis for the Assessment of Problem-Solving Skills in a Collaborative Online Game

Alejandro Andrade[1]([✉]), Bryan Maddox[2], David Edwards[1],
Pravin Chopade[1], and Saad Khan[1]

[1] ACTNext by ACT, Inc., 500 ACT Drive, Iowa City, IA, USA
alejandro.andrade@act.org
[2] Norwich Research Park, University of East Anglia, Norwich, Norfolk NR4
7TJ, UK
B.Maddox@uea.ac.uk

Abstract. We propose a novel method called Quantitative Multimodal Interaction Analysis to understand the meaning of interactions from a set of multimodal observable behavior. We apply this method for the measurement of collaborative problem-solving skills in a dyadic online game specially designed for this purpose. We outline our assumptions and describe the machine learning approach that help us tag multimodal behaviors connecting the theoretical construct with the empirical evidence.

Keywords: Interaction Analysis · Multimodal learning analytics · Collaborative Problem-Solving skills

1   Introduction

We propose a quantitative method for the analysis of multimodal data for the measurement of collaborative problem-solving (CPS) skills that builds upon the fields of Micro-Ethnography, Conversation Analysis, Interaction Analysis and Multimodal Discourse Analysis. For the purposes of this paper we call this approach Quantitative Multimodal Interaction Analysis. The purpose of our overarching study is summarized in the following statement: Given a set of completed tasks undertaken by a team, we need to estimate the CPS skills of each participating individual. To this end, we developed a novel methodology to help us understand patterns of action-oriented sequences of communication that took place while a group of participants played a collaborative online game. Put another way, our goal was to extract the meaning of these interactions from a set of multimodal observable behaviors in the context of a problem that requires team collaboration for its solution.

Our arriving at this methodological approach, however, did not go smoothly. Among all the various challenges we faced, three major issues seem worth mentioning. First, our methodological approach has been largely pragmatic in that it has been an abductive, iterative process. Rather than an attempt to fit a theoretical construct to empirical observations, or an attempt to extract grounded categories from observed

B. Eagan et al. (Eds.): ICQE 2019, CCIS 1112, pp. 281–290, 2019.
https://doi.org/10.1007/978-3-030-33232-7_24

phenomena, we have attempted to make best guesses from imperfect observations. A good example of our abductive reasoning took place in our discussions about units of analysis and the necessary conclusion to which we arrived—to observe multiple levels of units of analysis organizing the sequences of interaction. Second, the development of our methodological approach was challenging due to the interdisciplinary nature of our research team. Importantly, the center of tensions took place in the development of a coding scheme that was necessarily generalizable across participants but also sufficiently interpretative. This coding scheme with which we could annotate multimodal data for CPS skills was at the center of many exchanges between the quantitative-qualitative camps within our team. Third, a big hurdle for us has been the interdependence of multiple modes of communication. Multimodal data, by nature, cannot be characterized as a collection of parallel streams but rather as a synchronous collection of co-expressive modes of communication [1]. For instance, as we observed our participants play the collaborative game and their attempts at developing rapport among themselves, they used gestures, facial expressions, and verbal utterances. It quickly became apparent that evidence for collaborative skills comes as the coevolution of various communicative tiers. In what follows, we introduce our theoretical and methodological assumptions, the empirical context of measurement (i.e., the online collaborative game), an overview of the steps we have taken for using supervised machine learning methods, and then move on to illustrate our methodology with a case study.

2 Theoretical and Methodological Assumptions

Interaction Analysis was developed to understand empirical evidence from video recordings of the interaction of human beings with each other and with objects in their environment [2]. Interaction Analysis is related to ethnomethodology and it is interested in examining several themes related to human activity, such as the structure of events (e.g., beginnings, endings, segmentation), the temporal and spatial organization of activities (e.g., turn-taking, trouble and repair, use of tools and representations), and participation structures (e.g., roles and status). Multimodal Discourse Analysis was developed to understand human action as a function of language and gesture, and the organization of the social, cultural, material and sequential structures in which it takes place [3, 4]. Multimodal Discourse Analysis builds upon sociolinguistics and is interested in studying how participants interact with each other through talk while including larger semiotic frameworks created by nonverbal communication as well as relevant surrounding elements in the environment. Both Interaction Analysis and Multimodal Discourse Analysis address the issue of *how* students learn in an authentic context, and not just *what* is it that they have learned.

As we dive deep into the measurement of CPS, the overarching questions of IA and MDA become increasingly important. For instance, when we observe a dyad trying to succeed in a series of tasks, we should consider a series of assumptions for the analysis of multimodal data. In particular, these assumptions refer to the unit of analysis, the spatial and temporal structure of the activity, and the use of verbal and nonverbal elements to create a shared understanding of the problem and the required steps for its solution. First, we define our units of analysis as a hierarchical aggregate of low-level moment-by-moment turns, medium-level sequences of actions, and high-level skills. Second, it is assumed that a collaborative activity is constituted by a social and material framework that constraints the possible meanings that participants' verbal and non-verbal actions can take. Thus, a participant's gaze, prosody, or gesture are always interpreted within a social and material context. Specifically, each task a group attempts to solve is constrained by the objects it is constituted by (e.g., buttons, labels) and, in turn, the task affords participants with certain actions (e.g., press a button, read off a label). In what follows, we flesh out these assumptions before introducing our machine learning approach.

2.1   Units of Analysis

Conversational Turns. We conceived of as the most meaningful unit of analysis the attempt by a participant to create a meaningful exchange of information. For instance, a participant communicates a message that has a text and a subtext. That is, a participant utters some words that contain one or more ideas, and the meaning of these ideas is also framed by the underlying nonverbal and paralinguistic elements of the communicative act. A person might say "Move it to the right" where the "it" of the utterance is bounded by the materials within the environment and the "right" is bounded by the spatial organization of the objects and people in the room. Furthermore, the idea can be communicated with a tone of urgency—e.g., the satellite is going to crash with an asteroid unless an action is taken-, or it can be said with a hesitant tone. In this sense, a conversational turn is similar to a speech act, where language is understood as action-oriented and not just as a passive means for the representation of abstract concepts.

Action-Oriented Sequences. A task is constituted temporally by a *patterned sequence* of turns and is bounded by a beginning and an end of an activity. It is patterned because a task shows regularities that can be more or less expected in similar social situations. For instance, to solve a problem, participants first gather and share information, then analyze that information around some plausible conjectures, then execute some actions, and then make a final decision or conclusion. In addition, an activity can have clear limits (e.g., it has five minutes to be completed from start to end) or it can be bounded by a transition to a different activity, such as when participants reorient their attention to a different goal. Of course, these limits are arbitrary, and, in practice, people can move seamlessly in and out of an activity.

High-Level Skills. As participants move from task to task, they complete events that collectively reveal the overall object of the activity. The major difference between tasks and high-level skills is that tasks might have goals that do not directly reveal the motive of the group's performance. In a classical example from Activity Theory [5], Leontiev describes a primeval collective hunt, where an individual task is to beat a drum. However, an individual's goal can only be understood when examining the motive of the whole activity (i.e., a hunt), where the drum's beat would scare animals that would run into an ambush where the hunters are hiding. Although the drummer's object is to beat a drum, the group's motive is to feed themselves.

2.2   Context of the Activity

Social Organization. When humans communicate with each other, they are attuned to the particular social circumstances wherein that communication takes place. We are sensitive to social hierarchies and differences among speakers, as well as the timing and tone of the communicative efforts. Also, the number of participants in a conversation can affect the volume and level of communication, such as in large events (e.g., addressing an audience) versus two-people encounters (e.g., small talk or chit chat).

Material and Spatial Organization. Spatial organization of an action fundamentally frames its meaning. Proximity to participants or objects can convey important pieces of information, and the relative importance of different objects different people have access to can alter the communicative efforts of some speakers to address different goals. For instance, some form of division of labor takes place when participants have access to different pieces of information, materials, or other kinds of resources.

Observing Mutual Understanding. It is possible to determine whether participants achieve shared understanding of a task as they display observable cues during their communicative endeavors. For instance, Bavelas, Gerwing and Healing [6] describe an observable three-step micro-process: first, a speaker introduces new information; second, the receiver responds accordingly to this new information; third, the speaker acknowledges the response.

3   Empirical Context: Measuring CPS Skills in an Online Game

3.1   Collaborative Jigsaw Game: Crisis in Space

Crisis in Space is an online collaborative game where two participants take turns as the Operator and Engineer roles to solve missions in order to save the International Space Station (ISS) from a myriad of challenges [7]. Video, audio, eye gaze, and log data are captured for each participant. We use as a running example the task called Keypad. To solve the Keypad task the engineer needs to provide the correct order in which the operator has to press a series of buttons (see Fig. 1).

3.2   Collaborative Problem-Solving Skills

Our first step was to situate the CPS construct into the affordances of the CIS game. We focus our analysis on seven CPS skills (see Table 1). According to Fiore, Smith-Jentsch, Salas, Warner and Letsky [8], macro-cognition in collaborative teams consists of four cognitive processes of (1) individual information gathering, (2) team information exchange, (3) individual information synthesis, and (4) team knowledge sharing. In this account, information becomes knowledge when decision to take action (or actionable information) is developed. Additionally, we include three socio-emotional skills (leadership, rapport, and resilience).

Engineer's Symbol Sets

Operator's Symbol Buttons

Fig. 1. Crisis in Space (CIS) online game.

Table 1. Collaborative Problem-Solving Skills

Cognitive Team Task Skills

1. Individual information gathering: a player is tasked with a visual search for the relevant elements in the problem (number of wires, their color, their position on a circuit, etc.)
2. Team information exchange: communication about task parameters with the other team member. This communication includes reading off values, properties of game elements, suggesting, soliciting, and issuing of statements
3. Individual information synthesis: task parameters are to be organized and made sense of. This operation usually includes the use of the CIS manual and rule look up (e.g., find the correct wire to cut or the right sequence of symbols)
4. Knowledge building and decision to take action: when parameter and rule information have been located, steps need to be taken to complete the task. This move from information to action is conceived of as a collaborative endeavor–rule extraction and game actions are distributed/accessible across different team members

Socio-Emotional Team Skills

5. Leadership: takes risks in developing and maintaining relationships
6. Rapport: communicates with empathy, provides emotional support, and displays disposition to work collaboratively
7. Resilience: confidence and self-regulation in the face of obstacles

Game Tasks as CPS Evidence Eliciting Items. In the CIS game, a task is to our assessment of CPS skills like an item is to a multiple-choice test. A task (e.g., circuit panel, keypad, repair, asteroids) is an opportunity for the participants to demonstrate their CPS skills, and for us to evaluate the quality and attributes of their collaboration. A team task analysis of the Keypad task reveals the presence of the steps outlined in Table 2.

Table 2. Keypad Sequential Task Analysis

Cognitive skill	Sequence of elicited behaviors
	1. Mission starts

Info gathering and exchange	2. Location of Keypad task
	  a. Operator states verbal description of symbols or uses the word "symbols" or "shapes"
	    i. Click event on Keypad window or gaze on Keypad window
	  b. Engineer navigates to Keypad tab
	    i. Gaze over and Click event on Keypad tab. Might include verbal acknowledgment

	3. Location and naming of symbols
	  a. Operator describes four symbols with talk and gesture
	    i. Gaze on a particular symbol
	  b. Engineer acknowledges or asks for repeat/clarify

Info synthesis and take action	4. Map symbols to Symbol Set
	  a. Engineer visually identifies Symbol Set
	  b. Engineer communicates order of to-be-pressed buttons verbally and with gestures
	  c. Operator might ask for repeat/clarify
	  d. Operator presses symbol buttons on symbols window
	    i. Operator might verbally acknowledge clicks or Engineer might solicit acknowledgement

	5. Evaluation box:
	  a. If mission success,
	    i. Operator acknowledges task is completed successfully and Operator clicks on Back button and moves onto next task
	    ii. Engineer and Operator increase their positive valence
	  b. If not mission success,
	    i. Operator acknowledges task is not completed successfully
	    ii. Engineer and Operator increase their negative valence
	    iii. Repeat from Step 3

Socio-emotional skill	Observable behaviors
Leadership	I. Proportion of speech turns and words, new information in utterances, prosodic elements in talk indicating confidence, open body posture
Rapport	II. Number of smiles (positive valence), laughs, gaze on partner, positive and empathy words
Resilience	III. Positive words and valence given unsuccessful task or clarification statements

4   Supervised Machine Learning Approach

To analyze multimodal data such as video, audio, eye gaze, and log data, we train machine learning algorithms by manually annotating salient parts of the interaction, which in turn guide our measurement of CPS skills. After synchronizing the multiple data streams and stitching the individual videos together into a "dyadic" video, we use the annotation software ELAN [9] to create multimodal tagging of the content (see Fig. 2). As there are multiple facets to annotate, ELAN contains several tiers with specific "controlled vocabularies"—a.k.a. coding schemes. In particular, speech, gaze data, and other prosodic elements that reveal emotions are tagged for its cognitive (e.g., describing the task, describing the elements within the task) and socio-emotional content (e.g., acknowledgments, confusion, apologies). Then, we use deep learning algorithms to detect facial expressions (valence and arousal), gestures, laughter, semantic similarity, attention to areas of interest, and click-stream patterns, to automatically predict these behavioral tags. For instance, detecting gestures uses a deep learning algorithm to first predict the likelihood a hand is present in a video frame and then the likelihood of whether that hand is producing a communicative gesture [10]. To detect laughter, we use a pretrained laughter recognition algorithm and then use our manual annotations to improve the precision of the algorithm. In the same vein, a pretrained network detects and predicts the likelihood of facial expressions and affect states for a participant, and the annotated data is used as context to refine the predictions. Furthermore, word and sentence embeddings are used to represent verbal data and a distance metric is used to detect certain key words and phrases that represent tags such as "asking questions" and "describing symbols".

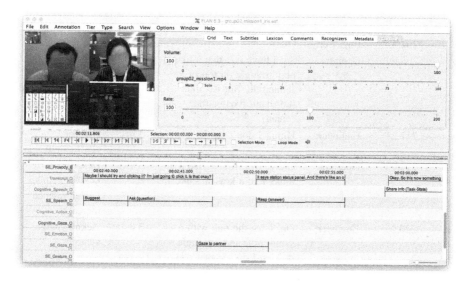

Fig. 2. Multimodal Annotation of Cognitive and Socio Emotional Behaviors using ELAN.

5   Case Study: Automated Assessment of Keypad Task

A combination of manual annotations and supervised learning has allowed us to develop an assessment of the specific steps present in a dyad's interaction in the Keypad task. An example is presented in Table 3. A high-level interpretation of this case study is as follows. Click-stream data help us segment tasks and particular task

Table 3. Excerpt 1. A dyad solving the keypad task.

Turn	Time	Speaker	Utterance	Task	Evidence
1	177.32	Operator:	So, now, there should be a bunch of shapes somewhere.	1	Click on keypad window
				2.a	State "shapes"
2	174.771	Engineer:	Okay. So, should I just press quit?	II	Facial valence increase
3	183.88	Operator:	Uh... No.	II	Facial valence increase
4	185.48		So... There should be a section with a bunch of shapes.	2.a	State "shapes"
5	184.49	Engineer:	Okay. Yes. I do see it. Okay.	2.b	Gaze on Keypad tab and verbal acknowledgement
6	191.8	Operator:	So, I have four shapes. I'm not entirely sure what I need to do with it but I should be able to tell you them, and you have to tell me which one to click on. Does that make sense?		
7	204.57	Engineer:	Okay.		
8	206.64	Operator:	So, I have one that's like a play triangle on top of a semicircle. But it's like... [left hand crosses the screen]	3.a	Verbal description Gesture a keypad symbol
9	217.21	Engineer:	Hmm. I don't see it.	3.b	Ask for clarification
10	219.96	Operator:	Okay. Um, I have one that looks like a key. It's like...	3.a	Describe a keypad symbol
				III	Facial valence increase
11	220.25	Engineer:	Oh, I see the one with a semi-circle and a triangle on it. Yes, I do see that one.	3.b	Acknowledge symbol
12	229.84	Operator:	Okay. I have one that's a key. So... It's a circle and then there's an F coming off of it. Looks like a key. [gesture]	3.a	Describe a keypad symbol
13	238.57	Engineer:	Okay.	3.b	Acknowledge symbol
14	237.96	Operator:	I have one that looks like a lightning bolt. It has a circle and then like the Harry Potter squiggle. [gesture]	3.a	Describe a keypad symbol

(continued)

Table 3. (*continued*)

Turn	Time	Speaker	Utterance	Task	Evidence
15	245.85	Engineer:	Ah? You said, Bolt?	3.b	Ask for clarification
16	249.4	Operator:	Bolt, lightning bolt.		
17	253.41	Engineer:	Okay. I do see it. Okay.	3.b	Acknowledge symbol
18	253.6	Operator:	I think this is the android symbol. It's like a weird, uh, like two donuts combined. [two hand gesture]	3.a	Describe a keypad
19	265.93	Engineer:	Okay.	3.b	Acknowledge symbol
20	262.92	Operator:	So... those are my four and I think you should somehow be able to tell me which one to click on.		
21	272.73	Engineer:	Ah... I have no clue. But just click on the one that you watched.	4.b	Communicate order of buttons
				II	Facial valence increase
22	285.84	Operator:	Oh, okay. Okay. Let's go with this one.	4.d	Click on symbol buttons
23	289.4		Oh, no, that was wrong!	5.b.i	Acknowledge task not successful
				5.b.ii	Facial valence increase

steps such as when a mission has begun (see Line 1). Although click data also determines which window the operator has opened, the operator needs to verbally communicate this information to the engineer. To detect this, we use natural language processing (NLP) techniques such as sentence embeddings and cosine distances to compare the utterances to a set of key phrases. For instance, we detect the description of a task (see Lines 1 and 4) and keypad symbols (e.g., see Lines 8, 10, and 12). Also, we detect "acknowledge of the message" tag by examining both gaze patterns over certain areas of interest and the associated verbal reply and clarification statements (e.g., see Lines 5, 9, and 11). In addition, we are able to locate "task completion" (i.e., successful or unsuccessful) using facial recognition techniques by processing the valence of facial expressions (see Line 23).

6   Conclusion and Discussion

Our goal has been to examine the overlap between qualitative (IA and MDA) and quantitative methodologies (machine learning) to assess CPS skills in an authentic setting. Quantitative Multimodal Interaction Analysis has been valuable in collecting evidence connecting the theoretical construct of CPS with empirical evidence stemming from audiovisual recordings and other multimodal data. As this approach is still in its initial stages, more evidence is required to understand its affordances and limitations. In particular, further work will add information from other CIS tasks and data from a different population.

References

1. McNeill, D.: Gesture and Thought. University of Chicago press, Chicago (2008)
2. Jordan, B., Henderson, A.: Interaction analysis: Foundations and practice. J. Learn. Sci. **4**, 39–103 (1995)
3. Goffman, E.: Forms of Talk. University of Pennsylvania Press, Philadelphia (1981)
4. Goodwin, C.: Action and embodiment within situated human interaction. J. Pragmat. **32**, 1489–1522 (2000)
5. Kaptelinin, V., Nardi, B.A.: Acting with Technology: Activity Theory and Interaction Design. MIT press, Cambridge (2006)
6. Bavelas, J., Gerwing, J., Healing, S.: Doing mutual understanding. Calibrating with microsequences in face-to-face dialogue. J. Pragmat. **121**, 91–112 (2017)
7. Chopade, P., Edwards, D., Khan, S.: Designing a digital jigsaw game based measurement of collaborative problem-solving skills. In: Cunningham, J., et al. (eds.) Companion Proceedings of the 9th International Learning Analytics and Knowledge Conference (LAK 2019), Tempe, Arizona, pp. 26–31. Society for Learning Analytics Research (SoLAR) (2019)
8. Fiore, S.M., Smith-Jentsch, K.A., Salas, E., Warner, N., Letsky, M.: Towards an understanding of macrocognition in teams: developing and defining complex collaborative processes and products. Theor. Issues Ergon. Sci. **11**, 250–271 (2010)
9. Wittenburg, P., Brugman, H., Russel, A., Klassmann, A., Sloetjes, H.: ELAN: a professional framework for multimodality research. In: 5th International Conference on Language Resources and Evaluation (LREC 2006), pp. 1556–1559. (2006)
10. Chopade, P., Khan, S., Edwards, D., Davier, A.A.V.: Machine learning for efficient assessment and prediction of human performance in collaborative learning environments. In: 2018 IEEE International Symposium on Technologies for Homeland Security (HST), pp. 1–6 (2018)

On the Equivalence of Inductive Content Analysis and Topic Modeling

Aneesha Bakharia[(✉)]

The University of Queensland, Brisbane, Australia
aneesha.bakharia@gmail.com

Abstract. Inductive content analysis is a research task in which a researcher manually reads text and identifies categories or themes that emerge from a document corpus. Inductive content analysis is usually performed as part of a formal qualitative research methodology such as Grounded Theory. Topic modeling algorithms discover the latent topics in a document corpus. There has been a general assumption, that topic modeling is a suitable algorithmic aid for inductive content analysis. In this short paper, the findings from a between-subjects experiment to evaluate the differences between topics identified by manual coders and topic modeling algorithms is discussed. The findings show that the topic modeling algorithm was only comparable to the human coders for broad topics and that topic modeling algorithms would require additional domain knowledge in order to identify more fine-grained topics. The paper also reports issues that impede the use of topic modeling within the quantitative ethnography process such as topic interpretation and topic size quantification.

Keywords: Topic modeling · Inductive content analysis

1 Introduction

The field of topic modeling has focused on developing algorithms that are able to discover themes or topics within a textual corpus. Topic modeling is an unsupervised machine learning technique and is often assumed to be equivalent to inductive content analysis techniques; as opposed to supervised text classification models used for directed content analysis that can be trained to identify predefined codes or categories (see Table 1). Topic modeling has been applied to datasets of varied size and origin, being particularly suited to large volumes of textual data where it is impractical for teams of researchers to perform inductive content analysis manually. The mathematical motivation behind topic modeling algorithms also provides a basis for quantifying and interpreting the derived topics, potentially making topic modeling a useful tool to aid quantitative ethnography.

There are two main types of topic modeling algorithms namely Non Negative Matrix Factorization (NMF) [5], and Latent Dirichlet Allocation (LDA) [2]. Both algorithms stem from a different mathematical basis but discover the latent topics within a corpus in an unsupervised manner. NMF uses principles from linear algebra and optimization while LDA is a probabilistic graphical model. NMF and LDA are not hard clustering algorithms and naturally allow for topic overlap (i.e., documents are

© Springer Nature Switzerland AG 2019
B. Eagan et al. (Eds.): ICQE 2019, CCIS 1112, pp. 291–298, 2019.
https://doi.org/10.1007/978-3-030-33232-7_25

able to belong to multiple topics). Topic modeling algorithms require a term-document matrix (i.e., bag of words representation) and produce matrices mapping latent topics to words and latent topics to documents. The weights produced that map words to latent topics are rarely used to aid interpretation or to quantify the magnitude of discovered topics. Many visualizations and topic modeling browsers also only show the top words in a topic without including functionality that allows the user to view the top documents that belongs to a topic [7].

Table 1. Coding differences between the three approaches to content analysis, adapted from Hsieh and Shannon [4]

Coding approach	Study	Code derivation
Summative	Keywords	Keywords identified before and during analysis
Inductive	Observation	Categories developed during analysis. Unsupervised algorithms: Topic Modeling (i.e., NMF, LDA) and clustering algorithms such as k-means
Directed	Theory	Categories derived from pre-existing theory prior to analysis. & Supervised classification algorithms: Support Vector Machines, Decision Trees and Deep Neural Networks

The research presented in this paper addresses two common issues. Firstly, topic modeling has been assumed to be an appropriate aid for inductive content analysis even though very few studies have compared the output of manual human coding with the derived topics from a topic modeling algorithm. Secondly the mathematical output of topic modeling algorithms, mapping latent topics to words and documents is rarely used as an aid to help researchers interpret and quantify topics. This paper details a between subjects study that compares the topics derived by manual coders with the topics derived with the aid of a topic modeling algorithm namely Non Negative Matrix Factorization (NMF). The paper seeks to address the following research questions:

1. Are topics derived via the NMF topic modeling algorithm comparable to the topics identified manually by qualitative researchers?
2. Are qualitative researchers able to interpret and quantify the output of derived topics (i.e., the mapping between topics and top terms; and the mapping between topics and corpus documents) from NMF?

2   Methodology

This section details the between-subjects experiment conducted to compare the topics that were manually derived by qualitative content analysts with the topics derived using the NMF algorithm. Twenty (20) participants were recruited and split into two (2) groups. Participants in Group A were required to read the provided corpus and manually find topics. No participant consensus process was included because Group A participants were required to perform inductive content analysis. In directed content analysis studies, analyst consensus in terms of the codes/categories and rationale for code assignment is usually decided upon before the coding process begins. Within the study presented in this paper, inductive content analysis was chosen to more closely model the process individual researchers take to develop a theory (i.e., methodologies closely related to Grounded Theory).

Participants in Group B were provided with a simple user interface that displayed the output of the NMF topic modeling algorithm. Participants in Group B were provided with a tutorial on using the user interface and a basic introduction to the topic modeling algorithm. Participants with domain knowledge related to the chosen dataset (i.e., Teaching and Learning background) and qualitative content analysis experience were recruited. In Group A, 4 participants had a Masters degree and 2 had a PhD. In Group B, 3 participants had a Masters degree and 4 had a PhD.

The chosen dataset contained responses to a single question namely "What types of support does your school provide to beginning science or mathematics teachers?" from a survey for First Year Mathematics and Science Teacher Professional Development. The corpus contained 100 responses, taking up approximately four (4) A4 size pages. The collective word count of the corpus was 1397 with individual responses varying between 1 and 4 lines of text. A small dataset was chosen to restrict the amount of time manual coders required to complete the coding task. The NMF algorithm was used in the experiment because it was able to provide topics of a higher quality than LDA for smaller datasets.

Participants in Group A were also provided with blank topic templates, asked to label the identified topics and add the response IDs of all related text documents to the templates. Participants in Group A were instructed to provide a reason for grouping the responses together in a topic.

Group B was provided with a user interface for the NMF algorithm. A more recent NMF algorithm, projected gradient descent NMF [6] was chosen because convergence was shown to be faster (i.e., reduced time to run the algorithm). Participants in Group B were required to specify the number of topics, a parameter required by the NMF algorithm. The instructions and user interface are shown in Fig. 1. The top weighted words in a topic and the top text responses in a topic were both shown to participants. Keyword-in-context functionality was provided to aid with topic interpretation and evidence gathering. The inclusion of functionality to allow participants to highlight the top words within the text responses was a required and included because prior studies and algorithm evaluation studies neglect to show the documents that map to a topic [3].

Participants in Group B were instructed to continue to change the number of generated topics until they were satisfied with the generated topics. Once the participants settled on a specific number of topics, participants were required to label and rate each topic. At the end of the research task, participants were presented with a questionnaire. The questionnaire was designed to gauge whether the derived topics were able to be interpreted. User interaction with the interface was tracked and the derived topics (i.e., word and document weightings within a topic at each iteration saved for analysis).

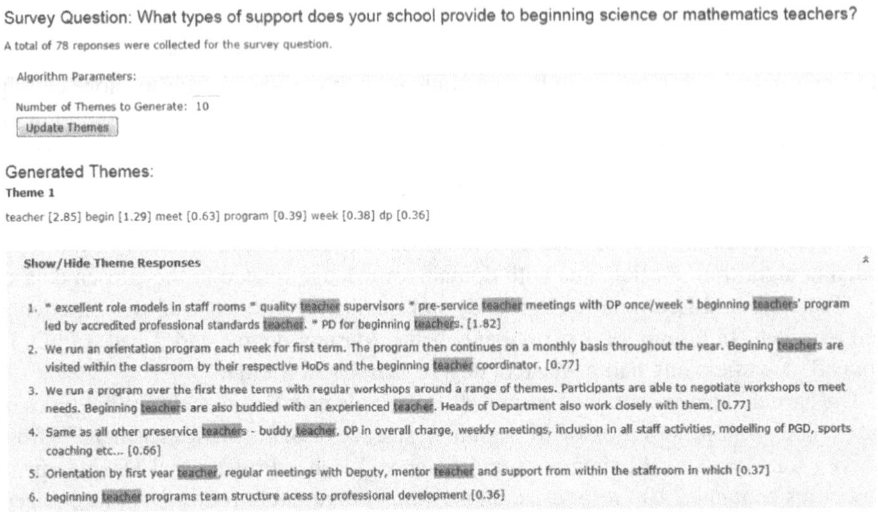

Survey Question: What types of support does your school provide to beginning science or mathematics teachers?

A total of 78 reponses were collected for the survey question.

Algorithm Parameters:

Number of Themes to Generate: 10

Update Themes

Generated Themes:

Theme 1

teacher [2.85] begin [1.29] meet [0.63] program [0.39] week [0.38] dp [0.36]

Show/Hide Theme Responses

1. " excellent role models in staff rooms " quality teacher supervisors " pre-service teacher meetings with DP once/week " beginning teachers' program led by accredited professional standards teacher. " PD for beginning teachers. [1.82]
2. We run an orientation program each week for first term. The program then continues on a monthly basis throughout the year. Begining teachers are visited within the classroom by their respective HoDs and the beginning teacher coordinator. [0.77]
3. We run a program over the first three terms with regular workshops around a range of themes. Participants are able to negotiate workshops to meet needs. Beginning teachers are also buddied with an experienced teacher. Heads of Department also work closely with them. [0.77]
4. Same as all other preservice teachers - buddy teacher, DP in overall charge, weekly meetings, inclusion in all staff activities, modelling of PGD, sports coaching etc... [0.66]
5. Orientation by first year teacher, regular meetings with Deputy, mentor teacher and support from within the staffroom in which [0.37]
6. beginning teacher programs team structure acess to professional development [0.36]

Fig. 1. The user interface provided to Group B for analyzing the output of the NMF algorithm.

3   Analysis

In this section the topics found manually by participants in Group A will be compared with the topics that were derived with the aid of the NMF algorithm. Additional data collected for Group B, including the use of interpretation tools embedded within the user interface, the ratings of the final topics and the questionnaire results are also discussed.

3.1   Analysis of the Topics Manually Discovered by Group A Participants

The analysis of the completed topic templates revealed that there was no consensus between participants in Group A on the exact number of topics. The number of discovered topics varied between five (5) to eleven (11). Significant overlap between the discovered topics was present in the completed templates for multiple Group A participants. Overlap is measured by counting documents that have been mapped to multiple topics. This finding clearly substantiates a requirement to use soft clustering computational aids (i.e. algorithms that allow overlap) such as the NMF topic modeling algorithm.

A common strategy employed by Group A participants was to group text responses using similar words together which resulted in broad topics such as Support, Mentoring, Induction and Professional Development. These broad topics were found by all participants in Group A. Group A participants were able to uncover additional fine-grained topics where the documents were grouped together because they contained related words. The related words in some cases were simple synonyms but were mainly related via complex domain relationships. As an illustrative example, the Workload Balance topic was created by a participant with the following rationale: "referred to reduction of timetable, efforts to lighten teaching load for beginning teachers, release time for professional development".

3.2   Analysis of the Topics Derived Using NMF as a Computational Aid

Similarly, to Group A, there was no consensus between the number of topics that Group B participants were able to derive using NMF. A greater variation in the number of derived topics was present in the derived topics - between five (5) and nineteen (19). It is important to note that the user interface allowed the participants to change the number of topics to be generated and review the resulting topics until they were able to settle on what they thought was a good approximation of the number of topics. Figure 2 includes step plots for each participant showing the changes in the number of topics and the time taken to review derived topics. Completion of the research task took between 15 and 90 min. Participants in Group B started with a lower number of topics before increasing the number of topics and finally settling on a lower number of topics. The selection of an ideal number of topics was a trial and error process for participants in Group B.

The main NMF derived topics were Support, Mentoring, Induction and Professional Development. These simple topics were also discovered by the manual coders in Group A and provides evidence that the NMF algorithm can successfully group text responses with common words together. More advanced and finer grained topics (i.e., Communication, Resources and Workload Balance) that were found by Group A participants were however not able to be derived by NMF. The NMF algorithm did not have the appropriate background or domain knowledge to map together related words and concepts. There were also topics that were generated by the NMF algorithm such as "Assessment and feedback of beginning teacher performance", "Personalized Beginning Teacher Learning Plans" and "Responsibility of HOD and support" that were not present in any of the topics found by the manual coders. These topics were only derived when the number of topics exceeded 10.

Participants in Group B were required to complete a questionnaire after they found an appropriate number of topics. The questionnaire included questions relating to the quality of the derived topics and whether the display of the topics (i.e., including both top words and top documents) supported topic interpretation. All participants in Group B completed the questionnaire. Question 1a asked participants to rate on a scale of 1 to 7 whether it was very difficult (1) or very easy (7) to use the top weighted words in the topic to interpret and understand a topic. All participants gave a rating or 4 or above, with seven participants giving a rating of 5 or above. Question 1b was used to determine if the documents within a topic were meaningfully grouped together. Question 1b was included to give an indication of the quality or coherence of the generated topics. A total of nine participants gave a rating of 4 or above, with five participants giving a rating of 5 on a scale of 1 (very difficult) to 7 (very easy).

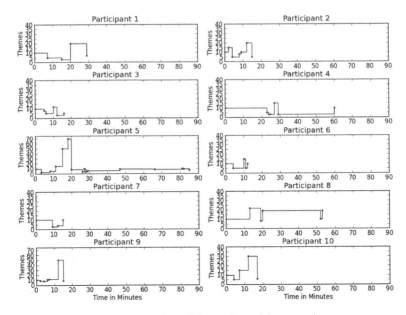

Fig. 2. Visualization of Group B participant sessions.

The inclusion of tools that facilitate in-context analysis such as keyword-in-context tools also led to improved theme interpretation, facilitating an enhanced understanding of the theme and providing evidence for validity. The responses to Question 4, "Please provide general feedback/comments relating to the automatically derived topics?" that relate to quantifying the derived topics are summarized below:

- **Display of Numeric Weightings**
  Participants were confused by what the weights (of both the top weighted words and top weighted documents) meant.
  "i didn't know what the numbers meant"
  "The numbers in brackets were at first misleading because I thought they might be averages of some sort"
  "Not sure how to compare the word weightings"
- **Topic Overlap and Size**
  Participants require topic overlap visualization.
  "Is there a way I can see the overlap between topics?"
  "Only the top 10 documents are shown in each theme. How big is a theme? Am I seeing all the survey responses in a theme?"
- **Theme Quality**
  The quality of the derived topics varied.
  "Not all returned themes are of the same quality."
  "Generally the derived themes worked excellently."
  "They were a good trigger - even when they were not obvious, they prompted thinking/new ideas."

4 Discussion

In this section, the study findings will be used to answer the research questions:

1. Are topics derived via the NMF topic modeling algorithm comparable to the topics identified manually by qualitative researchers?
   NMF can find the simple and broad topics that human coders are able to discover but was unable to discover the fine-grained topics that human coders were able to identify. This finding has implications for qualitative researchers currently using topic modeling to get an overview of the topics within a corpus. Topic modeling still needs to be used in combination with manual coding used to identify fine grained topics based upon complex word relationships. A follow up experiment was conducted that allowed researchers to interact with the algorithm and provide additional domain knowledge. The research findings showed that the research participants were able to use the interactivity to improve their ability to address their research questions [1].

2. Are qualitative researchers able to interpret and quantify the output of derived topics (i.e., the mapping between topics and top terms; and the mapping between topics and corpus documents) from NMF?
   Participants found the display of both the words and the documents that belonged to a topic to be a useful aid for topic interpretation. Participants however found the display of the word and document weightings to be confusing. One reason for this is that the NMF weightings were not presented as a percentage (or probability). Participants also wanted to view topic overlap and topic size. The required functionality could easily be included in a topic modeling browser. LDA-Vis for example is a topic modeling visualization tool that includes an algorithm for determining topic size and overlap [7]. The findings from this study suggest that topic modeling can be a valuable aid for topic discover, interpretation and statistical quantification with enhancements to the display of algorithm output. The required enhancements include keyword-in-context functionality (i.e., allowing analysts to review the location of words within their containing documents), visualizations to help the analyst determine the size and overlap between topics and the display of top word and document weightings that are a percentage value. With these suggested enhancments topic modeling browsers will become valuable aids within the quantitative ethnography process.

5 Future Directions

Future research will focus on evaluating the use of topic modeling as a starting point for quantitative ethnography and the design an algorithm to perform epistemic network analysis on the output of topic modeling algorithms.

6   Conclusion

The results presented within this paper provides valuable insight into the differences between the topics discovered by a topic modeling algorithm and those that have been identified by manual coders. As datasets are increasing in size it is imperative that further research is undertaken to help qualitative researchers quantify derived topics (i.e., understanding topic size and overlap) and make the output of topic modeling comparable to human coders.

Acknowledgement. The experiments described within this paper were conducted as part of my doctorate degree at Queensland University of Technology. I would like to thank and acknowledge my supervisors Peter Bruza, Jim Watters, Bhuva Narayan and Laurianne Sitbon.

References

1. Bakharia, A., Bruza, P., Watters, J., Narayan, B., Sitbon, L.: Interactive topic modeling for aiding qualitative content analysis. In: Proceedings of the 2016 ACM on Conference on Human Information Interaction and Retrieval, pp. 213–222. ACM (2016)
2. Blei, D.M., Ng, A.Y., Jordan, M.I.: Latent dirichlet allocation. J. Mach. Learn. Res. **3**, 993–1022 (2003)
3. Chang, J., Gerrish, S., Wang, C., Boyd-Graber, J.L., Blei, D.M.: Reading tea leaves: how humans interpret topic models. In: Advances in Neural Information Processing Systems, pp. 288–296 (2009)
4. Hsieh, H.F., Shannon, S.E.: Three approaches to qualitative content analysis. Qual. Health Res. **15**(9), 1277–1288 (2005)
5. Lee, D.D., Seung, H.S.: Algorithms for non-negative matrix factorization. In: Advances in Neural Information Processing Systems, pp. 556–562 (2001)
6. Lin, C.J.: Projected gradient methods for nonnegative matrix factorization. Neural Comput. **19**(10), 2756–2779 (2007)
7. Sievert, C., Shirley, K.: LDAVis: a method for visualizing and interpreting topics. In: Proceedings of the Workshop on Interactive Language Learning, Visualization, and Interfaces, pp. 63–70 (2014)

Using Recent Advances in Contextual Word Embeddings to Improve the Quantitative Ethnography Workflow

Aneesha Bakharia[1(✉)] and Linda Corrin[2]

[1] The University of Queensland, Brisbane, Australia
Aneesha.bakharia@gmail.com
[2] Swinburne University, Melbourne, Australia
lcorrin@swin.edu.au

Abstract. The qualitative content analysis process has traditionally been reliant on human researchers to read and code data, with limited use of automation. However, recent advances in Natural Language Processing (NLP) offer new techniques to improve the reliability and usefulness of content analysis, especially in the area of quantitative ethnography. In this paper we propose a new qualitative content analysis workflow that utilizes techniques such as contextual word embeddings and semantic search. Each of the design principles that inform this workflow are outlined and potential NLP solutions are discussed. This is followed by the description of a new prototype, currently in development, that implements elements of the workflow. The paper concludes with an outline of two proposed research studies to evaluate the effectiveness of the workflow and prototype as well as directions for future research.

Keywords: Quantitative ethnography · Qualitative analysis · Content analysis

1 Introduction

Most qualitative researchers are familiar with the analysis workflow that is dictated by software such as NVivo. The basis workflow involves reading the text corpus, highlighting key terms or phrases and either assigning these to a predefined category (i.e. Directed Content Analysis) or creating a new category (i.e., Inductive Content Analysis). Simple Boolean search techniques can also be used to identify additional documents that match a category. Search and retrieval functionality is provided by the majority of qualitative software applications available (e.g. NVivo, ATLAS.ti, MAXQDA, QualRus, QDAMiner, nCODER and DiscoverText). Scaling the process to larger datasets has usually involved the use of supervised machine learning where example documents that match a category are used to train a machine learning model. The machine learning model can then be used to classify unseen documents with reasonable accuracy. Machine learning, however, comes with its own caveats, particularly for qualitative content analysts and researchers. Qualitative researchers may not trust the algorithms and there are new processes, terminology, best-practice and algorithms to learn in terms of model training and validation.

© Springer Nature Switzerland AG 2019
B. Eagan et al. (Eds.): ICQE 2019, CCIS 1112, pp. 299–306, 2019.
https://doi.org/10.1007/978-3-030-33232-7_26

In this paper, we proposed a rethinking of both the human and algorithmic involvement in the qualitative content analysis process. The paper focuses on defining a set of design principles for a new qualitative content analysis workflow. Recent advances in Natural Language Processing (NLP), made possible by contextual word embeddings and semantic search, are then proposed as potential solutions to complete the workflow. Finally, an early prototype of an application that implements the workflow is introduced along with proposed research experiments and directions for future research.

2 Design Principles

In this section, we propose four design principles that aim to change the fundamental way in which qualitative researchers can use machine learning and deep learning algorithms for content analysis.

2.1 Design Principle 1: Promote Reading, Interpretation and Selection of Recommended Text Responses

Current workflows used by qualitative researchers involve the reading and tagging of words or phrases. These steps are also required on a subset of the data when supervised learning algorithms are used to create a dataset for training a classification model. Reading and interpretation are key steps followed in both directed and inductive content analysis [3]. These steps are also key to the new proposed workflow. It is, however, impractical when working with large datasets for the whole dataset to be read. Therefore, semantic search algorithms will be used to suggest text responses that are similar to identified words or phrases that the qualitative researcher has tagged. Semantic search in the proposed workflow will serve as a replacement for the simple search and retrieval techniques (i.e. Boolean search and regular expression usage) currently used by qualitative researchers. The proposed semantic search will also use recent advances in word embeddings [1, 4].

2.2 Design Principle 2: Seamless Integration of Classification Algorithms

The second design principle aims to remove the need for qualitative researchers to be familiar with applying supervised machine learning techniques. The process of selecting an algorithm, choosing algorithm meta-parameters and creating training and validation datasets will still be part of the process, but occur in the background. Knowledge of how to train a deep learning or machine learning text classification model should not be required by the qualitative researcher. It is important to note that Design Principle 2 does not mean that the algorithm is completely hidden from the researcher, rather it refers to the simplification of the supervised machine learning experiment lifecycle. Algorithmic transparency is extremely important and required by a qualitative researcher to gain trust in the algorithm, prevent bias and understand the capabilities and decisions being made by the algorithm.

2.3   Design Principle 3: Include Quantitative Metrics

An integral element of quantitative ethnography is to combine statistical and qualitative analysis [7]. It is therefore essential to allow analysts to quantify the size of classification groups (i.e. the number of responses matching a code) as well as review cross-rater correlations and accuracy metrics. These metrics help to demonstrate the validity and reliability of how the analysis method was employed and provide information for the reporting of research findings.

2.4   Design Principle 4: Support Multiple Analysts and Collaboration

In qualitative studies with multiple researchers, intercoder reliability is important to the quality of the analysis and a metric often required for publication. The focus of Design Principle 4 is to ensure that the proposed workflow is able to support research collaboration. A key proposed addition to the qualitative workflow is the addition of semantic search and recommendation after a researcher has tagged or coded a text response. As multiple researchers will be exposed to this new workflow, the selection of recommended items (i.e. the mapping of additional text responses to a code found via semantic search) must be included in intercoder reliability measures.

3   Recent Advances in Natural Language Processing

In 2018 several breakthroughs occurred in Natural Language Processing (NLP) allowing for each design principle to be addressed and for the proposed quantitative ethnography workflow to be implemented. In this section each breakthrough along with the details of how it can improve qualitative interpretation and recommendation are discussed.

3.1   Sentence Pieces

In NLP the predominant method for feeding text into algorithms has been the 'Bag-of-Words' matrix. The 'Bag-of-Words' matrix contains a count of each word in a document. However, the use of the 'Bag-of-Words' matrix has several shortcomings which include the inability to work with unseen words (i.e. words not part of the initial training corpus) and the inability to differentiate between word senses. The sentence pieces [4] research has been key in promoting the idea of using a technique known as byte-pair-encoding to efficiently group the combinations of characters found in words together. When used in combination with Word Embeddings (discussed in the next subsection), the meaning of words not present in the training corpus can be inferred using combination of characters found in the word.

3.2   Contextual Word Embeddings

A word embedding matrix maps each word to a large vector (between 300 and 500 dimensions) which allows for semantic word comparisons and vector algebra (e.g., King - Women = Queen). Word embeddings have existed for many years in NLP however the first large scale model used in downstream tasks was Word2Vec [5] followed by Glove [6]. Word embeddings take advantage of the premise that words that occur close to each other in a sentence should have a similar representation. Word embeddings however still map whole words to vectors, so are unable to deal with words not in a corpus or with multiple word senses.

In 2018, two contextual word embedding models namely ELMO [4] and BERT [1] were released. Neither provide an embeddings matrix, as they are deep neural learning language models, instead taking a word and its surrounding words and then provide a unique contextual vector representation. BERT uses sentence pieces (i.e. character word groupings) [4] and is able to provide a contextual vector for out of vocabulary words. BERT and ELMO models are also able to deal with multiple word senses. BERT is not based on difficult to train neural architectures such as Recurrent Neural Networks or Long Short-Term Memory models, rather it uses Convolution Neural Networks and attention mechanisms (known as a Transformer model).

Contextual word embeddings address two of the main shortcomings of the 'Bag-of-Words' model identified above. The proposed workflow for quantitative ethnography aims to incorporate contextual word embeddings to provide polysemy word filtering. Both models are open source with pre-trained models available for use in downstream tasks.

3.3   Semantic Search Using Contextual Word Embeddings

The contextual word embedding models, ELMO and BERT have an additional advantage in that they provide a vector space where multi-word expressions (i.e. phrases and sentences) can be compared. This feature enables semantic search. As an illustrative example, when a qualitative researcher highlights a word, phrase or sentence; the embeddings vector space would be searched for similar responses (i.e. via cosine similarity) and a list of recommended text responses to review and map to the category or code would be displayed. While simple cosine similarity will initially be used in the prototype described below, more advanced neural ranking algorithms are also available to be explored and evaluated at a later stage.

3.4   Finetuning, Classification and Transfer Learning

Transfer learning involves using a large dataset to train a model and then using the pre-trained model as the starting point to finetune the model for a more specific task on a smaller domain dataset. Transfer learning is largely responsible for the increase in accuracy and reduction in training time for many image and vision deep learning classification tasks. In image classification tasks, a pre-trained model such as ImageNet or ResNET is loaded and then finetuned. Up until 2018, transfer learning was rarely applied in NLP. UMLFit was one of the first NLP models to emerge that utilised transfer [2]. UMLFit uses a language model pre-trained on a large text corpus as the starting point for a classification neural network. A language model is a neural network trained to predict the next word in a sentence. The UMLFit model is then trained using additional examples from the domain specific task, a process known as finetuning. In several classification tasks, UMLFit was shown to produce higher classification accuracy and reduce training time [2].

Once a subset of text responses has been classified by a qualitative researcher with additional responses automatically identified via semantic search, UMLFit will then use the assigned labels to finetune the classifier.

4   Proposed Workflow Using Contextual Embeddings, Semantic Search and Transfer Learning

Figure 1 details the proposed quantitative ethnography workflow and the integration of key algorithms including contextual embeddings, semantic search and transfer learning.

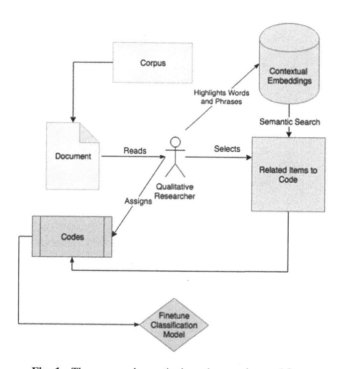

Fig. 1. The proposed quantitative ethnography workflow.

5   A Prototype Implementation

A key challenge has been implementing the workflow within a software product that is able to execute the required algorithms given that compute power and numerous deep learning libraries are required. As we are in the early stages of the proposed research, workflow design and user interface design, the software could not be distributed as a full executable download. We have rather sought to take advantage of the free compute resources (including TPU access) provided by Google Collaboratory to any Google

Drive account holder. Google Collaboratory, also known as Google Colab, is an enhanced version of Jupyter Notebook. The prototype implementation is available as a Google Colab notebook that can be cloned and used. As the project is open source, the notebook and associated widgets will also be able to be executed on other cloud Jupyter Notebook services such as AWS SageMaker.

The Colab notebook allows datasets stored on Google Drive to be loaded and incorporates custom built Jupyter widgets for text response review, coding (i.e. tagging or matching to categories) and the display of semantic search results. In Fig. 2, once a

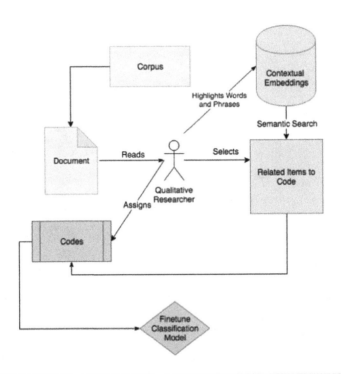

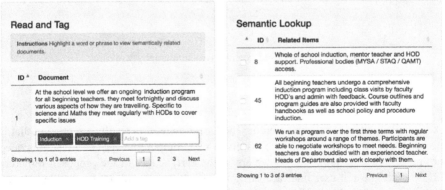

Fig. 2. Prototype implementation in Google Colab.

document has been read and coded (i.e. tags added), and a word or phrase is highlighted, the results from the semantic search using the contextual embedding are displayed. The researcher can then select additional documents that should be mapped to the code.

5.1 Planned Research Experiments

While four design guidelines have underpinned the design of the prototype tool, the current planned research studies will initially focus on "Design Principle 1: Promote Reading, Interpretation and Selection of Recommended Text Responses". Evaluation will focus on the inclusion of semantic search as an aid to assist the researcher in selecting additional documents that should be assigned to a code and whether using semantic search helps the researcher in gaining a deeper understanding of similar text responses and reduced the time required to complete the coding task. User interface usability experiments will not be conducted initially. The following two experiments have been designed:

5.2 Experiment 1: Multiple Researchers Collaborating on a Supplied Corpus

Qualitative researchers will be recruited, supplied a corpus and a predefined coding scheme. The recruited researchers will participate in a between-subjects study. Researchers in Group A will be required to code the corpus and review recommendations of additional items to assign to a code within the prototype tool. Participants in Group B will be required to manually read each response and assign a code. Group B participants will be allowed to perform simple Boolean searches and use regular expression syntax to locate additional responses that must be assigned to a code. Documents mapped to a code and time taken to complete the coding exercise by Group A and B will be compared. Both groups will be required to complete a survey.

5.3 Experiment 2: Research Teams Providing Their Own Data

Research teams will be recruited, given access to the prototype tool and allowed to upload their own dataset. Experiment 2 will allow for a variety of datasets to be coded with the assistance of semantic search. All participants in Experiments 2 will also be required to complete a survey and interviewed.

6 Future Directions

An interesting future research direction involves applying semantic search and recommendation within Epistemic Network Analysis (ENA). ENA is the process of identifying the connections between codes and uses dynamic network models to visualize and quantify code relationships [8]. In order to support ENA, the tool would need to facilitate the highlighting of multiple words or phrases in document and recommend related documents that match the combination of concepts. A network model will then be able to be built and displayed.

7   Conclusion

In 2018, numerous breakthroughs occurred in NLP which have the potential to transform the quantitative ethnography workflow. In this paper, contextual word embeddings, semantic search and related item recommendation are incorporated into the quantitative ethnography workflow with the aim of replacing current simplistic search and retrieval techniques used in the coding practices. The user interface for a prototype tool available as a Google Colab notebook was also presented along with proposed research experiments. In future publications, the implementation and evaluation of all the key design principles for rethinking the qualitative workflow for directed content analysis will be discussed.

References

1. Devlin, J., Chang, M.W., Lee, K., Toutanova, K.: Bert: Pre-training of deep bidirectional transformers for language understanding. arXiv preprint arXiv:1810.04805 (2018)
2. Howard, J., Ruder, S.: Universal language model fine-tuning for text classification. arXiv preprint arXiv:1801.06146 (2018)
3. Hsieh, H.F., Shannon, S.E.: Three approaches to qualitative content analysis. Qual. Health Res. 15(9), 1277–1288 (2005)
4. Kudo, T., Richardson, J.: Sentencepiece: A simple and language independent subword tokenizer and detokenizer for neural text processing. arXiv preprint arXiv:1808.06226 (2018)
5. Mikolov, T., Sutskever, I., Chen, K., Corrado, G.S., Dean, J.: Distributed representations of words and phrases and their compositionality. In: Advances in Neural Information Processing Systems, pp. 3111–3119 (2013)
6. Pennington, J., Socher, R., Manning, C.: Glove: global vectors for word representation. In: Proceedings of the 2014 Conference on Empirical Methods in Natural Language Processing (EMNLP), pp. 1532–1543 (2014)
7. Shaffer, D.W.: Quantitative Ethnography. Lulu.com (2017)
8. Shaffer, D.W., et al.: Epistemic network analysis: a prototype for 21st century assessment of learning. Int. J. Learn. Media 1(2), 33–53 (2009)

The Dynamic Interaction Between Engagement, Friendship, and Collaboration in Robot Children Triads

Yanghee Kim[(⊠)] and Michael Tscholl

College of Education, Northern Illinois University, Gabel Hall 155,
Dekalb, IL 60115, USA
{Ykim9,mtscholl}@niu.edu

Abstract. Grounded in child/robot interaction and inclusive education, this research has designed a small socio-technical community of a robot and two children where children play and learn equitably together while they help the robot learn. This designed community was implemented in a school media lab twice a week over three weeks, each session taking about 20 min. We ethnographically observed and video recorded children's participation in the triadic interaction naturally. The phenomena of interest include friendship development, collaborative communication, and engagement with the community. Data collection is still ongoing, and analysis will occur over this summer. This paper presents the theoretical frameworks and data analytic scheme. We expect to report the findings at the ICQE conference in October.

Keywords: Child-robot interaction · Collaboration · Engagement · Friendship

1 Research Problem

As the demographics of classrooms in public schools both in the U.S. and worldwide becomes increasingly diverse, public education faces the urgent challenge of finding ways to promote effective learning environments in which all children develop and learn equitably. We propose that supporting equitable friendships and collaboration at an early age can be an effective way to address this challenge. Also, recent research on humanoid sociable robots indicates that children develop social and affectionate relations with the robots, voluntarily engage in the interactions with the robots, and mimic the robots' behaviors [1–3]. This research project titled Inclusive Design for Engaging All Learners (IDEAL) aims to design a socio-technical learning community in which a social robot fosters friendship building and collaboration among kindergarten-aged children who come from diverse backgrounds. The research also examines the efficacy of this design by observing children's engagement, friendships, and collaboration as they interact with the robot and with their peer in a natural setting.

There has been a dearth of research on systematically assessing young children's verbal and nonverbal behaviors in learning contexts. Examining the efficacy of our design whose evidence is gleaned from children's interactive behaviors poses us a great challenge. We will use ENA (Epistemic Network Analysis), a statistical modelling tool

© Springer Nature Switzerland AG 2019
B. Eagan et al. (Eds.): ICQE 2019, CCIS 1112, pp. 307–314, 2019.
https://doi.org/10.1007/978-3-030-33232-7_27

which identifies, quantifies and represents (visualizes) connections among three phenomena of interest: friendship, collaboration, and engagement [4]. ENA is designed to analyze a large number of segments (snapshots of an ongoing activity), and may therefore allow us to systematically analyze and interpret the patterns and evolution of children's engagement in the triadic learning community and their friendship building and collaborative behaviors while they engage in the community.

2   Theoretical Frameworks

The design of a robotic learning community is grounded in playful learning theory [29] and culturally-sustaining pedagogy [28]. While they play with peers, children develop intellectually and socially; therefore, learning and play could be integrated fluidly when designing for children. Also, in their learning processes, children as cultural beings should be encouraged to share their personal experiences grounded in home language and culture. The learning community of a robot and children can offer a kind of third space where children can develop a sense of agency and comradery as they play and learn together. The robot verbally invites the children to tell their stories, providing opportunities for participation and demonstrating empathy and appreciation for children's contribution. In this type of community, children's diverse experiences are positioned as assets rather than deficits; children become fully engaged participants rather than marginalized. This way the robot acts as a cultural broker that mediates equitable interactions among children regardless of their cultural and linguistic backgrounds.

Data collection and analysis are grounded in a few theoretical traditions (e.g., ethnography, phenomenology, and symbolic interactionism). We ethnographically observe children's natural participation in the socio-technical community. As children are immersed in the triadic interaction with the peer and with the robot, we will pay attention to what new patterns and protocols of engagement, friendships, and collaboration as *lived experience* will emerge, what existing theories and practices of those phenomena will be replicated, and/or what new shared meanings and experiences come out of their interactions. Two main research questions involve (1) in what way and to what extent aspects of children's experiences in the robotic community (engagement, friendship, and collaboration) evolve over time and (2) in what way such experiences of children interact with each other.

2.1   Friendship

Friendships are characterized by companionship (seeking proximity and spending time together), intimacy (closeness and self-disclosure), and affection in which reciprocity and mutuality plays a core role [5]. It is well established that mutual friendship has a crucial influence on the cognitive, social, and emotional development and well-being of children [6]. Five-year-olds with a mutual friend significantly outperformed their friendless peers on a comprehensive social and cognitive development battery, after controlling for socio-economic status, group popularity, and language skill [30]. Friendship at early age also has a lasting impact on individuals' well-being in that negative friendship experience affects individuals' mental health adversely throughout life [7]. In recent decades, developmental psychologists and clinicians have implemented friendship training programs to coach young children to develop socially valid behaviors, the core of which include cooperation with peers, active listening, and having fun together equally (Frankel and Myatt 2003). In our robotic community, children are asked to collaborate with each other to help the robot learn while they play together. The robot models active listening to children and solicits equal participation.

According to classical intergroup contact theory [8], interacting with other ethnic groups may help reduce cognitive biases against outgroups. A volume of subsequent research confirms this theory, reporting that intergroup interactions enhances sociocognitive skills of children. Having cross-ethnic friendships in childhood has also been associated with positive intergroup attitudes in adolescence and adulthood [9, 10]. Particularly in ethnically heterogeneous contexts, cross-ethnic friendship is considered powerful in developing positive intergroup attitudes, such as equal status and cooperation [8, 11]. Children show less cognitive biases when they have more cross-ethnic companions and high-quality cross-ethnic friendships than when they do not have such relationships. Cross-ethnic friendship is also related to positive change in trust and sympathy toward other ethnic groups. These growing trust and sympathy in turn predict adolescents' inclusive attitudes [12]. In the context of inclusive schooling, direct dyadic friendship is more effective in changing intergroup attitudes than extended friendship – i.e., being aware of others' friendship [13]. Especially for young children, cross-ethnic friendships seem to be associated with other positive developmental outcomes of children judged by teachers [14] such as improved social adjustment, inclusive relationships, prosocial behaviors, and leadership skills. In this research, the robotic learning community sets physical space for direct friendship of two children coming from different backgrounds, where the robot model constant positive regard and appreciation for the information and help contributed by each child.

One important challenge is that some characteristics of friendships can be subtle and are difficult to identify particularly among young children whose thoughts, language, and emotions are still developing. For young children, therefore, friendship characteristics are typically inferred from their behaviors during interactions and play [15].

2.2 Collaboration

With solid curricular efforts, kindergartens can be the context where young children not only learn early academic skills but also develop such social skills that are necessary to be successful in school [16]. Studies on kindergartners' social skills – defined as the ability to resolve conflicts, to collaborate and to understand social cues – and academic development in the first years of school suggest a positive association (e.g. [17]). Mirroring this finding, Welsh, Parke, Widaman, and O'Neil [18] report that children rated at high risk of failure in school demonstrate less than average social competence already in kindergarten. Activities and task designs which foster collaboration are essential to help develop social skills. Through joint play, for example, children learn to share objects, how to resolve peer conflicts, and what it means to work together with others. Indeed, programs designed to help young children learn social skills are found to yield significant positive effects in the targeted competencies [19].

A promising means to foster collaboration among children is digital technology although the design and development of advanced technology to support collaborative interaction among young children are rare. Given the early stage of social and intellectual development of children, design can be focused on fostering effective collaborative communication, rather than collaborative problem solving. Our design of robotic triads seeks to elicit forms of collaborative communication which occur with increasing frequency in the real world. Referring to intercultural communication theory [20], we identify three core constructs of collaborative communication particularly for inclusion and diversity: common ground, equitable partnership, and co-cultural schemas. While two children and a robot play and engage in learning together, the robot can act as a mediator to draw both children to achieving the three communication goals.

2.3 Engagement

There is broad agreement that being engaged in learning means learners are participating actively in learning, persist when facing difficulties, and maintain a strong interest in resources available in the learning environment [21]. Renninger [22] emphasizes that strongly engaged learners appropriate resources for the purpose of learning, including to answer questions they themselves developed, in contrast to less engaged learners who simply carry out an assigned task or follow prescribed rules. Similar to the characterisation of engagement in older children and teens, engagement in younger children is described as demonstrating curiosity, enthusiasm, initiative and effort [23]. For many scholars, engagement is a description of learners' relation to the environment, not a psychological construct [24]. Psychological constructs such as motivation and interest however are directly related to engagement. It is presumed to be malleable, responsive to contextual features, and amenable to environmental change, making it an important measure to evaluate learning design.

While the positive correlation between high levels of engagement and learning achievements is well established for school grades 6–12 (e.g. [25]), much less is known on the impact of engagement on learning for young children. McClelland, Morrison, and Holmes [26] found that children demonstrating high task engagement (including with playful tasks) during kindergarten outperform their peers in academic tasks when in first and second grade. Brock, Rimm-Kaufman, Nathanson and Grimm [23] obtained a similar result for cognitive (task) engagement of kindergartners, but additionally found that high emotional engagement does not affect their academic achievement in the first school years. Blair, Denham, Kochanoff and Whipple [27] argue that social competence affects positively children's level of emotional engagement but may negatively influence their on-task behavior. Overall, there is agreement that different types of engagement affect children's learning differently, but very few studies have examined all components of engagement concurrently to identify the unique contribution of each to children's learning. Furthermore, there are additional challenges in examining the efficacy of this new type of robotic interaction community where children's engagement is multi-layered (i.e., engagement with the robot, with the peer, and with the task).

3 Method

3.1 Participants and Context

Participants were ten kindergarten-aged children (six girls and four boys) in a rural elementary school neighboring Northern Illinois University. Five groups were formed with two children per group (one child with the native-English speaking background and the other child from the Spanish speaking background). Four groups were mixed genders and one group was girls only. Each group participated in six interaction sessions (each taking 15 to 20 min). The interaction activities were implemented during an afterschool program run by the school two days per week over three weeks. The activities were video recorded and transcribed for analysis.

3.2 Intervention: Robotic Interaction Triads

In an interaction triad of robot and children, we personified the robot, *Skusie*, as a new friend who just arrived from another planet and did not know much about life on earth. In this learning community, Skusie needed children's help in order to learn about animal, birthday, school, and family. Skusie spoke both Spanish and English but its speech was not always perfect. Children were asked to work together to teach Skusie. We adopted a Wizard of Oz method to control Skusie, where a hidden researcher remotely controlled its pre-scripted utterances and bodily movement while children interacted with it. Skusie asked open-ended questions as prompts to initiate and then extend engaging conversations between pairs of children, e.g., *what are animals? What do you do on your birthday? Why do you come to school?*

4 Plan for Data Analysis and Interpretation

We will analyze three sets of data: video recording, transcripts, and ethnographic observation notes. Table 1 presents our initial analytic scheme. To examine friendship development, we will assess children's behaviors in terms of three core constructs of friendship (sharing, togetherness, and parity). Friendship in essence involves sharing physical space, tasks, ideas, and experiences. Friendship is inferred by proximity as friends sit together, draw together, and play together. Parity involves being equal while children exchange views, negotiate, and agree/disagree with each other. For our observation of children's engagement, we will start with the widely accepted categories of engagement (behavioral, emotional and cognitive) [21] as a tentative conceptual guide.

   Although we start with theory-based categories for each phenomenon of interest, we fundamentally will take a grounded approach to analyzing children's experiences while they are working on tasks and interacting with the robot and each other. In this grounded analysis, observation data will be evaluated qualitatively to determine the presence or absence of potentially meaningful behaviors. We expect that these data from children's natural interactions will enable us to produce genuine elements of each phenomena.

Table 1. Data analytic scheme

Phenomenon	Core categories	Behavioral indicators (BI)
Engagement	Cognitive engagement	Taking the initiative at task, voluntary elaboration
	Behavioral engagement	Immediate responses to events, initiating new actions without being prompted
	Emotional engagement	Strong emotional expressions (verbal, facial, and bodily)
Friendship	Sharing	Whispering, helping, being nice
	Togetherness	Mutual gaze, leaning toward
	Parity	Taking turns, agreeing
Collaboration	Common grounds	Talking about personal experiences, mindful listening, understanding the other's stories and symbols
	Equitable partnerships	Yielding turns, allowing autonomy, being nice
	Co-cultural schema	Agreeing, co-construction of experiences and artifacts

5 Significance of This Work

Being able to work collaboratively and equitably in diverse groups is an essential skill to succeed in schooling and career development. We view some urgent challenges that public education faces currently (e.g., high dropout rates of minority youths) through the lenses of equitable collaboration, friendships, and engagement. The provision of constructive contexts assisted by humanoid robots might offer a solution, where all students engage in collaborative learning of STEM topics and develop positive relationships. Importantly, the way in which collaboration, engagement and friendship interact and potentially strengthen each other has not yet been studied before. In our designed interaction setting, the three phenomena can be studied systematically. Using ENA will allow us to gain a deeper understanding of how the phenomena co-occur and evolve over time.

Acknowledgements. This work was funded by NSF IIS #1839194.

References

1. Breazeal, C.L.: Emotion and sociable humanoid robots. Int. J. Hum. Comput. Stud. **59**(1–2), 119–155 (2003). https://doi.org/10.1016/S1071-5819(03)00018-1
2. Crompton, H., Gregory, K., Burke, D.: Humanoid robots supporting children's learning in an early childhood setting. Br. J. Educ. Technol. Spec. Issue Digit. Devices Internet-Enabled Toys Digit. Games: Changing Nat. Young Children's Learn. Ecol. Exp. Pedagogies **49**(5), 911–927 (2018)

3. Kim, Y., Smith, D.: Pedagogical and technological augmentation of mobile learning for young children. Interact. Learn. Environ. **25**(1), 4–16 (2017)
4. Shaffer, D.W., Ruiz, A.R.: Epistemic network analysis: a worked example of theory-based learning analytics. In: Lang, C., Siemens, G., Wise, A.F., Gasevic, D., (eds.) Handbook of Learning Analytics, pp. 175–187. Society for Learning Analytics Research (2017)
5. Howes, C.: The earliest friendships. In: Bukowski, W.M., Newcomb, A.F., Hartup, W.W. (eds.) The Company They Keep: Friendship in Childhood and Adolescence, pp. 66–86. Cambridge University Press, Cambridge (1996)
6. Bukowsky, W., Sippola, L.: Friendship and development: Putting the most human relationship in its place. New Dir. Child Adolesc. Dev. **109**, 91–98 (2005)
7. Pedersen, S., Vitaro, F., Barker, E.D., Borge, A.I.: The timing of middle-childhood peer rejection and friendship: linking early behavior to early-adolescent adjustment. Child Dev. **78**, 1037–1051 (2007). https://doi.org/10.1111/j.1467-8624.2007.01051.x
8. Allport, G.W.: The Nature of Prejudice. Addison-Wesley, Cambridge (1954)
9. Ellison, C.G., Powers, D.A.: The contact hypothesis and racial attitudes among Black Americans. Soc. Sci. Q. **75**, 385–400 (1994)
10. Jackman, M.R., Crane, M.: "Some of my best friends are Black…": interracial friendship and Whites' racial attitudes. Public Opin. Q. **50**, 459–486 (1986)
11. Pettigrew, T.F.: Intergroup contact theory. Annu. Rev. Psychol. **49**, 65–85 (1998)
12. Grütter, J., Gasser, L., Zuffianò, A., Meyer, B.: Promoting inclusion via cross-group friendship: the mediating role of change in trust and sympathy. Child Dev. **89**(4), e414–e430 (2018)
13. Feddes, A.R., Noack, P., Rutland, A.: Direct and extended friendship effects on minority and majority children's interethnic attitudes: a longitudinal study. Child Dev. **80**(2), 377–390 (2009)
14. Kawabata, Y., Crick, N.R.: The role of cross-racial/ethnic friendships in social adjustment. Dev. Psychol. **44**(4), 1177–1183 (2008)
15. Papadopoulou, M.: The 'space' of friendship: young children's understandings and expressions of friendship in a reception class. Early Child Dev. Care **186**(10), 1544–1558 (2016). https://doi.org/10.1080/03004430.2015.1111879
16. Logue, M.E.: Early childhood learning standards: tools for promoting social and academic success in kindergarten. Child. Sch. **29**(1), 35–43 (2007)
17. Arnold, D.H., Kupersmidt, J.B., Voegler-Lee, M.E., Marshall, N.A.: The association between preschool children's social functioning and their emergent academic skills. Early Child. Res. Q. **27**(3), 376–386 (2012)
18. Welsh, M., Parke, R.D., Widaman, K., O'Neil, R.: Linkages between children's social and academic competence: a longitudinal analysis. J. Sch. Psychol. **39**, 463–481 (2001)
19. Durlak, J.A., Weissberg, R.P., Dymnicki, A.B., Taylor, R.D., Schellinger, K.B.: The impact of enhancing students' social and emotional learning: a meta-analysis of school-based universal interventions. Child Dev. **82**(1), 405–423 (2011)
20. Gudykunst, W.B. (ed.): Theorizing About Intercultural Communication, 10th edn. Sage Publications Inc., Thousand Oaks (2005)
21. Fredricks, J.A., Blumenfeld, P.C., Paris, A.H.: School engagement: potential of the concept, state of the evidence. Rev. Educ. Res. **74**, 59–109 (2004)
22. Renninger, K.A.: Working with and cultivating interest, self-efficacy, and self-regulation. In: Preiss, D., Sternberg, R. (eds.) Innovations in Educational Psychology: Perspectives on Learning, Teaching and Human Development, pp. 158–195. Springer, New York (2010)
23. Brock, L.L., Rimm-Kaufman, S.E., Nathanson, L., Grimm, K.J.: The contributions of "hot" and "cool" executive function to children's academic achievement, learning-related behaviors, and engagement in kindergarten. Early Child. Res. Q. **24**, 337–349 (2009)

24. Järvelä, S., Renninger, K.A.: Interest, motivation, engagement. In: Saywer, K. (ed.) The Handbook of the Learning Sciences, 2nd edn, pp. 668–685. Cambridge University Press, Cambridge (2014)
25. Marks, H.M.: Student engagement in instructional activity: patterns in the elementary, middle, and high school years. Am. Educ. Res. J. **37**, 153–184 (2000)
26. McClelland, M., Morrison, F., Holmes, D.: Children at-risk for early academic problems: the role of learning-related social skills. Early Child. Res. Q. **15**, 307–329 (2000)
27. Blair, K.A., Denham, S.A., Kochanoff, A., Whipple, B.: Playing it cool: temperament, emotion regulation, and social behavior in preschoolers. J. Sch. Psychol. **42**(6), 419–443 (2004)
28. Paris, D.: Culturally sustaining pedagogy: a needed change in stance, terminology, and practice. Educ. Res. **41**(3), 93–97 (2012)
29. Kangas, M.: Creative and playful learning: learning through game co-creation and games in a playful learning environment. Think. Skills Creativity **5**(1), 1–15 (2010)
30. Fink, E., Begeer, S., Peterson, C., Slaughter, V., de Rosnay, M.: Friendlessness and theory of mind: a prospective longitudinal study. Br. J. Dev. Psychol. **33**(1), 1–17 (2015). https://doi.org/10.1111/bjdp.12060

Effects of Perspective-Taking Through Tangible Puppetry in Microteaching and Reflection on the Role-Play with 3D Animation

Toshio Mochizuki[1]([envelope])[ORCID], Hiroshi Sasaki[2], Yuta Yamaguchi[1],
Ryoya Hirayama[1], Yoshihiko Kubota[3], Brendan Eagan[ORCID],
Takehiro Wakimoto[ORCID], Natsumi Yuki[1], Hideo Funaoi[6],
Hideyuki Suzuki[7], and Hiroshi Kato[8]

[1] Senshu University, 2-1-1 Higashi-mita, Tama-ku,
Kawasaki, Kanagawa 214-8580, Japan
educeboard@mochi-lab.net
[2] Kyoto University, 54 Kawaharacho, Syogoin, Sakyo-ku,
Kyoto, Kyoto 606-8507, Japan
[3] Tamagawa University, 6-1-1 Tamagawagakuen,
Machida, Tokyo 194-8610, Japan
[4] University of Wisconsin-Madison, Madison, WI 53706, USA
[5] Yokohama National University, 79-1 Tokiwadai, Hodogaya-ku,
Yokohama, Kanagawa 240-8501, Japan
[6] Soka University, 1-236 Tangi-machi, Hachioji, Tokyo 192-8577, Japan
[7] Ibaraki University, 2-1-1 Bunkyo, Mito, Ibaraki 310-8512, Japan
[8] The Open University of Japan, 2-11 Wakaba, Mihama-ku,
Chiba, Chiba 261-8586, Japan

Abstract. Perspective-taking of a wide variety of pupils or students is fundamental in designing a dialogic classroom. As a vehicle of perspective-taking, a tangible puppetry CSCL can create a learning environment that reduces the participants' anxiety or apprehension toward evaluation and elicits various types of pupils or students, allowing them to learn various perspectives. The CSCL also provides a 3D animation that records the puppetry for prompting perspective-taking of a variety of pupils in mutual feedback discussions. A comparative experiment, which comprised of a self-performed, a puppetry, and a second self-performed microteachings, showed a relatively stable impact of the puppetry microteaching in the mutual feedback discussions on the second self-performed. This paper discusses the potential effectiveness of puppetry as a catalyst of perspective-taking to learn a variety of pupils' viewpoints through their possible reactions in undergraduate teacher education.

Keywords: CSCL · Perspective-taking · Puppetry · 3D reflection animation

B. Eagan et al. (Eds.): ICQE 2019, CCIS 1112, pp. 315–325, 2019.
https://doi.org/10.1007/978-3-030-33232-7_28

1 Introduction

Designing an effective lesson leveraging dialogic pedagogy is an essential skill for schoolteachers [1]—but it is difficult even for experienced teachers to operationalize in a classroom. In the dialogic classroom, teachers need to design a dialogue to stimulate the students' thinking and advance their learning and understanding through structured and cumulative questioning and discussion, without monologic knowledge transmission. To prepare for designing a dialogue that ensures various students' participation, the teachers need to imagine a wide variety of voices of their students and possible reactions and questions [2].

Microteaching is one way to practice the implementation of dialogic pedagogy in teaching; however, it is not easy to achieve. One of the reasons discussed in the "apprenticeship of observation" framework [3] is that student teachers and novices had experienced monologic teaching as students themselves. However, we argue that there is another difficulty – excessive self-consciousness [4] or evaluation apprehension [5] during microteaching sessions. The role-play requires (student) teachers to act out young pupils roles in a realistic way in which they may find difficult, which creates a tendency to play "honest students" who follow the teacher's instructions without question.

Our past studies indicated that puppetry can serve as a powerful device for allowing people to overcome emotional or interpersonal obstacles in face-to-face role-plays and for eliciting reactions including inner emotions or unconscious experiences that they had in a problematic situation [6]. Then we developed a tangible puppetry CSCL system to help microteaching role-play in a puppetry format [7]. The system records the actions and conversations of the participants (hereinafter, the "character") on top of a transparent table (Fig. 1(a)). In Fig. 1, photo (a) shows the system ready to be implemented. Each puppet or prop is attached to a transparent box with an AR marker on the bottom. Each character can express his or her puppet's conditions (such as distracted or concentrated) by manipulating a switch to change the color of the LED in the box to either red or blue (Fig. 1(b)). These functions allow participants to elicit a variety of voices from possible pupils even in the self-performed role-play after the puppetry role-play (Fig. 1(c)) [8]. After the role-play, the participants can view the recorded puppetry to inspire reflection (Fig. 1(d)). This function provides a 3D animation movie of the recorded role-play. This 3D animation function was developed to foster deep perspective-taking by completely shifting a person's viewpoint, based on Lindgren's [9] argument that experiencing a first-person perspective in a virtual world can generate a person-centered learning stance and perspective-taking. This process enables the learner to see through the avatar's point of view and as a result blurs the boundaries between the self and the other; hence, the learner can gain novel perspectives. Thus, this animation movie allows participants to reflect upon their role-play by combining their wide and thorough (bird's-eye) view for all the dialogues and the various participant views (character points of view); the participants can examine the overall situation from the bird's-eye view, whereas, from the character points of view, they can consider the possible reactions (communication and behavior) of specific characters. The participants can switch the interface, while watching the role-play animation; as a result, they can consider the first-person perspective of each character, when necessary.

The present study aims to examine the effectiveness of 3D animation for reflection that the system generates to foster perspective-taking. We demonstrate a preliminary evaluation of the system by comparing mutual feedback discussions with the 3D animation and those with normal video recording of the puppetry (i.e. as similar to the self-performed microteaching). Then we discuss how an immediate transfer of perspective-taking training emerges.

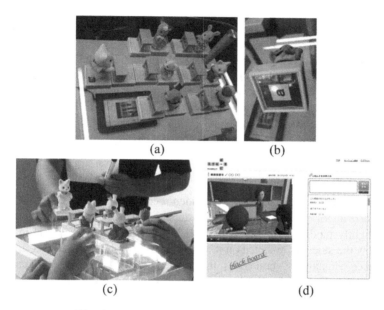

(a)   (b)

(c)   (d)

Fig. 1. The CSCL system for tangible puppetry.

2   Method

2.1   Participants and Design

We conducted a comparative experiment with participants that totaled 30 undergraduate student teachers (normal video only: 24; 3D animation introduced after the puppetry microteaching: 6; Female 86.7%) at a private university in Japan studying to become elementary school teachers and taking a pedagogy course. The participants in each condition participated in a part of the course in different years. Those students in each condition were randomly assigned to groups of three forming 8 and 2 triads, respectively.

Each microteaching session included a role-play and a mutual feedback discussion for reflection. Each participant in both conditions conducted a self-performed microteaching role-play or a puppetry microteaching for 10 min. To examine the effectiveness of perspective taking in the puppetry role-play and its 3D animation reflection, each participant enrolled in one puppetry microteaching and two self-performed microteachings; the first and third participants played the teacher in the self-performed role-plays, and the second participant played the teacher in the puppetry role-play. The rest of the participants played the pupil's role in every session in the same way (i.e., puppetry or self-performance) as the student teacher. Regardless of the form of microteaching, students playing the pupil's role were asked to act realistically, as though they were in an actual classroom. Thus, the first session was designed as the pretest, the second as the intervention, and the third as the posttest to examine the immediate transfer of the puppetry microteaching. Then the participants had a mutual feedback discussion for reflection according to instructions saying that the students needed to consider how to improve the lesson from the pupils' viewpoints, lasting for 20 min.

All the students in each group were video-recorded during the self-performed microteaching, as well as during the puppetry microteaching in the normal video only condition, then the triad students reviewed the corresponding video in each mutual feedback discussion. The 3D animation described above was used for recording and reflecting on the puppetry microteaching in the 3D animation condition; the triad students reviewed it instead of the video in the mutual feedback discussion for the puppetry microteaching.

2.2 Assessment

All the mutual feedback discussions were video-recorded and transcribed. Two of the authors coded all of the utterances in the student discussions for mutual feedback, adapting slightly modified Rosaen et al. [10] 's coding scheme (Table 1) in order to examine how the students reflected on their role-playing in both conditions ($\kappa = .729$). If an utterance contained several codes, the coders coded the corresponding categories. We did not code the microteachings utterances because previous studies showed that the puppetry changed the discourse patterns of the microteaching; the puppetry elicited a variety of informal discourse that is rarely used in self-performance, and those positive effects were also seen in the self-performance when made just after the tangible puppetry (see Mochizuki et al. [7] and Wakimoto et al. [11] for more details).

Table 1. Definition of codes for utterances in the mutual feedback discussions.

Code	Definition
Focus on Teacher-Management (TM)	Managing students' behavior, role in organization for a smooth lesson flow
Focus on Teacher-Instruction (TI)	Instructional strategy that facilitates the cognitive and social interaction around the goals of the lesson; focuses on the teacher's role
Focus on Student-Management (SM)	Managing students' behavior, organization for a smooth lesson flow; focuses on the children's behavior or attitudes
Focus on Student-Instruction (SI)	Instructional strategy that facilitates the cognitive and social interaction around the goals of the lesson; focuses on how the students responded to the instruction

2.3 Analysis

In this study, we applied Epistemic Network Analysis [12, 13] to our data using the ENA1.5.2 Web Tool [14]. We defined the units of analysis as all lines of data associated with a single value of session IDs (IDs for each microteaching session in each condition such as Video1, Video2, Video3, 3D1, 3D2, or 3D3), subsetted by group IDs (triad's IDs) and student IDs (participant's IDs).

The ENA algorithm uses a moving window to construct a network model for each line in the data, showing how codes in the current line are connected to codes that occur within the recent temporal context [15]. The moving window in this study was defined as four lines (each line plus the three previous lines) within a given conversation. The resulting networks are aggregated for all lines for each unit of analysis in the model. In this model, we aggregated networks using a binary summation in which the networks for a given line reflect the presence or absence of the co-occurrence of each pair of codes.

Our ENA model included the following codes: TM, TI, SM, and SI shown in Table 1. We defined conversations as all lines of data associated with a single value of group IDs subsetted by the session IDs, turn numbers in a conversation, and follow numbers within each turn.

The ENA model normalizes the networks for all units of analysis before they are subjected to a dimensional reduction, which accounts for the fact that different units of analysis may have different amounts of coded lines in the data. For the dimensional reduction, we used a singular value decomposition, which produces orthogonal dimensions that maximize the variance explained by each dimension.

ENA visualizes networks using network graphs where nodes correspond to the codes, and edges reflect the relative frequency of co-occurrence, or connection, between two codes. The result is two coordinated representations for each unit of analysis: (1) a plotted point, which represents the location of that unit's network in the low-dimensional projected space, and (2) a weighted network graph. The positions of the network graph nodes are fixed, and those positions are determined by an optimization routine that minimizes the difference between the plotted points and their corresponding network centroids. Because of this co-registration of network graphs and projected space, the positions of the network graph nodes—and the connections they define—can be used to interpret the dimensions of the projected space and explain the positions of plotted points in the space. Our model had co-registration correlations of 0.95 (Pearson) and 0.95 (Spearman) for the first dimension and co-registration correlations of 0.97 (Pearson) and 0.97 (Spearman) for the second. These measures indicate that there is a strong goodness of fit between the visualization and the original model.

ENA can be used to compare units of analysis in terms of their plotted point positions, individual networks, mean plotted point positions, and mean networks, which average the connection weights across individual networks. The networks may also be compared using network difference graphs. These graphs are calculated by subtracting the weight of each connection in one network from the corresponding connections in another.

To test for differences, we applied Mann-Whitney tests to the location of points in the projected ENA space for units in the first sessions in both conditions, those in the second sessions in both conditions, and those in the third sessions in both conditions.

3 Results

3.1 Results of the ENA

Table 2 shows epistemic networks of each session in each condition, as well as comparison plots between the conditions. There were significant differences between the video-only condition and the 3D animation condition in the first session (along the Y axis, the video-only: $Mdn = -0.19$, $N = 24$; the 3D: $Mdn = -0.77$, $N = 6$; $U = 116.00$, $p = .02$, $r = -.61$), as well as in the second and third sessions. We interpret the first session's differences to have been caused by the new curriculum standard that emphasizes student-centered teaching introduced in the video-only class condition. However, in the second session, the mean of the plotted points of the 3D animation condition moved up dramatically on the Y axis compared to the video-only condition where we did not observe a significant change on the Y axis. The strength of the connections between TI-SI was higher in the second session in the 3D animation condition, while that of TM-SM was higher in the video-only condition. There are significant differences along the X axis with a fairly large effect size (for the X axis, the video-only: $Mdn = 0.37$, $N = 24$; the 3D: $Mdn = -0.82$, $N = 6$; $U = 11.00$, $p = .00$, $r = .85$). These results indicate that the participants in the 3D animation condition made more instruction-centered connections during their reflection on the puppetry microteaching; that is, the participants tended to discuss how they should teach pupils who showed unexpected reactions (for example, they discussed the problem of pupils not understanding the instruction) in the puppetry microteaching in the 3D animation condition, while participants in the video-only condition discussed how they should use utterances of pupils and how they should ask pupils to do something (such as how to take a note, how many characters the pupils should take notes for preventing irregular actions such as chatting, and the like). This suggests that the 3D animation condition elicited more student-centered utterances that considered pupils' learning from the viewpoint of their understanding, which is important for achieving teaching objectives [16].

Furthermore, we also observed a significant difference between conditions in the third session along the X axis with a fairly large effect size (the video-only: $Mdn = 0.18$, $N = 24$; the 3D: $Mdn = -1.41$, $N = 6$; $U = 20.00$, $p = .01$, $r = .72$). The comparison plot between the two conditions in the third session shows that the co-occurrence connections of TI-SI-SM in the 3D animation condition are stronger than those in the video-only condition, which implies that the participants used more instruction-centered utterances as well. In addition, along the X axis, a Mann-Whitney test showed that the first session ($Mdn = 0.91$, $N = 6$) was statistically significantly different from the third session ($Mdn = -1.29$, $N = 6$; $U = 3.00$, $p = 0.02$, $r = 0.83$) with a substantive effect size when we examined within the 3D animation condition. This suggests that the 3D animation's effect that elicits utterances with student-centered connections remains even in the feedback discussion the third session.

Table 2. Epistemic networks of the mutual feedback discussions in each session in each condition and comparison plots between two conditions in each session.

	Video only	3D animation introduced in the 2nd (puppetry) session	Comparison plots between two conditions
1st			
2nd			
3rd			

3.2   Content Analysis for the Third Session

The ENA results described above show that the positive effects of perspective-taking persisted even in the mutual feedback discussion after the second self-performance when the student teachers watched the 3D animation for their reflection on their puppetry. In order to examine the characteristics of the differences between the video-only condition and the 3D animation condition, we qualitatively examined the reflective discourse. The ENA webtool extracted eight pieces of discourse in SI-TI and two pieces of discourse in TI-SM.

3.2.1   SI-TI: Discussion on How Possible Pupils Felt the Instructions

All eight discourse excerpts extracted by the ENA webtool for the stronger connection of SI-TI in the third session in the 3D animation condition contain the same pattern of participants discussing how pupils could feel in response to their instruction during the self-performed microteaching. The following is one representative excerpt:

C (pupil role):   I think it was a good idea to provide 3 × 5 at this point. [TI]
                    In some ways, the teacher asked, "how many?" and wanted to answer in a multiplication. [TI]
                    However, even without numbers, pupils who are good at math can answer it, but those who do not understand students are not making any sense. So only by providing 3 × 5. [SI]
                    Yeah. [SI]
                    I thought it would be easy to answer because pupils would probably understand that you are asking for a 3 × 5. [SI]
A (teacher role):   I wanted to do this in the other way, but I wrote 3 × 5 (on the blackboard) at first. [TI]
                    Because the topic of the class was division. [TI]
                    After confirming the multiplication, I will delete a number of one side, but I forgot to do so and left it for a while. [TI]

Before participant A, in the role of the teacher, explained their original intention of instruction as a teacher, participant C, in the role of a pupil, said that the flow of A's instruction was good, imagining that various pupils would exist in an actual classroom. Other episodes also showed that the participants discussions were based on the assumption of various pupils' perspectives and focused on improving their instruction from the viewpoints of possible pupils.

3.2.2   TI-SM: Discussion on Their Possible Instructions with Imagining Possible Slow Pupils

Two excerpts extracted by the ENA webtool for the TI-SM connection showed that the student teachers discussed how they should manage the pupils who are not good at math. The following is a representative example of the TI-SM connection:

E (teacher role):   Honestly, I do not know how to cope with pupils who are not good at math. [SM]
                    I did not say anything but "that's true". [SM]
F (pupil role):   I think it's difficult. [SM]
                    What should the teacher do for pupils who cannot understand? [SM]
                    I think they would probably keep going uneasy for a long time. [SM]
                    So what a kind of message should the teacher talk to? [SM]
                    To the pupils who do not understand our instruction [SM]
                    There are so many children who don't like math … [SM]
                    To pupils feeling uneasiness [SM]

D (pupil role):	Anyway, was it the first time for the pupils to learn division? [TI]
E:	Yup. [TI]
	The unit has just started and it's the first class for the unit. [TI]
D:	I thought that it would be easy for pupils to understand even if the instruction started from a small number from 4 or 6 or so. [SI]
	I thought there would be two pupils, and it would seem like it would be easy to understand if you put the example, as it would be like dividing the four bonds into two. [SI]

The example began what participant E, who played a teacher role, sharing their anxiety surrounding their teaching in light of pupils' viewpoints, such as whether their students are participating in their class with feeling of uneasiness or low confidence, and how to respond to them. Participant F, who played a pupil role, showed sympathy, and participant D was trying to propose an actual solution which considered pupils' perspectives. Another excerpt also showed that the participants imagined how pupils would use mathematical manipulatives, their textbook, or other provided resources. As such, the student teachers considered how pupils would behave while learning math in their classroom during their reflective discussion of their performance.

4   Discussion and Conclusion

This study shows how the use of puppets—as transitional objects that elicit the projection of self (puppeteer) to non-self (puppet)—elicited a variety of informal utterances, enabling student teachers to achieve perspective-taking of a variety of possible pupils in actual classrooms even when the student teachers reflected on their performance. We introduced a 3D animation that records the puppetry to prompt perspective-taking of a variety of pupils in the mutual feedback discussion when student teachers reflected on their microteaching performances. The comparative experiment revealed that the positive effects of perspective-taking were maintained even in the mutual feedback discussion after the second self-performance when the student teachers watched the 3D animation for their reflection on their puppetry. The qualitative analysis of the discourse in the third reflective discussion showed that the student teachers maintained pupils' perspective while they discussed their performance.

Our past study, which introduced 2D animation for reflection on puppetry [7], showed that the effects were lost in the mutual feedback discussions after the second self-performed microteaching. The current study showed that the 3D animation, which allows a first-person view [9], is powerful enough to elicit student teachers' discussion of diverse perspectives. This may enhance the student teachers' perspectives in imagining possible pupils' voices for achieving dialogic teaching.

Further research is needed to investigate the effectiveness of the perspective-taking that the system and its 3D reflection movies prompted in this study, by examining dialogues in a more qualitative manner, after getting more data in additional experiments. In addition, other contexts such as nursing or disaster prevention should also be studied [6] to generalize the effectiveness of the tangible puppetry CSCL.

Acknowledgements. This work was supported in part by JSPS KAKENHI Grants-in-Aids for Scientific Research (B) (Nos. JP26282060, JP26282045, JP26282058, JP15H02937, & JP17H02001) from the Japan Society for the Promotion of Science, as well as the National Science Foundation (DRL-1661036, DRL-1713110), the Wisconsin Alumni Research Foundation, and the Office of the Vice Chancellor for Research and Graduate Education at the University of Wisconsin-Madison. The opinions, findings, and conclusions do not reflect the views of the funding agencies, cooperating institutions, or other individuals.

References

1. Mutton, T., Hagger, H., Burn, K.: Learning to plan, planning to learn: the developing expertise of beginning teachers. Teach. Teach. Theory Pract. **17**(4), 399–416 (2011)
2. Bakhtin, M.: Discourse in the novel. In: Holquist, M. (ed.) The dialogic imagination, pp. 259–422. University of Texas, Austin (1981)
3. Lortie, D.: Schoolteacher: A Sociological Study. University of Chicago Press, Chicago (1975)
4. Ladrousse, G.P.: Role Play. Oxford University Press, Oxford (1989)
5. Cottrell, N., Wack, D., Sekerak, G., Rittle, R.: Social facilitation of dominant responses by the presence of an audience and the mere presence of others. J. Pers. Soc. Psychol. **9**(3), 245–250 (1968)
6. Mochizuki, T., Wakimoto, T., Sasaki, H., Hirayama, R., Kubota, Y., Suzuki, H.: Fostering and reflecting on diverse perspective-taking in role-play utilizing puppets as the catalyst material under CSCL. In: Lindwall, O., Häkkinen, P., Koschmann, T., Tchounikine, P., Ludvigsen, S.R. (eds.) Exploring the Material Conditions of Learning: The Computer Supported Collaborative Learning (CSCL) Conference 2015, vol. 2, pp. 509–513 (2015)
7. Sasaki, H., et al.: Development of a tangible learning system that supports role-play simulation and reflection by playing puppet shows. In: Kurosu, M. (ed.) HCI 2017. LNCS, vol. 10272, pp. 364–376. Springer, Cham (2017). https://doi.org/10.1007/978-3-319-58077-7_29
8. Mochizuki, T., et al.: Effects of perspective-taking through tangible puppetry in microteaching role-play. In: Smith, B.K., Borge, M., Mercier, E., Lim, K.Y. (eds.) Making a Difference: Prioritizing Equity and Access in CSCL, 12th International Conference on Computer Supported Collaborative Learning (CSCL), vol. 2, pp. 593–596 (2017)
9. Lindgren, R.: Generating a learning stance through perspective-taking in a virtual environment. Comput. Hum. Behav. **28**, 1130–1139 (2012)
10. Rosaen, C.L., Lundeberg, M., Cooper, M., Fritzen, A., Terpstra, M.: Noticing noticing. How does investigation of video records change how teachers reflect on their experiences? J. Teach. Educ. **59**(4), 347–360 (2008)
11. Wakimoto, T., et al.: Student teachers' discourse during puppetry-based microteaching. In: Misfeldt, M., Eagan, B. (eds.) Proceedings of the First International Conference on Quantitative Ethnography (2019). (in printing)
12. Shaffer, D.W.: Quantitative Ethnography. Cathcart Press, Madison (2017)
13. Shaffer, D.W., Collier, W., Ruis, A.R.: A tutorial on epistemic network analysis: analyzing the structure of connections in cognitive, social, and interaction data. J. Learn. Anal. **3**(3), 9–45 (2016)
14. Marquart, C.L., Hinojosa, C., Swiecki, Z., Eagan, B., Shaffer, D.W.: Epistemic network analysis (Version 1.5.2) (Software). http://app.epistemicnetwork.org. Accessed 16 June 2019

15. Siebert-Evenstone, A., Arastoopour Irgens, G., Collier, W., Swiecki, Z., Ruis, A. R., Shaffer, D.W.: In search of conversational grain size: modelling semantic structure using moving stanza windows. J. Learn. Anal. **4**(3), 123–139 (2017). https://doi.org/10.18608/jla.2017.43.7
16. Mochizuki, T., Kubota, Y., Suzuki, H.: Cartoon-based teaching simulation for reflective improvement of lesson plans in pre-service teacher training. In: Gibson, D., Dodge, B. (eds.) Proceedings of Society for Information Technology & Teacher Education International Conference 2010, pp. 1983–1990. Association for Advancement of Computing in Education, Chesapeake (2010)

A Socio-Semantic Network Analysis of Discourse Using the Network Lifetime and the Moving Stanza Window Method

Ayano Ohsaki[1,2]([⊠]) [iD] and Jun Oshima[2] [iD]

[1] Advanced Institute of Industrial Technology, 1-10-40, Higashi-Ooi,
Shinagawa-ku, Tokyo 1400022, Japan
ohsaki-ayano@aiit.ac.jp
[2] Research and Education Center for the Learning Sciences, Shizuoka
University, 836 Ohya, Suruga-ku, Shizuoka-Shi, Shizuoka 4228529, Japan

Abstract. This study proposes a new temporal Socio-Semantic Network Analysis (SSNA) of discourse by using the network lifetime and the moving stanza window method to analyze idea improvement in learning as knowledge-creation. The procedure of our proposed method has four steps. The first step entails making a discourse analysis unit. One discourse analysis unit is composed of discourses depending on the set numbers at a size of the moving stanza window method. The second step is calculating the total value of degree centrality for each discourse analysis unit with periods of the network lifetime by using SSNA. The third step involves calculating the difference value between discourse analysis units to define the candidates for the pivotal points. The last step is tracing the discourse back from the candidates for the pivotal points to identify segments for in-depth dialogical discourse analysis. To evaluate the proposed method, we analyzed discourse data in collaborative learning using different methods with and without the network lifetime and moving stanza window. As a result, new pivotal points were detected by implementing both the network lifetime and the moving stanza window method. An in-depth dialogical discourse analysis of a new pivotal discourse segment confirmed the appropriateness of the detection. Based on the results, it is concluded that our proposed method is better in detecting pivotal points of learning as knowledge-creation compared to the previous approach.

Keywords: Visualization · Socio-Semantic Network Analysis · Temporal analysis · Discourse analysis

1 Theoretical Background and Research Purpose

This study proposes the use of temporal analysis methodology to examine how learners engage in learning as knowledge-creation [1]. Over the past few years, Socio-Semantic Network Analysis (SSNA) has been recognized as a valuable method for analyzing idea improvement [2, 3]. The total value of degree centralities has been used to evaluate idea improvement. A recent study revealed the usefulness of a combination of SSNA and in-depth dialogical discourse analysis as a mixed method approach [4]. However,

© Springer Nature Switzerland AG 2019
B. Eagan et al. (Eds.): ICQE 2019, CCIS 1112, pp. 326–333, 2019.
https://doi.org/10.1007/978-3-030-33232-7_29

existing SSNA research has mostly focused on aggregative discourse analysis and has not paid much attention to the nature of temporary interactions. Thus, the appropriate implementation of dialogical and temporal analysis techniques in SSNA needs to be further examined in the quantitative approach to learning as knowledge-creation. Our study attempts to pursue the issue by suggesting a new analysis technique.

We propose a new temporal method of analysis by coordinating two important ideas: (1) the moving stanza window method, and (2) the network lifetime. Figure 1 shows how the aggregative discourse analysis is conducted in SSNA. The cooccurrence of vocabularies is calculated an exchange by another. It shows the temporal nature of how the network structure develops but not how the structure is restructured.

Figure 2 shows how SSNA is conducted with the moving stanza window method. The difference between the aggregative analysis and the analysis with the moving stanza window method is shown in the difference in network structures in Figs. 1(a) and 2(a). The unit of analysis is comprised of multiple exchanges based on the assumption of dialogism of human conversation. Dyke, Kumar, and Rosé [5] suggested analyzing a small amount of discourse using a sliding window to examine the discourse that is in close temporal proximity. Siebert-Evenstone et al. [6] have further proposed moving a stanza window method model within a conversation by dividing the activity into multiple overlapping stanzas. We integrated these two features into our current SSNA application to evaluate the temporal nature of discourse in collaborative learning.

Figure 3 shows how SSNA is conducted with the moving stanza window method and the network lifetime. The network lifetime is a concept concerning the calculation for the period of the number of discourse analysis units (Fig. 3) from the network science field. The concept of the lifetime contributes to making a temporal network for analysis of the interactions because most interactions in a network do not continue, and the network has a finite duration [7]. Barabási [8] has suggested the burst model i.e., the human activity patterns involve two types as follows: long periods of rest and short periods of intense activity. To visualize this model, we used the concept of network lifetime in that the effect of links should be sustained within a limited time. The network of keywords is constructed of all of the discourses from the discourse analysis unit 0 to the discourse analysis unit 2, when we do not set the network lifetime (Fig. 2 (b)). Thereby, the total value of degree centrality in the keyword network is calculated the discourse analysis unit 0 to the discourse analysis unit 2. In the case of setting the network lifetime, the keyword network is constructed of the limited periods, and the sum total value of degree centrality of keywords is reckoned for only periods of the set numbers at a size of the moving stanza window method (see Fig. 2). The keywords **a**, **b**, **c**, **d**, **e**, and **f** are connected to the network from the discourse analysis unit 0 and the discourse analysis unit 1 in the first calculation (see Fig. 3(a)). Following this, the keyword network is constructed of keywords **b**, **c**, **d**, **f**, and **g**. In the second calculation, keyword **a** and keyword **e** are missing from the keyword network (Fig. 3(b)).

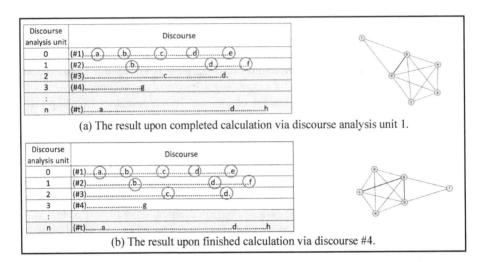

(a) The result upon completed calculation via discourse analysis unit 1.

(b) The result upon finished calculation via discourse #4.

Fig. 1. Examples of the aggregative SSNA method. (Discourse analysis unit and keywords on the left; keyword network on the right)

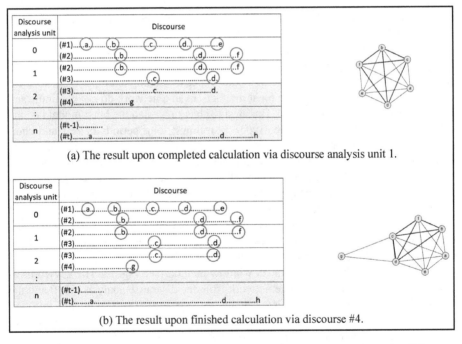

(a) The result upon completed calculation via discourse analysis unit 1.

(b) The result upon finished calculation via discourse #4.

Fig. 2. Examples of SSNA with the moving stanza window method (size = 2). (Discourse analysis unit and keywords on the left; keyword network on the right)

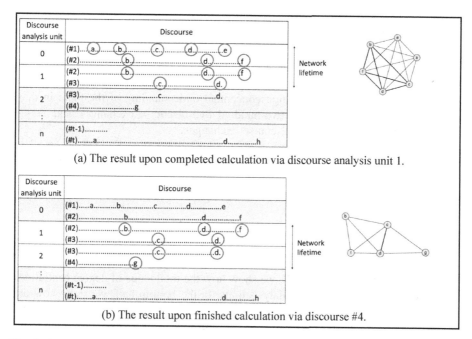

(a) The result upon completed calculation via discourse analysis unit 1.

(b) The result upon finished calculation via discourse #4.

Fig. 3. Examples of SSNA with the moving stanza window method (size = 2) and the network lifetime (period = 2). (Discourse analysis unit and keywords on the left; keyword network on the right)

2 Method

Our proposed method has four steps. Firstly, we concatenated discourses of the set numbers at a size of the moving stanza window method as a discourse analysis unit (see Fig. 2). Secondly, we calculated the total value of degree centrality (C_d) for each discourse analysis unit with periods of the network lifetime by using SSNA. Thirdly, we calculated the differences value ΔC_d between the total value of degree centrality at a discourse analysis unit n and the total value of degree centrality at a discourse analysis unit n-1 in formula (1):

$$\Delta C_d = \sum C_{dn} - \sum C_{dn-1} \tag{1}$$

We defined the candidates of the pivotal points based on the differences value ΔC_d. Finally, to define the pivotal point, we traced the discourse back from the candidates for the pivotal points to identify segments for in-depth dialogical discourse analysis.

To evaluate our proposed temporal analysis, we conducted a comparative study by analyzing the same set of discourse data with using the three different methods: (1) the aggregative SSNA; (2) SSNA with the moving stanza window method (size = 2); and, (3) SSNA with the moving stanza window method (size = 2) and the network lifetime (period = 2).

We used the data set of tenth grade students' collaborative discourses to solve the problem related to the human immune system in their regular biology classes. Thirty-nine students participated. There were twelve groups of three or four students. The problem they discussed was "Can you explain how vaccinations protect us from infections?" We selected the transcribed conversation from two successful groups where all the students attained a high conceptual understanding of the topic in their post-test [9]. Target Group 1 had two-hundred five discourse exchanges, and Target Group 2 had two-hundred seven discourse exchanges.

Following Oshima et al. [2], we used an application called KBDeX to calculate how the total value of degree centralities of 23 nodes (words representing their conceptual understanding of the human immune system) in the network transitioned across discourse exchanges.

3 Results and Discussion

Figures 4 and 5 show the transitions of values of degree centralities in the high-performance groups. In the top half, we show how the total value of degree centralities transitioned across the discourse analysis units. In the bottom half, we show how the difference value (ΔC_d) of the total value of degree centralities transitioned. Line 1 (gray line) shows the result of the aggregative SSNA, and Line 2 (black line) shows the result of SSNA with the moving stanza window method (size = 2). Line 3 (red dotted line) shows the result of SSNA with the moving stanza window method (size = 2) and the network lifetime (period = 2). The data of a discourse analysis unit is changed as per the size of the moving stanza window method. However, the same discourse analysis unit number includes the same discourse in each method (see Figs. 1, 2 and 3).

Line 1 and Line 2 had similar patterns although Line 2 showed the differences more clearly because the same discourses had been calculated twice. For example, discourse #2 is used for a calculation at discourse analysis units 0 and 1 (see Fig. 2). Line 3 had a different pattern from Lines 1 and 2. It was found that the addition of the moving stanza window method and the network lifetime demonstrated the different temporal nature of discourse in SSNA.

Discourse #70 and discourse #71 were detected as the candidates of the pivotal points when we focused on the difference value (ΔC_d) at Target Group 1 (see Fig. 4). These candidates of pivotal points are detected in Line3. Line 2 and Line1 did not demonstrate the changes in network structure at these points. To examine whether these points are "pivotal," we conducted dialogical analysis and confirmed that discourses #68, #69, #70, #71, and #72 were pivotal points for improving the group's idea in high-performance Target Group 1 as follows. The original discourse was in Japanese and translated into English by the first author (keywords in SSNA is shown in bold). In this conversation, the question from Student B (#68), the comment of Student C (#70) and a statement of Student A (#71) contributed the group's understanding from the discourse #68 to the discourse #72. The conversation sequence started from discourse #68: Student B expressed doubt for previous one utterance of Student A. Student A explained own new understanding about the human immune system with discourse #71 as a reaction for discourse #68. Student C supported Student A, and Student B, in

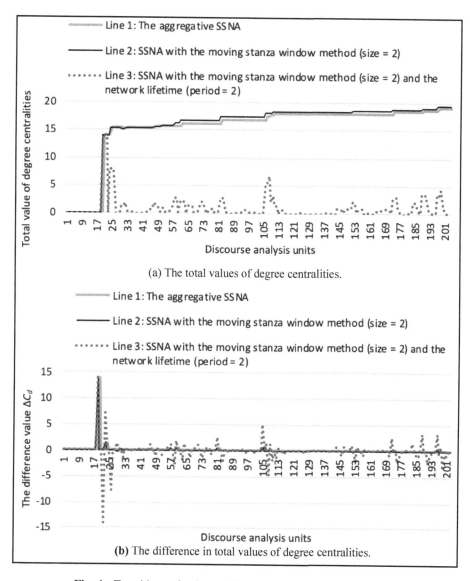

(a) The total values of degree centralities.

(b) The difference in total values of degree centralities.

Fig. 4. Transitions of values of high-performance Target Group 1.

discourse #70. This is a result of in-depth dialogical discourse analysis by tracing the discourse back from the candidates for the pivotal points. Our previous SSNA did not detect a rise in total degree centrality at this point.

Student B (#68):   So, really?

Student A (#69):   **The B cell** is also slow.

Student C (#70):   **The humoral immunity** doesn't work the first time. It starts working after the second time after memorizing its [virus's] the information. So, **the B cell** is slow.

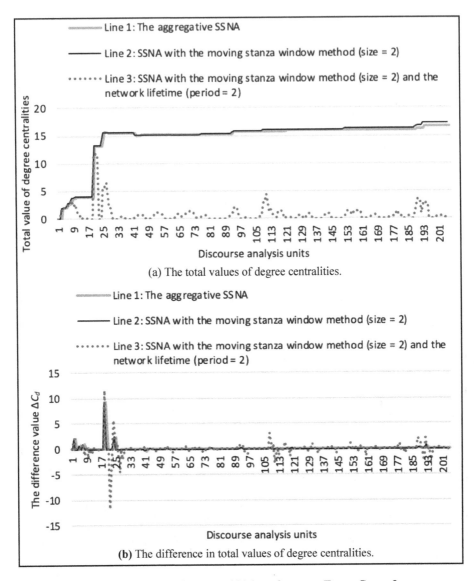

Fig. 5. Transitions of values of high-performance Target Group 2.

Student A (#71):   Slow. No, I made a mistake. The first, only the natural [immunity] works for protection. For the first time. For the virus of that vaccination. A handful of **T cells** and **B cells** receive it [the virus of that vaccination] at that time, then **the B cells** makes **antibodies**. The number of **T cells** increases. So, the second time, when it [virus] arrives really, **the B cells** put **antibodies** and **the T cells** attack it [virus]. Because it [T cells] increases. Maybe so.

Student B (#72):   I got it.

The purpose of this study was to develop a new method of analysis to analyze idea improvement. SSNA is a well-known analytical method for learning as knowledge-creation. However, a more dialogical and temporal analysis method is required for a quantitative approach to knowledge-creation. We proposed SSNA with coordinating the network lifetime and the moving stanza window as a new analytical method, and conducted a comparative study of three methods: (1) the aggregative SSNA, (2) the aggregative SSNA with the moving stanza window method, and (3) SSNA with the moving stanza window method and the network lifetime. As a result of the examination of three methods, we confirmed the effects of the implementation both the network lifetime and the moving stanza window method as follows: (1) the moving stanza window method was effective in more greatly calculating a difference value, and (2) the network lifetime contributed to analysis for details by calculating the data of the short periods. In conclusion, our results suggest the usefulness of our proposed method. On the other hand, as the current research was conducted with high-performance groups, only, and thus further study is required on other groups of varying capabilities.

Acknowledgements. This work was supported by JSPS KAKENHI Grant Numbers 16H0187, JP18K13238, 19H01715.

References

1. Paavola, S., Lipponen, L., Hakkarainen, K.: Models of innovative knowledge communities and three metaphors of learning. Rev. Educ. Res. **74**(4), 557–576 (2004)
2. Oshima, J., Oshima, R., Matsuzawa, Y.: Knowledge building discourse explorer: a social network analysis application for knowledge building discourse. Educ. Technol. Res. Dev. **60**, 903–921 (2012)
3. Scardamalia, M., Bereiter, C.: Knowledge building and knowledge creation: theory, pedagogy, and technology. In: Sawyer, K. (ed.) The Cambridge Handbook of the Learning Sciences, 2nd edn, pp. 397–417. Cambridge University Press, New York (2014)
4. Oshima, J., Oshima, R., Fujita, W.: A mixed-methods approach to analyze shared epistemic agency in jigsaw instruction at multiple scales of temporality. J. Learn. Anal. **5**(1), 10–24 (2018)
5. Dyke, G., Kumar, R., Ai, H., Rosé, C.P.: Challenging assumptions: using sliding window visualizations to reveal time-based irregularities in CSCL processes. In: Proceedings of the 10th ICLS, vol. 1, pp. 363–370 (2012)
6. Siebert-Evenstone, A.L., Irgens, G.A., Collier, W., Swiecki, Z., Ruis, A.R., Shaffer, D.W.: In search of conversational grain size: modeling semantic structure using moving stanza windows. J. Learn. Anal. **4**(3), 123–139 (2017)
7. Barabási, A.: Network Science. Cambridge University Press, Cambridge (2016)
8. Barabási, A.: Bursts: The Hidden Patterns Behind Everything We Do, From Your E-mail to Bloody Crusades. Plume, New York (2011)
9. Oshima, J., Ohsaki, A., Yamada, Y., Oshima, R.: Collective knowledge advancement and conceptual understanding of complex scientific concepts in the jigsaw instruction. In: Smith, B.K., Borge, M., Mercier, E., Lim, K.Y. (eds.). Making a Difference: Prioritizing Equity and Access in CSCL, 12th International Conference on Computer Supported Collaborative Learning (CSCL) 2017, vol. 1, pp. 57–64 (2017)

Designing an Interface for Sharing Quantitative Ethnographic Research Data

Zachari Swiecki[1](✉) ⓘ, Cody Marquart[1], Arjun Sachar[1],
Cesar Hinojosa[1], Andrew R. Ruis[1] ⓘ,
and David Williamson Shaffer[1,2] ⓘ

[1] University of Wisconsin, Madison, WI 53706, USA
swiecki@wisc.edu
[2] Aalborg University Copenhagen, A. C. Meyers Vænge 15,
2450 Copenhagen, Denmark

Abstract. Recently, there have been growing calls to make research data more widely available. While the potential benefits of sharing research data are many, there are also many challenges, including the interpretability, attendability, and complexity of the data. These challenges are particularly salient for research data associated with quantitative ethnographic analyses, which often use relatively novel and sophisticated techniques. In this paper, we explore design considerations for an interface for sharing research data that attempts to address these challenges for quantitative ethnographic analyses. These considerations include: (a) maintaining the consistency of the interpretive space, (b) simplifying model details, (c) including example results and interpretations, and (d) highlighting key affordances in the user interface. To explore these considerations, we describe the design of an interactive visualization of the thematic networks present in the HBO television series, *Game of Thrones*.

Keywords: Sharing research data · Interface design · Epistemic network analysis · Quantitative ethnography

1 Introduction

Recently, there have been growing calls to make research data—the raw data, field notes, models, software and so on that are integral to scientific results—more widely available (see, e.g., Borgman 2012; Feldman & Shaw 2019; Tsai et al 2016). While the potential benefits of sharing research data are many, there are also many challenges, including the interpretability, attendability, and complexity of the data. These challenges are particularly salient for research data associated with quantitative ethnographic analyses (Shaffer 2017), which often use relatively novel and sophisticated techniques such as epistemic network analysis (ENA). In this paper, we explore design considerations for an interface for sharing research data that attempts to address these challenges for quantitative ethnographic analyses. To do so, we describe the design of an interactive ENA-based visualization of the thematic networks present in HBO's television series, *Game of Thrones*.

© Springer Nature Switzerland AG 2019
B. Eagan et al. (Eds.): ICQE 2019, CCIS 1112, pp. 334–341, 2019.
https://doi.org/10.1007/978-3-030-33232-7_30

2 Theory

Calls for sharing research data have come from variety of audiences including funding agencies, publishers, the academic community, and the public. These calls may go beyond sharing the raw data used in a given analysis and extend to the constellation of information and tools that researchers use to produce scientific results, such as interview protocols, field notes, statistical models, code, and other software. As Borgman argues (2012), there are at least three rationales for sharing such research data. First, sharing research data may make it easier to reproduce or verify research results. Second, it may make publicly funded research more available and transparent to the public. And third, it may enable others (including researchers and the public) to ask new questions of the data and potentially advance the state of research.

While there may be good reasons to share research data, there exist many challenges to doing so. The processes researchers use to generate research data require expertise and are often complex, non-linear, and labor intensive (Borgman, 2012). Moreover, it is difficult, if not impossible, to separate research data from the interpretations of the primary researchers, especially when the data have qualitative components (Feldman & Shaw 2019; Tsai et al. 2016). Because audiences may lack the expertise of the primary researchers, and almost certainly were not participants in the production of the research data and its interpretation, an important challenge associated with sharing such data is *interpretability*, or the extent to which audiences can understand the meaning of the data and analyses.

A second challenge in data sharing is that quantitative analyses often use techniques that require both sophisticated mathematical reasoning and prior experience and training in statistical techniques. Even a relatively straightforward regression analysis presents a number of "results" which might or not be important in answering a question, as well as a range of a priori and post hoc tests that can (and in many cases should) be run to determine the validity of a result. Thus, there is an inherent complexity that needs to be addressed in any context of data sharing.

Finally, in a nascent field, by definition techniques and their applications are novel. In turn, any attempt at data sharing needs to address the problem of *attendability*, or the ability of audiences to focus on the salient affordances of the analyses being conducted. For example, many audiences are not (yet!) familiar with quantitative ethnographic techniques such as *Shaffer's Rho* and *epistemic network analysis*.

Sharing quantitative ethnographic research data thus faces all of these challenges.[1] Hence any approach to doing so needs to address to questions of interpretability, complexity, and attendability.

In this paper, we describe one attempt to address these issues by exploring design considerations for an interface for sharing quantitative ethnographic research data. We present some of the critical design principles of an interactive ENA-based visualization

---

[1] Of course, there are many other important challenges associated with sharing research data, including: practical challenges, such as the establishment of sharing protocols, epistemological challenges, such as what it means to reproduce results or whether it is possible at all, and ethical challenges, such as maintaining the confidentiality of participants involved in the data collection. However, these challenges are beyond the scope of this paper.

of the thematic networks present in HBO's television series, *Game of Thrones*. This dataset was chosen because it exemplifies the structural features of datasets commonly analyzed using quantitative ethnographic techniques, and because it comes from a context with which many audiences are likely to be at least peripherally familiar.

The design considerations we explore draw on prior work described by Herder and colleagues (2018) to design and test an ENA-based teacher dashboard. While the interface described here shares some affordances with the dashboard, they differ in important ways. The teacher dashboard was designed to scaffold the real-time assessment of students participating in virtual internships. The time-constraints associated with real-time assessment meant that the presentation of the research data needed to be greatly simplified. In addition to being designed for audiences that include but are not limited to educators, the lack of such time constraints allows the present interface to maintain a higher fidelity to the original research data. In the sections below, we describe the analysis of the data and design considerations for the interface in more detail.

3   Methods

Game of Thrones focuses on the efforts of four families to secure power over the fictional continent of Westeros. While the social network of the show's many characters has been analyzed before (Beveridge & Shan 2016), our goal was to model how their dialogue, actions, and interactions serve to advance the themes of the story.

To conduct this analysis, we obtained transcripts of all episodes from the first 7 seasons of the show using a web-scraping tool (Miller 2017). Next, we constructed a dataset by including dialogue and significant events from the transcripts, as well as metadata about families of characters, locations, seasons, episodes, and scene breaks.

Using a grounded approach, we developed codes that capture major themes of the show: Honor, Family, Sex, Love, Violence, Death, Religion, and Drinking. To code the data, we developed automated classifiers for each code using the ncodeR R package (Marquart et al. 2018). For each code, we achieved $\kappa > 0.76$ and ρ (0.65) < 0.05 between human raters and the automated classifier.

To analyze the data, we constructed three ENA models using the rENA R package (Marquart et al. 2018), each with different units of analysis: characters, episodes, and characters by season.

ENA uses a moving window to construct a network model for each line in the data that occurs within a given conversation, or collection of related lines. Here, we defined lines as lines in the transcripts, which could contain dialogue or events, and conversations as scenes. Connections in the network are defined as the co-occurrence between codes in the current line and codes within the window. For this analysis, we used an infinite moving window to identify connections between codes in the current line and any prior line in the conversation. The resulting networks are aggregated for all lines for each unit of analysis. ENA then normalizes these networks and performs a dimensional reduction via singular value decomposition to project the networks into a lower dimensional space.

Networks are visualized using two coordinated representations: (1) *projected points*, which represent the location of the units of analysis in the space created by the dimensional reduction, and (2) *weighted network graphs* in which the nodes correspond to codes, and the edges are proportional to the relative frequency of connection between two codes. These network graphs can be used to interpret the dimensions of the space, and thus the positions of the projected points: dimensions in the space distinguish units of analysis in terms of connections between codes whose network nodes are located at the extremes of the space. Network graphs for groups of units can be computed and plotted by taking the mean of the individual networks in the group. Moreover, network graphs for any two units (or any two groups of units) can be subtracted and plotted in the space to show the connections that are stronger in one unit or group relative to the other.

We used the Shiny application for R (Chang et al. 2019) to create an interactive web interface for exploring these models.

4   Results

The interface is divided into three main sections (see Fig. 1). On the left is the panel for selecting units or groups to plot. In the middle, network graphs and their associated projected points are plotted in real-time. And on the right is a record of the data used to generate the models. Interested readers can explore the interface by visiting https://got. epistemicnetwork.org/. In the sections below, we describe in more detail the design principles we used to create the interface.

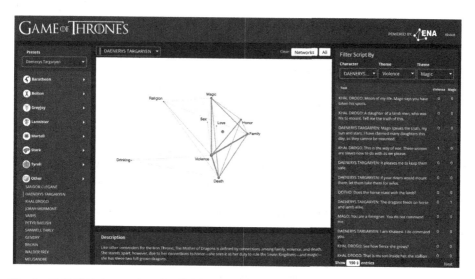

Fig. 1. GoT-ENA interactive visualization. Plotting selections are made on the right; network graphs and projected points appear in the middle; data is shown on the right.

4.1   Consistency of Interpretation

The three ENA models we created each used different units of analysis. In turn, they each produced different dimensional reductions, network node locations, and dimensional interpretations. For example, Fig. 2 shows the network layout for the model that used characters as units of analysis (left) and the layout for the model that used episodes as units (right). In the character model, the X dimension distinguishes units in terms of connections to Magic versus Family, Love, and Religion, while in the episode model, it distinguishes units in terms of connections to Drinking versus Family. In the character model, the Y dimension distinguishes units in terms of connections to Death, Honor, and Family versus Drinking, while in the episode model, it distinguishes units in terms of connections to Religion and Magic versus Death.

Fig. 2. Network layout for character model (left) and episode model (right).

Interpreting the dimensions from a dimensional reduction can be difficult (Shenet et al. 2008), however, we have found that spaces where nodes clearly distinguish the extremes of the two dimensions are easier to interpret. We thus attempted to reduce the interpretive challenge for users by selecting the set of dimensions that was easiest to interpret by this criterion—in this case, those from the episode model—and projecting the data points from the remaining models into that space. As a result, *users can change between different models without having to re-interpret the dimensions.*

4.2   Simplification of Model Details

ENA models are typically used by researchers to warrant qualitative claims using statistical methods. As such, standard ENA representations often include visual aspects that convey statistical information such as the amount of variance explained by each dimension and confidence intervals around the mean of the projected points for a given group. While these aspects are important for researchers, they can be distracting and confusing for other audiences. Thus, to make the ENA models in this tool more accessible, we *removed information that would be critical for a researcher in validating a model, but was not necessary to understand the results of a valid model.* This included removing variance explained metrics, confidence intervals, and outlier intervals. In future versions of the tool, we plan to make this feature optional so as to support audiences with different goals and expertise.

4.3   Example Results and Interpretations

Within the interface, we also included several example results. These included network graphs for key characters and network comparisons between key characters and families. We also provided a narrative interpretation for each example result to help users (a) make sense of the results, and thus (b) *scaffold their ability to understand the networks and comparisons for other units and models*. In addition to the presets, users can view networks and comparisons (without interpretations) for all units within a given model.

4.4   Highlighting Key Affordances

One of the most powerful affordances of ENA that distinguishes it from other network techniques is that it allows comparisons between networks in terms of both network graphs and network summary statistics. To focus users on this affordance and its benefits, we designed the user interface to facilitate network comparisons in several ways. For example, clicking a unit on the left panel plots its projected point and network. Clicking subsequent units adds their projected points to the plot and automatically plots the network subtraction between the first unit and the most recently added unit. Similarly, clicking a group of units in on the left panel—for example, a family or a season— plots all of the projected points associated with that group, the mean projected point position for the group, and the mean network for the group. Clicking subsequent groups adds their projected points and means to the plot and automatically plots the mean network subtraction between the first group and the most recently added group (see Fig. 3). In this way, the *interface focuses user attention on the key affordances of the analytical technique*: in this case allowing them to quickly and easily compare multiple networks simultaneously via there projected points and means, and to investigate the differences between any two networks in more detail using network subtractions.

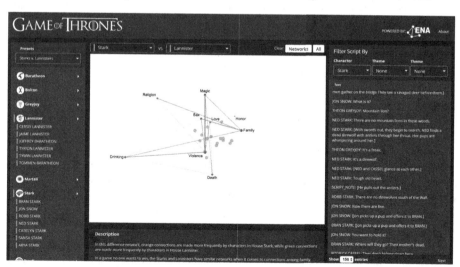

Fig. 3. Comparison between two families (green and orange) showing the projected points for characters in each family, mean points for each family, and the subtracted network for the families (Color figure online).

4.5   Linking Visualizations and Data

A critical component of ENA analyses and quantitative ethnographic methods more generally is investigating the alignment between the model and the data used to generate it—a process known as *closing the interpretive loop* (Shaffer 2017). Closing the interpretive loop is a critical step in checking the validity of the model and its interpretation. While this process is typically undertaken by the primary researcher, it can be an important process for external audiences as well, particularly if their aim is to better understand the model or to verify its accuracy. We *scaffolded the process of closing the interpretive loop by including both analytical visualizations and the underlying data to which they correspond*: in this case, by including network models and their corresponding data in the interface. Specifically, users can filter the data to show lines corresponding to particular units, codes, and connections. These affordances make the warrants for model interpretations explicit in ways that are infeasible with traditional methods of sharing research data.

5   Discussion

In this paper, we explored design considerations for an interface for sharing quantitative ethnographic research data. In particular, our design attempted to address the challenges of interpretability, attendability, and complexity.

The design presented here supported interpretability by (a) holding constant the dimensions of the visualization so that they did not have to be reinterpreted for different views of that data; (b) removing information about the models that was not central to interpreting the results; (c) providing a set of example results with corresponding interpretations, and (d) including the underlying data that corresponded to visualization. That is, we tried to *remove extraneous statistical information and include additional interpretive scaffolds*.

The design supports attendability by (a) removing information that was not central to interpreting the results, and also (b) designing the tool such that the affordances of the user interface were aligned with the affordances of the analytical technique. That is, we tried to *focus user attention on the key modeling choices and their associated results*.

Finally, all of these design decisions reduced the complexity of the analysis by (a) limiting the set of decisions that a user could make, (b) removing unnecessary details of the model. That is, we tried to *simplify the analytic space*.

These design considerations allow the presentation of research data in way that addresses the potential lack of expertise and experience of the audience. By making it possible for audiences to easily change models, see interpreted research data, and interact with the data to verify those interpretations and produce new insights, the gap between the primary researchers and the audience is made smaller.

Our results are of course limited by the specific data we analyzed and the processes we used to create the interface, which required a team of researchers, designers, and programmers. But while the design considerations are derived from one particular tool and interface, we believe that they have the potential to scaffold other attempts to communicate quantitative ethnographic analyses through data sharing. Our future work will include efforts to scaffold researchers in the creation of an interface for sharing their own data. Such a tool would make a variety of quantitative ethnographic analyses more widely usable, and in turn, would have the potential to grow the Quantitative Ethnography community and advance the state of its research.

Acknowledgments. This work was funded in part by the National Science Foundation (DRL-1661036, DRL-1713110), the Wisconsin Alumni Research Foundation, and the Office of the Vice Chancellor for Research and Graduate Education at the University of Wisconsin-Madison. The opinions, findings, and conclusions do not reflect the views of the funding agencies, cooperating institutions, or other individuals.

References

Beveridge, A., Shan, J.: Network of thrones. Math. Horiz. **23**(4), 18–22 (2016)

Borgman, C.L.: The conundrum of sharing research data. J. Am. Soc. Inform. Sci. Technol. **63**(6), 1059–1078 (2012). https://doi.org/10.1002/asi.22634

Chang, W., Cheng, J., Allaire, J., Xie, Y., McPherson, J.: shiny: Web Application Framework for R. (Version 1.3.2) (2019). https://CRAN.R-project.org/package=shiny

Feldman, S., Shaw, L.: The epistemological and ethical challenges of archiving and sharing qualitative data. Am. Behav. Sci. **63**(6), 699–721 (2019). https://doi.org/10.1177/0002764218796084

Herder, T., et al.: Supporting teacher's intervention in student's virtual collaboration using a network based model. In: Proceedings of the International Conference on Learning Analytics, Sydney, Australia, pp. 21–25 (2018)

Marquart, C.L., Swiecki, Z., Collier, W., Eagan, B., Woodward, R., Shaffer, D.W.: rENA: Epistemic Network Analysis (Version 0.1.3) (2018). https://cran.r-project.org/web/packages/rENA/index.html

Marquart, C.L., Swiecki, Z., Eagan, B., Shaffer, D.W.: ncodeR (Version 0.1.2) (2018). https://cran.r-project.org/web/packages/ncodeR/ncodeR.pdf

Miller, J.W.: Scraping song lyrics from Genius.com (2017). https://www.johnwmillr.com/scraping-genius-lyrics/

Shaffer, D.W.: Quantitative Ethnography. Cathcart Press, Madison (2017)

Shen, H., Huang, J.Z.: Sparse principal component analysis via regularized low rank matrix approximation. J. Multivar. Anal. **99**(6), 1015–1034 (2008)

Tsai, A.C., et al.: Promises and pitfalls of data sharing in qualitative research. Soc. Sci. Med. **169**, 191–198 (2016). https://doi.org/10.1016/j.socscimed.2016.08.004

Post-hoc Bayesian Hypothesis Tests
in Epistemic Network Analyses

M. Shane Tutwiler[⊠] [iD]

University of Rhode Island, Kingston, RI 02881, USA
shane_tutwiler@uri.edu

Abstract. Applied researchers are often forced to test an uninteresting (and unrealistic) hypothesis: that the mean difference between groups is zero in some imagined population. Misinterpretation of these common null hypothesis tests often obscure actual findings, and the testing process itself can result in inflated estimates over time. In this paper, we demonstrate the use of freely available software to conduct Bayesian hypothesis tests on ENA findings, in addition to traditional null hypothesis testing.

Keywords: Bayesian hypothesis testing · ENA · Quantitative methods

1 Introduction

The social and psychological sciences are currently in the throes of a replication crisis, resulting in revised perspectives on best practices regarding quantitative research design and analysis over the last half-decade [1]. The field of quantitative ethnography and its prime methodology, Epistemic Network Analysis (ENA), had its genesis at the beginning of the replication crisis, and has evolved rapidly along with the methodological best-practices in the social sciences. In an effort to avoid some of the analysis-related pitfalls identified in the replication crisis, we motivate the utility and demonstrate the use of a freely available statistical software package [2] to conduct post-hoc Bayesian hypothesis testing on common classes of t-tests that researchers using ENA might encounter.

2 Background

2.1 Limitations of Null Hypothesis Significance Testing

Most of the hypothesis testing mechanisms built into the extant ENA software are based on parametric and non-parametric null-hypothesis significance testing (NHST). In this inferential paradigm, applied researchers estimate the mean difference between two sampled groups, and use that information, along with estimates of a standard error, to determine an appropriate test statistic to summarize the magnitude of the observed difference. Then, based on assumptions about the sample and its relationship to a hypothesized population of interest, a p-value is calculated that communicates the probability that a test statistic of that magnitude or greater would have been observed if

© Springer Nature Switzerland AG 2019
B. Eagan et al. (Eds.): ICQE 2019, CCIS 1112, pp. 342–348, 2019.
https://doi.org/10.1007/978-3-030-33232-7_31

the data were sampled from a population in which no relationship exists. If this probability (p-value) is below a pre-designated, and often arbitrary, cut-point (e.g., 0.05), then the observed relationship is deemed "statistically significant" in the sample, and generalized to the population.

The limitations of this procedure are manifold, and have been highlighted in recent critical research [3]. Of chief concern for this paper are the facts that: (a) rejecting a null hypothesis does not "prove" that a particular hypothesized relationship exists [4]; (b) failing to reject a null hypothesis does not "prove" that a relationship does not exist [4]; and (c) NHST may result in errors in estimated magnitudes and directions of effects, a problem that is even more pronounced in small noisy data [5].

2.2 Introduction to Bayesian Parameter Estimation and Hypothesis Testing

Bayesian Parameter Estimation. Frequentist parametric statistical tests (such as t-tests) are computationally tractable and deceptively easy to interpret. This ease of estimation and interpretation come at the expense of strong assumptions, however. As highlighted above, analysts assume that data are independent and represent a random, adequately powered (i.e., large) sample from a population of interest. They then calculate some type of point-estimate that represents their best guess at a population-level difference given the data (usually via mean square error reduction or maximum likelihood estimation). Oft-misinterpreted 95% Confidence Intervals can then be estimated by adding/subtracting approximately two standard errors from the point estimate. This results in a range of probable values, given the data and associated analytic assumptions. The point estimate should not be privileged in this model, and all values within the range of the 95% Confidence Interval should be communicated in order to properly capture the uncertainty in the data.

In practice, this is rarely done. The point estimates are presented with their associated NHST results (p-values), and treated as "real" if the null is rejected. However, the results of frequentist t-tests can be used to generate post-hoc Bayesian estimates of effect size differences that are less vulnerable to the influence of spuriously large or small data points (especially important when dealing with small data sets) and that treats the estimated effects as a probability distribution, with the median being the most likely value (in essence, how most people interpret the point estimates from frequentist parametric models).

For example, assume that we wanted to estimate the mean difference (as an effect size, δ) between two groups. Using Bayes Theorem, we can express the probability of a certain value of δ given the data (D) as:

$$P(\delta|D) = \frac{P(D|\delta)}{P(D)} * P(\delta) \tag{1}$$

That is to say, our belief in the probability of a given effect size given the data, $P(\delta|D)$ (also known as the posterior) is a function of two things:

1. the degree to which the given data are consistent with a particular effect size (also known as the likelihood): $(P(D|\delta))/(P(D))$, and
2. our prior knowledge or belief about the distribution of plausible effect sizes: $P(\delta)$.

Because it is framed as a probability distribution, our interpretation of the area under the curve where 95% of the estimates are located (the 95% Credibility Interval) is much easier to interpret as a measure of uncertainty. Another important distinction is the fact that frequentist methods do not allow us to incorporate prior knowledge, as they have uniformly wide prior distributions that treat all potential values as equally plausible. As such, their estimates are often overly sensitive to the data at hand, an issue known as over-fitting [6]. The JASP software allows us to use default or user-specified prior estimates to regularize the findings [6], decreasing over-fitting the model to the data and increasing our out-of-sample fit (i.e., generalizability). The ability to specify a range of prior is especially important in small samples, such as those often found in ENA analyses, which can be particularly sensitive to abnormal data.

Model-Based Bayesian Hypothesis Testing. Unlike NHST, which can only test the hypothesis that the data were sampled from a population in which the relationship under consideration is exactly zero, Bayesian Hypothesis Testing allows analysts to compare the probability that a host of hypotheses are true, given the data. For example, consider two hypotheses, one in which the relationship under study is zero (H_0) and the other in which the relationship is non-zero (H_1). Under NHST, we would only be able to directly test H_0 and would have to qualitatively build the case that H_1 is the best alternative hypothesis, given our rejection of H_0. The Bayesian paradigm, on the other hand, allows us to directly compare the relative probabilities of the competing hypotheses given the observed data. Again utilizing Bayes Theorem, we can write the pertinent probabilities as:

$$P(H_0|D) = \frac{P(D|H_0)}{P(D)} * P(H_0) \tag{2}$$

$$P(H_1|D) = \frac{P(D|H_1)}{P(D)} * P(H_1) \tag{3}$$

In simple terms (using the null hypothesis as an example), we are stating that support for a hypothesis given observed data, $P(H_0|D)$ (the posterior) is a function of two things:

3. the degree to which a given data are supported by a particular hypothesis (the likelihood): $\frac{P(D|H_0)}{P(D)}$, and
4. our prior belief about the hypothesis under question: $P(H_0)$.

We can then compare the likelihoods directly by taking their ratio, called a Bayes Factor:

$$BF_{10} = \frac{P(D|H_1)}{P(D|H_0)} \tag{4}$$

which indicates the degree to which data are supported by the alternative hypothesis. Thus, we are able to directly quantify our preference for an alternative hypothesis in comparison to the null. Using heuristic guidelines based on the work of [7] in Table 1, we can then label the degree of support ranging from Anecdotal (BF_{10} = 1 to 3) to Extreme (BF_{10} = 100+).

Table 1. Interpretation of Bayes Factors in support of a given hypothesis.

Bayes factor	Magnitude of support
1–3	Anecdotal
3–10	Moderate
10–30	Strong
30–100	Very strong
100+	Extreme

The JASP software computes the posterior estimates of parameters, model-based likelihoods, and associated Bayes Factor based on information about the t-statistic and sample size from a traditional t-test, such as those conducted by the ENA software. Further details and examples of Bayesian estimation and hypothesis testing can be found in [8].

3   Analysis and Results

3.1   Hypothetical ENA t-test Example

To highlight the above process, we consider a t-test conducted in a hypothesized ENA study that shows a comparison of two groups of observations in a frame (8 in each group) at a difference of 0.20 and standard error of 0.09, which results in a t-statistic of 2.22 with 14 degrees of freedom. The associated two-tailed p-value with such an analysis would be 0.043. As such, one might be tempted to declare statistical significance and move on. However, as previously discussed, small and potentially noisy datasets are vulnerable to spurious findings under the NHST paradigm, and rejecting a null hypothesis does not prove the existence of a given alternative hypothesis [4].

To illustrate this, we use the *retrodesign* package [9] in R [10] to simulate the power, expected magnitude of inflation, and rate of sign-reversal in studies similar to the one described above, over time. In ENA studies similar to the one described above, we have a power to properly reject null hypotheses about 53% of the time (well below the preferred 80%), and the magnitude of estimated difference is expected to be inflated by a factor of about 1.3. That is to say, this would be a particularly noisy analysis, the results of which might be inconclusive.

3.2   Bayesian Estimation and Testing of ENA Results

A better approach would be to supplement the NHST with a post-hoc Bayesian hypothesis test. As shown in Fig. 1 (which compares the posterior and prior parameter estimates described above), the median estimated difference was 0.77 with a very wide 95% Credible Interval ranging from about −0.12 to +1.9. This indicates a large degree of uncertainty in our estimation. We also note in Fig. 2 that the Bayes Factor associated with these findings are robust across a range of assumed prior probabilities, and indicate that the data at best only anecdotally support a hypothesis that the observed relationship is non-zero.

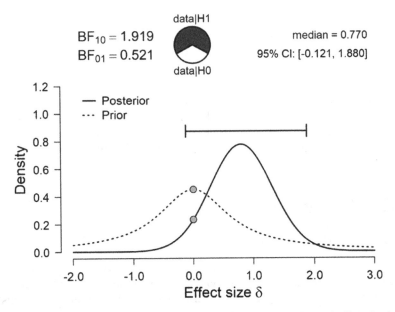

Fig. 1. Results of post-hoc Bayesian parameter estimation of the estimated effect-size in mean differences.

Given this analysis, we can now speak to our degree of confidence about the observed difference on the ENA graph. Instead of relying upon NHST as a heuristic decision-making guide, we can explicitly state that, while we do note a probable effect-size difference ranging between −.12 to +1.9, with a majority of estimated effect-sizes being greater than 0 and the most probable effect being 0.80, evidence for it is anecdotal at best and interpretations should be made cautiously.

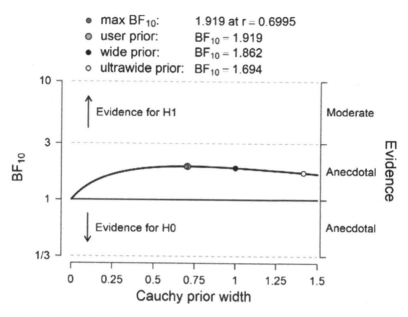

Fig. 2. Results of post-hoc Bayesian hypothesis testing robustness checks across a range of priors.

4 Discussion and Conclusion

Traditional parametric statistical analyses, such as t-tests, have historically relied upon NHST (i.e., statistical significance) to determine if sample-based differences might exist in a hypothesized population. The strength of these inferences depends largely on the quality and size of the samples being tested, and the degree to which a host of assumptions are tested and met. As such, critiques of their continued use across the social sciences have been widespread.

As demonstrated in the above analyses, post-hoc Bayesian hypothesis testing of the t-tests produced by the ENA tool offers researchers the ability to make sample-specific inferences in a framework that includes information about the uncertainty of estimation (via the 95% Credibility Interval of effect sizes) as well as an estimation of the degree of support the data provide for a given hypothesis (via the Bayes Factor). This is a vast improvement to traditional parametric analyses and NHST, which focus primarily on testing a sharp null hypothesis on a "point-estimate" of an effect via a common but often arbitrary cut-point ($\alpha = 0.05$).

Future work should seek to include Bayesian hypothesis testing, ideally at the parameter-level, in ENA software using methods such as those proposed by [11] or [12]. Until then, secondary analyses such as those described here can be used to directly test hypotheses of interests based on the results of t-test results reported in the ENA software.

References

1. Begley, C.G., Ioannidis, J.P.: Reproducibility in science: improving the standard for basic and preclinical research. Circ. Res. **116**(1), 116–126 (2015)
2. JASP Team: JASP (Version 0.9.2) [Computer software] (2019)
3. McShane, B.B., Gal, D., Gelman, A., Robert, C., Tackett, J.L.: Abandon statistical significance. Am. Stat. **73**(Suppl. 1), 235–245 (2019)
4. Leppink, J., Winston, K., O'Sullivan, P.: Statistical significance does not imply a real effect. Perspect. Med. Educ. **5**(2), 122–124 (2016)
5. Gelman, A., Carlin, J.: Beyond power calculations: assessing type S (sign) and type M (magnitude) errors. Perspect. Psychol. Sci. **9**(6), 641–651 (2014)
6. McElreath, R.: Statistical Rethinking: A Bayesian Course with Examples in R and Stan, vol. 122. CRC Press, Boca Raton (2016)
7. Jeffreys, H.: Theory of Probability, 3rd edn. Oxford University Press, Oxford (1961)
8. Wagenmakers, E.-J., Morey, R.D., Lee, M.D.: Bayesian benefits for the pragmatic researcher. Curr. Dir. Psychol. Sci. **25**, 169–176 (2016)
9. Timm, A.: retrodesign: Tools for Type S (Sign) and Type M (Magnitude) Errors. R package version 0.1.0 (2019)
10. R Core Team: R: A language and environment for statistical computing. R Foundation for Statistical Computing, Vienna, Austria (2019)
11. Kruschke, J.K.: Bayesian estimation supersedes the t test. J. Exp. Psychol. Gen. **142**(2), 573 (2013)
12. Gelman, A., Carlin, J.B., Stern, H.S., Dunson, D.B., Vehtari, A., Rubin, D.B.: Bayesian Data Analysis, 3rd edn. Chapman and Hall/CRC, Boca Raton (2013)

Applying Epistemic Network Analysis to Explore the Application of Teaching Assistant Software in Classroom Learning

Lijiao Yue[1(✉)], Youli Hu[1], and Jing Xiao[2]

[1] East China Normal University, 3663 North Zhongshan Road, Shanghai, China
51184108028@stu.ecnu.edu.cn
[2] University College London, Gower Street, London, UK

Abstract. With the rapid development of information technology, teaching assistant software has been constantly appearing in classroom learning. How to effectively apply this technical resource in classroom learning has become one of the focus of educational research and practice. In this study, the epistemic network analysis method was used to process the interview text of students using teaching AIDS, and the effect of the application of teaching AIDS in classroom learning was discussed. The results show that there is a significant difference between high-score students and low-score students in using instructional software in classroom learning, especially with regards to their learning motivation towards it. Additionally, the use of epistemic network analysis technology could improve the accuracy of decision-making reference for evaluating the effect of software use and implementing accurate teaching.

Keywords: Rain Classroom · Teaching aid software · Epistemic network analysis

1 Introduction

With the rapid development of information technology, the application of mobile internet and big data technology in the field of education has become more and more extensive. Computer network technology not only provides resources support for learning, but also promotes the reform and renewal of teaching ideas and forms. Most disciplines are currently facing the challenges like the increasing requirements for student individuality, diverse learning needs and teaching information resources of various aspects. How to make full use of the new era of information technology to effectively improve the students' classroom learning with the support of teaching software is pretty essential for education. As an intelligent teaching tool in higher education, "Rain Classroom" is widely used in different disciplines among Chinese universities. As of December 9, 2017, the active teaching teachers and students in the rain classroom have exceeded 2.2 million, covering 150,000 classes, and the monthly living number has reached 760,000. It is one of the most active intelligent teaching tool in China. Therefore, the research sample of "Rain Classroom" as teaching aid software is representative.

© Springer Nature Switzerland AG 2019
B. Eagan et al. (Eds.): ICQE 2019, CCIS 1112, pp. 349–357, 2019.
https://doi.org/10.1007/978-3-030-33232-7_32

The existing literature on the Rain Classroom mostly focuses on the interpretation and application of the functions of the tool itself, without in-depth exploration of the effect of this teaching aid on different aspects of college students' classroom learning. Simple qualitative or quantitative analysis is difficult to identify the use difference, engagement and thinking process of different students in the learning process of using auxiliary software. It is difficult to provide authentic and reliable evidence for the reasonable application of auxiliary software. Epistemic Network Analysis (ENA), as a data analysis method for quantifying ethnography, combines qualitative and quantitative analysis. It enables individuals (or groups) to associate with different abilities to be represented by coding in a networked form. In-depth analysis is made on the learning attitude and process of students at different levels to efficiently and conveniently evaluate the dynamic coupling relationship between different "objects", so as to compare the network structure characteristics and differences of students or learning groups [1]. Aiming at this problem, this study will focus on three important elements of students' learning in classroom, namely, learning motivation, learning strategy and learning effect, using ENA to reveal the differences and changes in the use of teaching assistant software among students of high-score and low-score learning level, so as to explore the relationship between the application of teaching assistant software and students' classroom learning.

2 Theoretical Basis

2.1 Rain Classroom

"Rain Classroom" is an online hybrid teaching tool, which integrates technology into PPT and micro-letters. By creating online virtual classroom, it creates an efficient learning environment to ensure effective interaction between teachers and students in pre-class preparation, classroom teaching and after-class review. One of the innovations of the concept of "Rain Classroom" is the realization of Digital-channel Teaching under the background of mobile Internet. Through information technology, two communication channels, Synchronous and Asynchronous, are established in the teaching process, so that all teaching contents and forms can be carried out in the most appropriate channel. Its basic functions can be roughly divided into: question feedback; classroom exercise response system; "mobile courseware" push; "bullet curtain" classroom discussion; data acquisition and analysis. Xiao Anbao confirmed that "Rain Classroom" helps to improve learning initiative in the flip teaching of Ideological and political lessons in Colleges and universities [2]. If the data recorded in rain classes are taken as part of the course assessment results, students will be more active in completing tasks. Wang Ruijuan and others have constructed a new teaching model based on rain classroom by analyzing the characteristics of rain classroom [3]. They have carried out practical research in the teaching of Computer Network course. It is found that the application of rain classroom in the reverse classroom can stimulate students' learning initiative and achieve better teaching results. At present, most researchers are still in the interpretation and application of the functions of the tools themselves, and there are few studies on the relationship between teaching aids software and scientific learning. This study can be further explored on the basis of existing research. From the perspective of subject learning, this paper reflects on the design and application of teaching assistant software.

2.2 Theoretical Definition of Classroom Learning

The three important elements of students' learning in class include learning motivation, learning strategy and learning effect. Among them, the learning motivation is related to self-regulation, which means individuals coordinate their internal requirements with the external incentives of learning behaviors, so as to form the dynamic factors to stimulate and maintain learning behaviors [4]. In order to control and maintain a good positive learning motivation, Keller (2008) proposed an ARCSV model of great practical value to improve the rate of knowledge absorption and the level of academic achievement, which includes: Attention, Relevance, Confidence, Satisfaction, and Volition [5].

The learning resources provided to students by teaching AIDS are often fragmentary and short. For better understanding and mastery of knowledge, one learning strategy is to establish and arrange a good knowledge structure. However, learners of different levels differ in the quantity, frequency and quality of learning strategies [6].

In terms of process data, the learning effect of students is closely related to their participation and engagement during class. Learning outcome has a strong correlation with students' academic performance. Some researchers have pointed out that the involvement of teaching AIDS directly affects the learner-instructor interaction, teaching effect and learning result in the classroom learning environment [7].

2.3 Theoretical Framework of ENA

Epistemic Network Analysis (ENA) is a data analysis tool, which identifies and quantifies the core elements in text (discourse) data. It can also simulate the connection structure between the core elements, and construct the visual dynamic network graph [8]. ENA is based on the cognitive framework theory and the "evidence-based" educational evaluation design model [9]. The cognitive framework theory holds that professional knowledge can be modeled as the understanding expressed through specific discourses. The model compares the differences between different cognitive networks through statistical data and visualized graphs reflecting the connections between various elements, and then presents the differences of cognitive development reflected by corresponding teaching activities. ENA provides reliable ideas and methods for measuring complex thinking and problem solving ability through quantitative connection model.

In recent years, researchers have used ENA to analyze and visualize cognitive phenomena in various fields and come to many meaningful conclusions. For example, ENA is used to evaluate the chat discourse in online classroom collaborative learning and analyze the thematic connection of students' performance in the discourse, so as to predict students' performance [10]. Beside this, ENA can be used as a tool to assist teachers in formative and summative evaluation, and visual analysis can be carried out on college students' compositions at different levels [11]. This study further suggests that ENA can be used as a method or tool to help demand objects to evaluate.

3 Experimental Research

3.1 Research Objects

This study takes the course Python Programming in the spring semester of a university in east China as the research context, and conducts in-depth interviews on the impact of the teaching mode based on Rain Classroom on students' learning. The total number of students in this course is 25, and the final number of participants in the in-depth interview is 17 (no. S1–S17), accounting for about 2/3 of the total number, so the questions reflected by the interviewees are representative to a certain extent. Research interview samples are distributed from the first-year to the third-year undergraduates, involving multiple majors, ensuring the diversity of data sources. As a rain classroom promotion lecturer, the teacher has a good command of the software, which avoids the possibility of poor teaching effect due to human and technical factors. In case analysis, this study collects and analyses data through in-depth interviews with relevant individuals.

3.2 Data Collection and Analysis

The research mainly adopts semi-structured interview and records the whole interview process in the form of recording. The interview time of each person is about 20–30 min. After the interview, the recordings were transcribed into words by the researchers. The total text is more than 100,000 words, providing abundant materials for qualitative research.

The data were collected using the traditional in-depth interview method, and the content of the interview was divided into the following five parts according to the theoretical basis: general perception, learning motivation, learning strategy, learning effect, and students' suggestions and opinions on it. The general perception is mainly based on the learners' previous experience of using the rain classroom to get the information about their overall feelings towards this software. Learning motivation is designed according to the ARCSV model proposed by Keller in 2008. The three dimensions of Confidence, Satisfaction and Volition are selected because of their close correlation with teaching aid. Learning strategies are divided into pre-class preparation, notes, feedback in class, and review after class according to chronological order. At last, the students' understanding of subject knowledge, particularly the change of their level of mastery were investigated. The data processing includes collecting, transcribing and coding the interview contents of different students, and finally quantitative analysis using the method of cognitive network modeling.

3.3 Coding Scheme

According to the interview structure, the study focused on three important elements in classroom learning: learning motivation, learning strategy and learning effect. Coding scheme was designed on the basis of these three dimensions to analyze the interview text data (Table 1).

Firstly, the audio of the interview was transcribed into text. The dialogue recorded in the transcript reflects the speakers' conversation rounds. And then encodes it

Table 1. Coding scheme.

Dimension	Code	Definition
Learning motivation	Confidence	The use of Rain Class, on the ease of learning and students to complete the learning task confidence
	Satisfaction	Students' learning expectation and Rain Class play a role in achieving the goal
	Volition	The use of mobile phones in class has an effect on students' attention
Learning strategy	Preview	Student preview survey and the value of Rain Class
	Note	Changes in students' note-taking habits
	Feedback	Student learning feedback
	Revision	Survey of students' review after class and the role of Rain Class
Learning effect	Atmosphere	Compared with the traditional classroom, the classroom atmosphere changes
	knowledge Master	The influence of Rain Class on students' learning effect

sentence by sentence, Table 2 shows the coding example of the part of the interview data in the learning strategy dimension. The 17 interview records were independently coded by two researchers and the coding results were found to be consistent (Kappa = 0.78). Differences in individual coding scores reached to consensus after consultations.

Table 2. The coding example of the part of the interview data in the learning strategy dimension.

Student name	Interview text	Preview	Note	Feedback	Revision
LYW	I usually preview, preview time mainly depends on the length or difficulty of the courseware, about ten minutes to half an hour	1	0	0	0
LYW	There will be some review, which is what I have talked about in this lesson. I may review it every other day or the day, or I will review it when I am doing my homework	0	0	0	1
WSL	I have clicked the "Understand" and "Do not understand" buttons	0	1	0	0
WSL	For the barrage and submission, I will send it if I can. And if it's about feelings, I'll send it	0	0	1	0
WYQ	Yes. I often take notes.	0	1	0	0
WYQ	I usually preview, and I think it's worth it	1	0	0	0
GXY	I like the feedback after class, and it helps me make a very good note so that I can understand what I have learned	0	0	1	0
GXY	I usually do a quick review according to the homework after class, and review what I have learned and what I have talked about in this lesson	0	0	0	1

4 Analysis Result

In the interview records, there are dialogues about the correct rate of homework. According to the correct rate, students are divided into high-score group and low-score group. The overall correct rate of high-score students in homework is more than 60%, while those below 60% are classified as low-score group. The final statistical results show that there are 11 high-score group students and 6 low-score group students. In order to illustrate the characteristics of using teaching assistant software in learning for high-score and low-score students, this study adopted epistemic network analysis and established a quantitative analysis model based on all the interview data to further show the differences between the two groups in the two dimensions of learning motivation and learning strategy.

The author drew the cognitive network graphs of two groups and each student according to the students interview text coding data. As shown in Fig. 1, each dot in the figure represents the center of mass of each student's cognitive network graph. Two squares represent the average value of the center of mass of each group member's network graph, which can also be understood as the center of mass of two groups' network graph. The surrounding rectangle represents the confidence interval of the center of mass of each group at 95% level.

According to the results of epistemic network analysis, the first dimension (X-axis) and the second dimension (Y-axis) in the generated two-dimensional projection account for 40.1% and 15.5% of the total variance of the data. It can also be seen from the figure that there is a significant difference in the data of cognitive network between

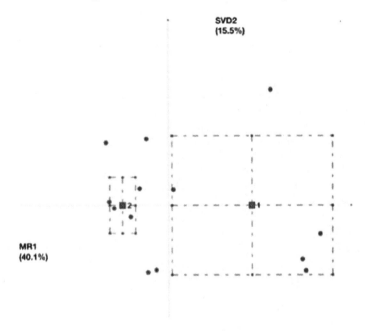

Fig. 1. Cognitive network map of high-score grouping (Blue) and low-score grouping (Red) students in projection space. (Color figure online)

the two groups in the first dimension (high grouping M = 1.26, low-score grouping M = −2.31, t = 4.07, p = 0.01 < 0.05), while there is no significant difference in the data in the second dimension.

In addition, ENA was used to draw the average cognitive network graph of the high-score and low-score groups as a whole (Fig. 2), so as to further analyze the differences between the cognitive network structures of the two groups. In the ENA network model, line thickness and saturation represent the strength of the connection between two elements. As shown in Fig. 2, the connection between elements in the high-score group is more complex, and the connection between volition and preview and review is stronger than that in the low-score group, as well as the connection between note-taking and feedback and confidence. This shows that the students with higher grades have stronger volition and self-control, and they can keep preview and review. When using the teaching assistant software, they still take notes and give feedback to the teacher in time, so they are full of confidence in learning. The students in low-score group are more likely to have a strong connection between confidence and review, preview and review. In addition, you can see from the two figures that both groups are closely linked to the other elements in the preview.

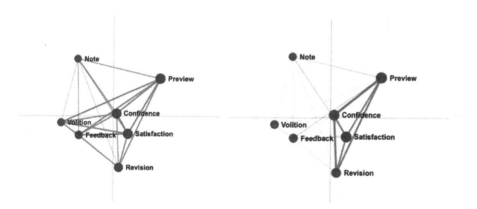

Fig. 2. Average network graph (left: high-group, right: low-group).

By subtracting two average network graphs, we can get the difference network graph as shown in Fig. 3. If there is overlap between the two groups of elements, the lines connecting the stronger groups will eventually appear, and the lines will be superimposed on each other, which can show the difference of average cognitive network between the high-score group and the low-score group students more clearly. As can be seen from the figure, the lower group focuses more on the connection of elements in the lower right corner, while the higher group focuses more on the connection of elements in the upper left corner. The difference between the left and right regions is more obvious, which just confirms that the difference between the two groups described above is significant in the first dimension, but not significant in the second dimension.

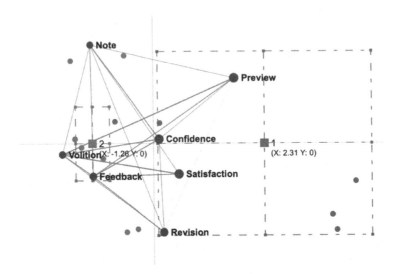

Fig. 3. Difference network figure.

5 Discussion and Suggestion

In this study, epistemic network analysis is used to analyze the interview data of students' use of teaching assistant software. By comparing the visual cognitive network structure figure, it is found that there are obvious differences in learning motivation and learning strategies between high-score group students and low-score group students. This difference brings new ideas and methods for teachers to predict the effect of software use and students' learning under the teaching mode of teaching assistant software. Through the learning analysis technology based on ENA and other tools and means, the feedback data of students using teaching aids software are processed and analyzed, and the different cognitive levels and structures of the evaluators about the use of teaching aids software can be obtained. Relevant analysis can help teachers understand more detailed and specific information about learning status and usage than examination results. It is able to help educators predict students' performance trends, thereby implementing more targeted and accurate teaching. Relevant differences can also inform learners their gap in learning effectiveness, and improve the software use method and learning method in a targeted way, so as to achieve a comprehensive grasp of the course knowledge.

What's more, some suggestions have emerged from the conclusions of the study for students' learning, teacher teaching, and development and design of teaching assistant software such as Rain Class.

Firstly, for students, when they utilize teaching assistant software in the classroom learning, certain methods should be adopted to maintain strong volition and self-control. In addition, students should also pay attention to the use of effective learning strategies, especially to timely feedback the learning situation to the teacher, adhere to preview review. Secondly, in the teaching process, teachers need to adjust the classroom mode according to the actual situation of the use of the software, such as strengthening the tracking and management of learners' after-class learning. In addition, prior to the decision to use the tool, teachers should make full preparations, familiar with the software features and functions, design appropriate teaching plan. Thirdly, there is still a large room for improvement in software functions. For example, teaching assistant software can set certain encouragement mechanism, such as points and rewards, so as to enhance students' learning motivation and confidence; provide more feedback mechanisms for teachers and students, such as analyzing students' problem solving and constructing students' knowledge map.

To sum up, existing studies have shown that epistemic network analysis is an extremely effective method for quantitative analysis of qualitative data, and this study further confirms the significant value of ENA in t exploring the use and learning of teaching assistant software. In future research, the combination of other information means and technologies can better improve the analysis effect and efficiency of ENA and give full play to its value.

References

1. Shaffer, D.W., Collier, W., Ruis, A.R.: A tutorial on epistemic network analysis: analyzing the structure of connections in cognitive, social, and interaction data. J. Learn. Anal. **3**, 9–45 (2016)
2. Xiao, A.-B., Xie, J., Gong, F.-Q.: The application of rain classroom on the flipped teaching of ideological and political theory course in universities. Mod. Educ. Technol. **05**, 47–53 (2017)
3. Wang, R.-J., Yang, Z.-H.: Application of rain classroom on flipped classroom. Mod. Comput. (2018)
4. Liu, M.-J., Xiao, H.-Y.: A review of the research on study motivation. J. Shanxi Datong Univ. (2009)
5. Keller, J.M.: First Principles of motivation to learn and e-learning. Distance Educ. **29**(2), 175–185 (2008)
6. Gu, P.Y.: Strategies in learning and using a second language (second edition). ELT J. **66**(2), 251–253 (2012)
7. Schroedermoreno, M.S.: Enhancing active and interactive learning online - lessons learned from an online introductory agroecology course. NACTA J. **54**, 21–30 (2010)
8. Shaffer, D.W., Hatfield, D.: Epistemic network analysis: a prototype for 21st century assessment of learning. Int. J. Learn. Media **2**, 33–53 (2009)
9. Shaffer, D.W., Hatfield, D.: Epistemic network analysis: a worked example of theory-based learning analytics. In: Handbook of Learning Analytics, pp. 175–187 (2017)
10. Cai, Z., Eagan, B., Dowell, N., et al.: Epistemic network analysis and topic modeling for chat data from collaborative learning environment. In: Proceedings of the 10th International Conference on Educational Data Mining, pp. 104–111 (2017)
11. Fougt, S.S., Siebert-Evenstone, A., Eagan, B., et al.: Epistemic network analysis of students' longer written assignments as formative/summative evaluation. In: Proceedings of the 8th International Conference on Learning Analytics and Knowledge, pp. 126–130. ACM (2018)

Author Index

Printed in the United States
By Bookmasters